Reiner Westermeier
Electrophoresis in Practice

Related title from Wiley-VCH

Reiner Westermeier, Tom Naven

Proteomics in Practice
A Laboratory Manual of Proteome Analysis

ISBN 3-527-30354-5

Reiner Westermeier

Electrophoresis in Practice

A Guide to Methods and Applications
of DNA and Protein Separations
Fourth, revised and enlarged Edition

in collaboration with

Sonja Gronau
Phil Becket
Josef Bülles
Hanspeter Schickle
Günter Theßeling

WILEY-
VCH

WILEY-VCH Verlag GmbH & Co. KGaA

Dr. Reiner Westermeier
GE Healthcare Bio-Sciences
Munzinger Str. 9
79111 Freiburg
Germany

in collaboration with
Sonja Gronau
Phil Becket
Josef Bülles
Hanspeter Schickle
Günter Theßeling

All books published by Wiley-VCH are carefully produced. Nevertheless, authors, editors, and publisher do not warrant the information contained in these books, including this book, to be free of errors. Readers are advised to keep in mind that statements, data, illustrations, procedural details or other items may inadvertently be inaccurate.

Library of Congress Card No.:
applied for
British Library Cataloguing-in-Publication Data
A catalogue record for this book is available from the British Library.
Bibliographic information published by Die Deutsche Bibliothek
Die Deutsche Bibliothek lists this publication in the Deutsche Nationalbibliografie; detailed bibliographic data is available in the Internet at
<http://dnb.ddb.de>.

© 2005 WILEY-VCH Verlag GmbH & Co. KGaA, Weinheim

Printed in the Federal Republic of Germany.

Printed on acid-free paper.

Composition Kühn & Weyh, Satz und Medien, Freiburg
Printing betz-druck GmbH, Darmstadt
Bookbinding Litges & Dopf Buchbinderei GmbH, Heppenheim

ISBN 3-527-31181-5

Contents

Foreword   *XI*

Preface   *XIII*

Abbreviations   *XVII*

Part I   **Fundamentals**   *1*

Introduction   *3*

1	**Electrophoresis** *9*	
1.0	General *9*	
1.1	Electrophoresis in non-restrictive gels *17*	
1.1.1	Agarose gel electrophoresis *17*	
1.1.2	Polyacrylamide gel electrophoresis of low-molecular weight substances *20*	
1.2	Electrophoresis in restrictive gels *21*	
1.2.1	The Ferguson plot *21*	
1.2.2	Agarose gel electrophoresis *22*	
1.2.3	Polyacrylamide gel electrophoresis of nucleic acids *24*	
1.2.4	Polyacrylamide gel electrophoresis of proteins *34*	

2	**Isotachophoresis** *45*	
2.1	Migration with the same speed *45*	
2.2	"Ion train" separation *47*	
2.3	Zone sharpening effect *47*	
2.4	Concentration regulation effect *47*	

3	**Isoelectric focusing** *51*	
3.1	Principles *51*	
3.2	Gels for IEF *53*	
3.3	Temperature *54*	
3.4	Controlling the pH gradient *54*	

Electrophoresis in Practice, Fourth Edition. Reiner Westermeier
Copyright © 2005 WILEY-VCH Verlag GmbH & Co. KGaA, Weinheim
ISBN: 3-527-31181-5

3.5   The kinds of pH gradients   *55*
3.5.1   Free carrier ampholytes   *55*
3.5.2   Immobilized pH gradients   *59*
3.6   Protein detection in IEF gels   *62*
3.7   Preparative isoelectric focusing   *62*
3.8   Titration curve analysis   *64*

4   **Blotting**   *67*
4.1   Principle   *67*
4.2   Transfer methods   *68*
4.2.1   Diffusion blotting   *68*
4.2.2   Capillary blotting   *68*
4.2.3   Pressure blotting   *68*
4.2.4   Vacuum blotting   *69*
4.2.5   Electrophoretic blotting   *70*
4.3   Blotting membranes   *72*
4.4   Buffers for electrophoretic transfers   *73*
4.5   General staining   *75*
4.6   Blocking   *75*
4.7   Specific detection   *76*
4.8   Protein sequencing   *78*
4.9   Transfer problems   *79*

5   **Interpretation of electropherograms**   *81*
5.1   Introduction   *81*
5.1.1   Purity control   *81*
5.1.2   Quantification prerequisites   *81*
5.2   Image analysis   *83*
5.2.1   Hardware for image analysis   *84*
5.2.2   Software for image analysis   *87*

6   **Proteome Analysis**   *91*
6.1   General   *91*
6.2   Sample preparation   *96*
6.3   Two-dimensional electrophoresis   *100*
6.4   Detection techniques   *102*
6.5   Image analysis   *107*
6.6   Protein spot identification   *109*
6.6.1   Mass spectrometry methods   *111*
6.6.2   Peptide mass fingerprinting   *116*
6.6.3   Protein characterization   *118*
6.7   Bioinformatics   *118*
6.8   Functional proteomics   *119*

7   **Instrumentation**   *121*
7.1   Current and voltage conditions   *121*
7.2   Power supply   *123*
7.3   Separation chambers   *124*
7.3.1   Vertical apparatus   *124*
7.3.2   Horizontal apparatus   *125*
7.4   Staining apparatus for gels and blots   *128*
7.5   Automated electrophoresis   *128*
7.6   Instruments for 2-D electrophoresis   *130*
7.6.1   Isoelectric focusing apparatus   *130*
7.6.2   Multiple slab gel apparatus   *131*
7.7   Safety measures   *131*
7.8   Environmental aspects   *132*

Part II   **Equipment and Methods**   *133*

Equipment for Part II   *135*
Instrumentation   *136*
Special laboratory equipment   *138*
Consumables   *139*
Chemicals   *140*

Method 1: PAGE of dyes   *145*
1   Sample preparation   *145*
2   Stock solutions   *145*
3   Preparing the casting cassette   *146*
4   Casting ultrathin-layer gels   *148*
5   Electrophoretic separation   *149*

Method 2: Agarose and immuno electrophoresis   *151*
1   Sample preparation   *151*
2   Stock solutions   *152*
3   Preparing the gels   *152*
4   Electrophoresis   *157*
5   Protein detection   *159*

Method 3: Titration curve analysis   *163*
1   Sample preparation   *163*
2   Stock solutions   *164*
3   Preparing the blank gels   *164*
4   Titration curve analysis   *167*
5   Coomassie and silver staining   *170*
6   Interpreting the curves   *172*

Method 4: Native PAGE in amphoteric buffers   *175*
1   Sample preparation   *176*
2   Stock solutions   *177*
3   Preparing the empty gels   *178*
4   Electrophoresis   *182*
5   Coomassie and silver staining   *185*

Method 5: Agarose IEF   *189*
1   Sample preparation   *189*
2   Preparing the agarose gel   *190*
3   Isoelectric focusing   *193*
4   Protein detection   *195*

Method 6: PAGIEF in rehydrated gels   *197*
1   Sample preparation   *198*
2   Stock solutions   *198*
3   Preparing the blank gels   *199*
4   Isoelectric focusing   *201*
5   Coomassie and silver staining   *204*
6   Perspectives   *208*

Method 7: Horizontal SDS-PAGE   *211*
1   Sample preparation   *211*
2   Stock solutions for gel preparation   *215*
3   Preparing the casting cassette   *216*
4   Gradient gel   *218*
5   Electrophoresis   *221*
6   Protein detection   *224*
7   Blotting   *227*
8   Perspectives   *227*

Method 8: Vertical PAGE   *231*
1   Sample preparation   *232*
2   Stock solutions   *233*
3   Single gel casting   *233*
4   Multiple gel casting   *237*
5   Electrophoresis   *241*
6   SDS electrophoresis of small peptides   *242*
7   Two-dimensional electrophoresis   *244*
8   DNA electrophoresis   *244*
9   Long shelflife gels   *245*
10   Detection of bands   *245*

Method 9: Semi-dry blotting of proteins *247*
1   Transfer buffers *248*
2   Technical procedure *249*
3   Staining of blotting membranes *253*

Method 10: IEF in immobilized pH gradients *257*
1   Sample preparation *258*
2   Stock solutions *258*
3   Immobiline recipes *259*
4   Preparing the casting cassette *262*
5   Preparing the pH gradient gels *264*
6   Isoelectric focusing *269*
7   Staining *271*
8   Strategies for IPG focusing *273*

Method 11: High-resolution 2-D electrophoresis *275*
1   Sample preparation *276*
2   Stock solutions *279*
3   Preparing the gels *280*
4   Separation conditions *285*
5   Staining procedures *293*

Method 12: PAGE of double stranded DNA *299*
1   Stock solutions *300*
2   Preparing the gels *301*
3   Sample preparation *305*
4   Electrophoresis *305*
5   Silver staining *310*

Method 13: Native PAGE of single stranded DNA *313*
1   Sample treatment *315*
2   Gel properties *316*
3   Buffers and additives *317*
4   Conditions for electrophoresis *318*
5   Strategy for SSCP analysis *318*

Method 14: Denaturing gradient gel electrophoresis *321*
1   Sample preparation *322*
2   Rehydration solutions *322*
3   Preparing the rehydration cassette *323*
4   Rehydration *325*
5   Electrophoresis *327*
6   Silver staining *329*

Method 15: Denaturing PAGE of DNA   *331*
1   Sample preparation   *332*
2   Solutions   *333*
3   Rehydration   *333*
4   Electrophoresis   *334*
5   Silver staining   *337*

Appendix   Trouble-shooting   *339*
A1   Isoelectric focusing   *339*
A1.1   PAGIEF with carrier ampholytes   *339*
A1.2   Agarose IEF with carrier ampholytes   *348*
A1.3   Immobilized pH gradients   *352*
A2   SDS electrophoresis   *359*
A3   Vertical PAGE   *369*
A4   Semidry blotting   *371*
A5   2-D electrophoresis   *379*
A6   DNA electrophoresis   *385*

References   *389*

Index   *399*

Foreword

The number of electrophoretic separation methods has increased dramatically since Tiselius' pioneer work for which he received the Nobel Prize. Development of these methods has progressed from paper, cellulose acetate membranes and starch gel electrophoresis to molecular sieve, disc, SDS, and immunoelectrophoresis and finally to isoelectric focusing but also to high resolution two-dimensional electrophoresis. Together with silver and gold staining, autoradiography, fluorography and blotting, these techniques afford better resolution, sensitivity and specificity for the analysis of proteins. In addition, gel electrophoresis has proved to be a unique method for DNA sequencing while high resolution two-dimensional electrophoresis has smoothed the fascinating path from isolation of the protein to the gene through amino acid sequencing and after gene cloning, to protein synthesis.

The spectrum of analytical possibilities has become so varied that an overview of electrophoretic separation methods seems desirable not only for beginners but also for experienced users. This book has been written for this purpose.

The author belongs to the circle of the bluefingers and experienced this in Milan in 1979 when he was accused of being a money forger when buying cigarettes in a kiosk after work because his hands were stained by Coomassie. Prof. Righetti and I had to extricate him from this tricky situation. According to Maurer's definition (Proceedings of the first small conference of the bluefingers, Tübingen 1972) an expert was at work on this book and he can teach the whitefingers, who only know of the methods by hearsay, fore example, how not to get blue fingers.

As it is, I am sure that this complete survey of the methods will not only help the whitefingers but also the community of the bluefingers, silverfingers, goldfingers etc. and will teach them many technical details.

Weihenstephan, February 1990

Prof. Dr. Angelika Görg
FG Proteomik,
Technische Universität München,
Freising-Weihenstephan

Electrophoresis in Practice, Fourth Edition. Reiner Westermeier
Copyright © 2005 WILEY-VCH Verlag GmbH & Co. KGaA, Weinheim
ISBN: 3-527-31181-5

Preface

German version

This book was written for the practician in the electrophoresis laboratory. For this reason we have avoided physico-chemical derivations
and formulas concerning electrophoretic phenomena.

The type of explanation and presentation stems from several years
of experience in giving user seminars and courses, writing hand-
books and solving user problems. They should be clear for technical
assistants as well as for researchers in the laboratory. The commen-
tary column offers room for personal notes.

In part I, an introduction – as short as possible – to the actual state
of the art will be given. The references are not meant to be exhaus-
tive.

Part II contains exact instructions for 11 chosen electrophoretic
methods, which can be carried out with one single piece of equipment. The sequence of the methods was planned so that an electrophoresis course for beginners and advanced users can be established
afterwards. The major methods used in biology, biochemistry, medicine and food science methods have been covered.

If – despite following the method precisely – unexplained effects
should arise, their cause and the remedies can be found in the trouble-shooting guide in the appendix.

The author would be thankful for any additional comments and solutions for the trouble-shooting guide which the reader could supply.

Freiburg, March 1990   *R. Westermeier*

English version, First Edition

The author is grateful to Dr. Michael J. Dunn, Senior Lecturer at the
National Heart and Lung Institute, Harefield, Middlesex, UK, for his
kind engagement of reading the manuscript, correcting the english
and for his excellent and informed advices.

Electrophoresis in Practice, Fourth Edition. Reiner Westermeier
Copyright © 2005 WILEY-VCH Verlag GmbH & Co. KGaA, Weinheim
ISBN: 3-527-31181-5

In this version, some updates have been made to methodological aspects, new experiences, applications, and the references. A new drawing program is used, which allows higher resolution in the explanatory figures.

Leonberg, February 1993   *R. Westermeier*

English version, Second Edition

The author thanks Professor Görg for her tips for the state of the art of high resolution two-dimensional electrophoresis, and Dr. Gabriel Peltre, Institute Pasteur, Paris, for valuable hints on the practice of immunoelectrophoresis, agarose isoelectric focusing, blotting, and titration curves.

This version has been updated in the wording, the way of quoting the references, and in the methodology. A few figures, hints for problem solving, and a few very important references have been added. The main differences to the previous issue, however, are constituted by the addition of the lately developed methods for DNA typing and the methodology for vertical gels. Thus, section II contains now 15 chosen electrophoretic methods.

Freiburg, November 1996   *R. Westermeier*

English version, Third Edition

Three years ago, just when the second issue reached the book stores, "Proteome analysis" became a buzz word. Proteomics seems to become the continuation of the genome sequencing approach with even greater efforts. Many molecular biology laboratories start to work with two-dimensional electrophoresis of proteins now. It was thus necessary to update the book again, this time with the further developed state of the art of 2-D electrophoresis, and a short overview on the proteome analysis methodology. Some space could be gained for this, because image analysis of electrophoresis gels has become much easier in the last few years. Thus its description became much shorter. Also the front picture has been renewed to express the importance of 2-D electrophoresis nowadays. Of course, all the other technical areas have been checked for new developments and updated as well.

Freiburg, July 2000   *R. Westermeier*

English version, Fourth Edition

During the last few years the Proteomics technologies have been further developed and became more robust. A very powerful new method for quantitative 2-D electrophoresis, difference gel electrophoresis (DIGE), has meanwhile been established. It is the only method, which allows sample multiplexing in 2-D gels and enables the use of an internal standard. Therefore its description was added to the Proteomics chapter. Also in some other areas of electrophoretic methodology progresses have been accomplished; they are now included. As with the other editions, the occasion of preparing a new edition was used to revise the entire manuscript again, and to fix some errors. Finally, the author highly appreciates the complete layout refurbishment of the book, which has been carried out to give it a more professional look and improve its readability.

Freiburg, July 2004   *R. Westermeier*

Abbreviations, symbols, units

2-D electrophoresis	Two-dimensional electrophoresis
A	Ampere
acc.	according
A,C,G,T	Adenine, cytosine, guanine, thymine
A/D-transformer	Analog-digital transformer
ACES	N-2-acetamido-2-aminoethanesulfonic acid
AEBSF	Aminoethyl benzylsulfonyl fluoride
AFLP	Amplified restriction fragment length polymorphism
API	Atmospheric pressure ionization
APS	Ammonium persulphate
ARDRA	Amplified ribosomal DNA restriction analysis
AU	Absorbance units
16-BAC	Benzyldimethyl-n-hexadecylammonium chloride
BAC	Bisacryloylcystamine
Bis	N, N'-methylenebisacrylamide
bp	Base pair
BSA	Bovine serum albumin
C	Crosslinking factor [%]
CA	Carbonic anhydrase
CAF	Chemically assisted fragmentation
CAM	Co-analytical modification
CAPS	3-(cyclohexylamino)-propanesulfonic acid
CCD	Charge-coupled device
CDGE	Constant denaturing gel electrophoresis
CHAPS	3-(3-cholamidopropyl)dimethylammonio-1-propane sulfonate
CE	Capillary electrophoresis
CID	Collision induced dissociation
conc	Concentrated

Electrophoresis in Practice, Fourth Edition. Reiner Westermeier
Copyright © 2005 WILEY-VCH Verlag GmbH & Co. KGaA, Weinheim
ISBN: 3-527-31181-5

CM	Carboxylmethyl
CMW	Collagen molecular weight marker
const.	Constant
CTAB	Cetyltrimethylammonium bromide
Da	Dalton
DAF	DNA amplification fingerprinting
DBM	Diazobenzyloxymethyl
DDRT	Differential display reverse transcription
DEA	Diethanolamine
DEAE	Diethylaminoethyl
DGGE	Denaturing gradient gel electrophoresis
DHB	2,5-dihydroxybenzoic acid
DIGE	Difference gel electrophoresis
Disc	Discontinuous
DMSO	Dimethylsulfoxide
DNA	Desoxyribonucleic acid
DPT	Diazophenylthioether
dsDNA	Double stranded DNA
DSCP	Double strand conformation polymorphism
DTE	Dithioerythreitol
DTT	Dithiothreitol
E	Field strength in V/cm
EDTA	Ethylenediaminetetraacetic acid
ESI	Electro spray ionization
EST	Expressed sequence tag
FT-ICR	Fourier transform - Ion cyclotron resonance
GC	Group specific component
h	Hour
HEPES	N-2-hydroxyethylpiperazine-N′-2-ethanane-sulfonic acid
HMW	High Molecular Weight
HPCE	High Performance Capillary Electrophoresis
HPLC	High Performance Liquid Chromatography
I	Current in A, mA
IEF	Isoelectric focusing
IgG	Immunoglobulin G
IPG	Immobilized pH gradients
ITP	Isotachophoresis
kB	Kilobases
kDa	Kilodaltons
KR	Retardation coefficient
LDAO	Lauryldimethylamine-N-oxide
LMW	Low Molecular Weight
M	mass
mA	Milliampere

MALDI	Matrix assisted laser desorption ionization
MEKC	Micellar electrokinetic chromatography
MES	2-(N-morpholino)ethanesulfonic acid
min	Minute
mol/L	Molecular mass
MOPS	3-(N-morpholino)propanesulfonic acid
m_r	Relative electrophoretic mobility
mRNA	messenger RNA
MS	Mass spectrometry
Ms^n	Mass spectrometry with n mass analysis experiments
MS/MS	Tandem mass spectrometry
MW	Molecular weight
NAP	Nucleic Acid Purifier
Nonidet	Non-ionic detergent
NEPHGE	Non equilibrium pH gradient electrophoresis
O.D.	Optical density
P	Power in W
PAG	Polyacrylamide gel
PAGE	Polyacrylamide gel electrophoresis
PAGIEF	Polyacrylamide gel isoelectric focusing
PBS	Phosphate buffered saline
PCR®	Polymerase Chain Reaction
PEG	Polyethylene glycol
PFG	Pulsed Field Gel (electrophoresis)
PGM	Phosphoglucose mutase
pI	Isoelectric point
PI	Protease inhibitor
pK value	Dissociation constant
PMSF	Phenylmethyl-sulfonyl fluoride
PPA	Piperidino propionamide
PSD	Post source dissociation (decay)
PTM	Post-translational modification
PVC	Polyvinylchloride
PVDF	Polyvinylidene difluoride
r	Molecular radius
RAPD	Random amplified polymorphic DNA
REN	Rapid efficient nonradioactive
R_f value	Relative distance of migration
RFLP	Restriction fragment length polymorphism
R_m	Relative electrophoretic mobility
RNA	Ribonucleic acid
RPA	Ribonuclease protection assay
s	Second
SDS	Sodium dodecyl sulfate

SNP	Single Nucleotide Polymorphism
ssDNA	Single stranded DNA
T	Total acrylamide concentration [%]
t	Time, in h, min, s
TBE	Tris borate EDTA
TBP	Tributyl phosphine
TBS	Tris buffered saline
TCA	Trichloroacetic acid
TCEP	Tris carboxyethyl phosphine
TEMED	N,N,N′,N′-tetramethylethylenediamine
TF	Transferrin
TGGE	Temperature gradient gel electrophoresis
TMPTMA	Trimethylolpropane-trimethacrylate[2-ethyl-2(hydrohymethyl) 1,3-propandiol-trimeth-acrylate]
ToF	Time of Flight
Tricine	N,tris(hydroxymethyl)-methyl glycine
Tris	Tris(hydroxymethyl)-aminoethane
U	Volt
V	Volume in L
v	Speed of migation in m/s
v/v	Volume per volume
VLDL	Very low density lipoproteins
W	Watt
w/v	Weight per volume (mass concentration)
ZE	Zone electrophoresis

Part I:
Fundamentals

Electrophoresis in Practice, Fourth Edition. Reiner Westermeier
Copyright © 2005 WILEY-VCH Verlag GmbH & Co. KGaA, Weinheim
ISBN: 3-527-31181-5

Introduction

Electrophoretic separation techniques are at least as widely distributed as chromatographic methods. With electrophoresis a high separation efficiency can be achieved using a relatively limited amount of
equipment. It is mainly applied for analytical rather than for preparative purposes. However, with the advent of new technology like
amplification of DNA fragments with Polymerase Chain Reaction
(PCR®), and highly sensitive and powerful mass spectrometry analysis of proteins and peptides, so called "analytical amounts" of electrophoretically separated fractions can now be further analysed.

The main fields of application are biological and biochemical
research, protein chemistry, pharmacology, forensic medicine, clinical investigations, veterinary science, food control as well as molecular biology. It will become increasingly important to be able to choose
and carry out the appropriate electrophoresis technique for specific
separation problems.

The monograph by Andrews (Andrews 1986) is one of the most
complete and practice-oriented books about electrophoretic methods.
In the present book, electrophoretic methods and their applications
will be presented in a much more condensed form.

Andrews AT. Electrophoresis, theory techniques and biochemical and clinical applications. Clarendon Press, Oxford (1986).

Principle: Under the influence of an electrical field charged molecules and particles migrate in the direction of the electrode bearing
the opposite charge. During this process, the substances are usually
in aqueous solution. Because of their varying charges and masses,
different molecules and particles of a mixture will migrate at different
velocities and will thus be separated into single fractions.

The electrophoretic mobility, which is a measure of the migration
velocity, is a significant and characteristic parameter of a charged
molecule or particle. It is dependent on the pK values of the charged
groups and the size of the molecule or particle. It is influenced by the
type, concentration and pH of the buffer, by the temperature and the
field strength as well as by the nature of the support material. Electrophoretic separations are carried out in free solutions as in capillary

Chrambach A. The practice of quantitative gel electrophoresis. VCH Weinheim (1985).

Mosher RA, Saville DA, Thormann W. The dynamics of electrophoresis. VCH Weinheim (1992).

Electrophoresis in Practice, Fourth Edition. Reiner Westermeier
Copyright © 2005 WILEY-VCH Verlag GmbH & Co. KGaA, Weinheim
ISBN: 3-527-31181-5

and free flow systems, or in stabilizing media such as thin-layer plates, films or gels. Detailed theoretical explanations can be found in the books by Chrambach (1985) and Mosher *et al.* (1992).

The relative mobility is abbreviated as m_r or R_m.

Sometimes the *relative* electrophoretic mobility of substances is specified. It is calculated relative to the migration distance of a standard substance, mostly a dye like bromophenol blue, which has been applied as an internal standard.

There is a fourth method: "Moving Boundary Electrophoresis", which is described on page 9. However this technique has no practical importance anymore.
"Electrophoresis" is a general term for all these methods. Blotting is not seen as a separation, but as a detection method.

Three basically different electrophoretic separation methods are employed in practice nowadays:

a)   Electrophoresis, sometimes called zone electrophoresis (ZE).
b)   Isotachophoresis or ITP
c)   Isoelectric focusing or IEF

The three separation principles are illustrated in Fig. 1.

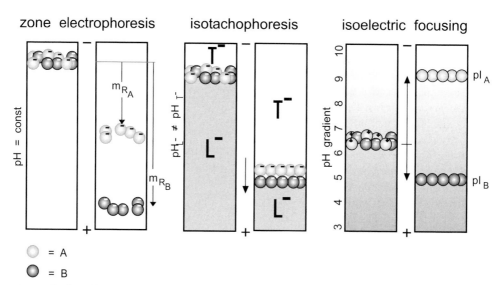

Fig. 1:   The three electrophoretic separation principles. Explanations in the text.
A and B are the components of the sample.

a) In *zone electrophoresis* a homogeneous buffer system is used over the whole separation time and range so as to ensure a constant pH value. The migration distances during a defined time limit are a measure of the electrophoretic mobilities of the various substances. It can be applied to nonamphoteric as well as amphoteric molecules. During the separation diffusion can lead to blurred zones, which reduces the sensitivity of detection and the resolution.

This is also valid for disc electrophoresis, because a discontinuous system exists only at the beginning of the separation and changes into a homogeneous one.

b) In *isotachophoresis* (ITP), the separation is carried out in a discontinuous buffer system. The ionized sample migrates between a leading electrolyte with a high mobility and a terminating – sometimes called trailing – ion with a low mobility, all of them migrating with the same speed. The different components are separated according to their electrophoretic mobilities and form stacks: the substance with the highest mobility directly follows the leading ion, the one with the lowest mobility migrates directly in front of the terminating electrolyte. In ITP there is a concentration regulating effect which works against diffusion.

In comparison to other electrophoretic and chromatographic separation methods, ITP is considered exotic because there are no spaces between the zones: the bands are not "peaks" (Gaussian curves) but "spikes" (concentration dependent bands). ITP is mostly applied for stacking of the samples during the first phase of disc electrophoresis.

c) *Isoelectric focusing* (IEF) takes place in a pH gradient and can only be used for amphoteric substances such as peptides and proteins. The molecules move towards the anode or the cathode until they reach a position in the pH gradient where their net charges are zero. This pH value is the *"isoelectric point"* (pI) of the substance. Since it is no longer charged, the electric field does not have any influence on it. Should the substance diffuse away, it will gain a net charge again, and the applied electric field will cause it to migrate back to its pI. This concentrating effect leads to the name *focusing*. Thus also with IEF there is no problem with diffusion.

In IEF it is important to find the correct place in the pH gradient to apply the sample, since some substances are unstable at certain pH values (see below).

Areas of applications: Mainly proteins, peptides, sugars, and nucleic acids are separated. Electrophoretic methods are used for the qualitative characterization of a substance or mixture of substances, for control of purity, quantitative determinations, and preparative purposes. The most prominent fields are the Genome and the Proteome analysis. The word "Proteome" was introduced by Mark Wilkins during a congress in Sienna 1994, in written form in the publication by Wasinger *et al.* one year later.

Wasinger VC, Cordwell SJ, Cerpa-Poljak A, Yan JX, Gooley AA, Wilkins MR, Duncan MW, Harris R, Williams KL, Humphery-Smith I. Electrophoresis 16 (1995) 1090–1094.

The scope of the applications ranges from whole cells and particles to nucleic acids, proteins, peptides, amino acids, organic acids and bases, drugs, pesticides and inorganic anions and cations – in short – everything that can carry a charge.

Sample application on gels which are immersed in buffer (e.g. vertical and submarine gels) is done with syringes into sample wells polymerized into the gel or into glass tubes, the sample density is raised with glycerol or sucrose.

The sample: An important criterion for the choice of the appropriate electrophoretic method is the nature of the sample to be analyzed. There must be no solid particles or fatty components suspended in the solution. Those interfere with the separation by blocking the pores of the matrix. Sample solutions are mostly centrifuged, sometimes also desalted, before electrophoresis.

Substances which are exclusively negatively or positively charged are easy to run: Examples of such anions or cations are: nucleic acids, dyes, phenols and organic acids or bases. Amphoteric molecules such as amino acids, peptides, proteins and enzymes have positive or negative net charges depending on the pH of the buffer, because they possess acidic as well as basic groups.

In "Proteome analysis", where complex mixtures of several thousand proteins have to be separated in one gel; the sample preparation procedure greatly influences the result.

Proteins and enzymes are often sensitive to certain pH values or buffer substances; conformational changes, denaturation, complex formation, and intermolecular interactions are possible. The concentration of the substances in the solution also plays a role. In particular, when the sample enters the gel, overloading effects can occur when the protein concentration reaches a critical value during the transition from the solution into the more restrictive gel matrix.

For open surfaces as in horizontal systems (e.g. cellulose acetate, agarose gels and automated electrophoresis) either sample applicators are used, or the sample is pipetted into sample wells with a micropipette. Capillary systems usually have an automated sample applicator.

For sodium dodecyl sulphate electrophoresis, the sample must first be denatured; which means it must be converted into molecule-detergent micelles. The method of selective sample extraction, particularly the extraction of not easily soluble substances often determines the nature of the buffer to be used. The nature of the stabilizing medium, e.g. a gel, is dependent on the size of the molecule to be analyzed.

The buffer: The electrophoretic separation of samples is done in a buffer with a precise pH value and a constant ionic strength. The ionic strength should be as low as possible so that both the contribution of the sample ions to the total current and their speed will be high enough.

Yet a minimum buffering capacity is required so that the pH value of the samples analyzed does not have any influence on the system.

During electrophoresis, the buffer ions are carried through the gel just like the sample ions: negatively charged ions towards the anode, positively charged ones towards the cathode. This should be achieved with as little energy as possible so that not much Joule heat is developed.

Elemental charge:
1.602×10^{-19} As;
Avogadro constant:
1.602×10^{23} elemental units per mol.

With the help of the Second Law of Electrolysis by Michael Faraday it is possible to calculate the amount of ions migrating in an electrophoresis experiment: The amounts of electricity are equal to the amount of substances, which are eliminated from different electrolytes. Taking the Avogadro constant and the elemental charge, this means: 1 mAh equals 36.4 µmol.

To guarantee constant pH and buffer conditions the supplies of electrode buffers must be large enough. The use of buffer gel strips or wicks instead of tanks is very practical, though only feasible in horizontal flatbed systems.

For anionic electrophoresis very basic, and for cationic electrophoresis very acidic buffers are used.

In vertical or capillary systems, the pH is very often set to a very high (or low) value, so that as many as possible sample molecules are negatively (or positively) charged, and thus migrate in the same direction.

In these systems the sample is loaded at one end of the separation medium.

When a gel matrix does not contain any ions from polymerization, amphoteric buffers can be applied, which do not migrate during electrophoresis. Such a buffer substance, however, must possess a high buffering capacity at its isoelectric point. For some applications, no buffer reservoirs are necessary with this method.

A polyacrylamide gel, which is covalently bound to a plastic film can be washed after polymerization. See method 4 in this book.

Electroendosmosis: The static support, the stabilizing medium (e.g. the gel) and/or the surface of the separation equipment such as glass plates, tubes or capillaries can carry charged groups: e.g. carboxylic groups in starch and agarose, sulfonic groups in agarose, silicium oxide on glass surfaces. These groups become ionized in basic and neutral buffers: in the electric field they will be attracted by the anode. As they are fixed in the matrix, they cannot migrate. This results in a compensation by the counterflow of H_3O^+ ions towards the cathode: electroendosmosis.

Electroendosmosis is normally seen as a negative effect, yet a few methods take advantage of this effect to achieve separation or detection results (see page 12: MEKC and page 19: counter immunoelectrophoresis).

In gels, this effect is observed as a water flow towards the cathode, which carries the solubilized substances along. The electrophoretic and electroosmotic migrations are then additive (see Fig. 2). The results are: blurred zones, and drying of the gel in the anodal area of flatbed gels.

In capillary electrophoresis mostly the term "electroosmotic flow" is applied, the term "electroendosmosis" is only used in gel electrophoresis.

When fixed groups are positively charged, the electro-osmotic flow is directed towards the anode.

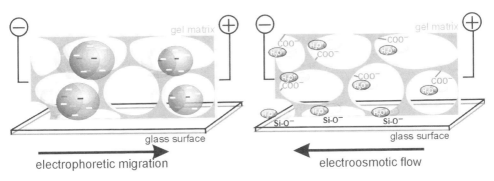

Fig. 2: Electroendosmosis: Negatively charged groups fixed to the gel matrix or to a surface cause the flow of water ions. This results in a water transport into the opposite direction of the electrophoretic migration of the sample ions, leading to blurred band pattern.

1
Electrophoresis

1.0
General

Electrophoresis in free solution

Moving boundary electrophoresis: Arne Tiselius (1937) developed the moving boundary technique for the electrophoretic separation of substances, for which, besides his work on adsorption analysis, he received the Nobel prize in 1948. The sample, a mixture of proteins for example, is applied in a U-shaped cell filled with a buffer solution and at the end of which electrodes are immersed. Under the influence of the applied voltage, the compounds will migrate at different velocities towards the anode or the cathode depending on their charges. The changes in the refractive index at the boundary during migration can be detected at both ends on the solution using Schlieren optics.

Tiselius A. Trans Faraday Soc. 33 (1937) 524–531.

Nowadays moving boundary electrophoresis in free solution is mainly used in fundamental research to determine exact electrophoretic mobilities.

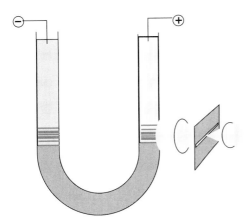

Fig. 3: Moving boundary electrophoresis in a U-shaped cell according to Tiselius. Measurement of the electrophoretic mobility with Schlieren optics.

Hannig K. Electrophoresis 3 (1982) 235–243.

Free flow electrophoresis: in this technique developed by Hannig (1982) a continuous stream of buffer flows perpendicular to the electrical field through a buffer film between two cooled glass plates which is 0.5 to 1.0 mm wide. At one end the sample is injected at a defined spot and at the other end, the fractions are collected in an array of tubes.

This is the only continuous electrophoretic separation method.

The varying electrophoretic mobilities perpendicular to the flow lead to differently heavy but constant deviations of the components so that they reach the end of the separation chamber at different though stable positions (see Fig. 4).

Wagner H, Kuhn R, Hofstetter S. In: Wagner H, Blasius E. Ed. Praxis der elektrophoretischen Trennmethoden. Springer Verlag, Heidelberg (1989) 223–261.

Unfortunately electrophoresis in free solution cannot yet be applied on an industrial scale. The upscaling of the instrumentation is limited by the thermal convection which results from the insufficient dissipation of Joule heat from the flowing electrolyte. Loading cannot be freely increased because highly concentrated samples begin to sediment. Both these limiting factors occur only under gravity. Since 1971, ever since Apollo 14, experiments have been conducted in space to try and develop production in an orbital station.

Besides the separation of soluble substances, this technique is also used for the identification, purification and isolation of cell organelles and membranes or whole cells such as erythrocytes, leukocytes, tissue cells, the causal agent of malaria and other parasites (Hannig 1982, Wagner *et al.* 1989).This method is very effective since even minimal differences in the surface charge of particles and cells can be used for separation.

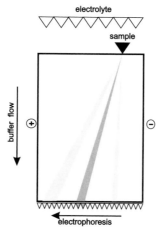

Fig. 4: Schemtic drawing of a continuous free flow electrophoresis system. According to Wagner *et al.* (1989)

Jorgenson JW, Lukacs KD. Anal Chem 53 (1981) 1298–1302

Capillary electrophoresis (CE): this technique is being used increasingly for analytical and micropreparative electrophoresis (Jorgenson and Lukacs, 1981; Hjertén, 1983): as for HPLC, the abbreviation HPCE for High Performance Capillary Electrophoresis is often used. Separation is carried out in a fused silica capillary 20 to 30 cm long and with an internal diameter of 50 to 100 μm. Both ends are immersed in a buffer container into which the electrodes are built (see Fig. 5).

Hjertén S. J Chromatogr. 270 (1983) 1–6.
Fused silica capillaries are otherwise used in gas chromatography.

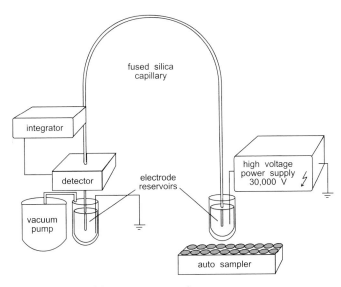

Fig. 5: Example of the instrumentation for capillary electrophoresis.

The amount of chemicals and sample needed is very low. The volume of injected material is usually not more than 2 – 4 nL, nanograms of sample material is required.

Field strengths of up to 1 kV/cm and currents of 10 to 20 mA are used; for this reason a power supply which can yield voltages up to 30 kV is needed. Joule heat can be dissipated very effectively from these thin capillaries with a fan.

CE separations typically take 10 to 20 min. There are many detection methods possible: UV/VIS, fluorescence, conductivity, electrochemistry etc. In most applications the fractions are detected by UV measurement at 280, 260 or in some cases even 185 nm directly in the capillary.

For some substances and applications the limit of detection can go as low as to the atto-mole level.

In general the results are then further processed by HPLC interpretation software on personal computers.

To prevent adsorption of components on the surface of the capillary and electro-osmotic effects, the inside of the capillary can be coated with linear polyacrylamide or methyl cellulose. Capillary electrophoresis instruments can be used for all three of the separation methods: electrophoresis, isotachophoresis and isoelectric focusing. Even an additional new method, a hybrid of electrophoresis and chromatography, has been developed:

The buffer used depends on the nature of the separation: e.g. 20 to 30 mmol/L sodium phosphate buffer pH 2.6 for electrophoresis of peptides.

Terabe S, Otsuka K, Ichikawa K, Tsuchiya A, Ando T. Anal Chem.64 (1984) 111–113
Terabe S, Chen N, Otsuka K. In Chrambach A, Dunn M, Radola BJ. Eds. Advances in Electrophoresis 7. VCH Weinheim (1994) 87–153.

Micellar electrokinetic chromatography (MEKC) introduced by Terabe *et al.* (1984). It is the only electrophoretic method, which can separate neutral as well as charged compounds. Surfactants are used at concentrations over the the critical micelle concentration. The charged micelles migrate in the opposite direction to the electro-osmotic flow created by the capillary wall. The electro-osmotic counter-flow is faster than the migration of the micelles. During migration, the micelles interact with the sample compounds in a chromatographic manner through both hydrophobic and electrostatic interactions.It has become one of the most widely used CE methods. More details on this method are found in a review by Terabe *et al.* (1994).

However, the investment for such an instrument is by far higher compared to a gel electrophoresis equipment.

One great advantage of capillary electrophoresis lies in its automation. Every step can be controlled by semiautomatic or full automatic instrumentation. An autosampler is a standard part of this equipment.

Another advantage is the possibility of linking with other analytical instruments either before electrophoresis: HPLC/HPCE or after: HPCE/MS.

In contrast to Reversed Phase Chromatography proteins are not damaged during HPCE and, in addition, the resolution is better.

For preparative separations a fraction collector is attached to the UV detector. The identification of the individual substances is done by the relative mobility or the molecular weight, or else the collected fractions are analyzed.

Cohen AS, Karger BL. J Chromatogr. 397 (1987) 409–417.

See also page 27

For molecular weight separations of proteins, peptides,and nucleic acids capillaries filled with linear (non crosslinked) polyacrylamide gel are used (Cohen *et al.* 1987).

The most successful application of capillary electrophoresis is the separation of DNA fragments. Because of the possibility of automation and the repeated use of a high number of capillaries in parallel, this technique is ideal for high throughput DNA sequencing. It had been predicted, that the complete knowledge of the human genome would be available in the year 2005. But the introduction of the new multi-capillary sequencers has speeded up the Human Genome Project considerably.

Electrophoresis in supporting media

The instructions in the second part are limited to electrophoresis in supporting media since these techniques only require minimal equipment.

Compact material such as paper, films or gels are used. So as to monitor the progress of the separation and to recognize the end of the run, dyes with a high electrophoretic mobility are applied together with the sample.

For separation of proteins in anodal direction Bromophenol Blue, Xylenecyanol or Orange G are used, in the cathodal direction Bromocresol Green, Pyronine or Methylene Blue.

Detection of the separated zones can either be done directly in the medium by positive staining with Coomassie blue or silver, negative staining with zinc-imidazole, spraying with specific reagents, enzyme substrate coupling reactions, immuno precipitation, autoradiography, fluorography, or indirectly by immunoprinting or blotting methods. A comprehensive survey on enzyme staining methods has been published by Rothe (1994).

Blotting: transfer to immobilizing membranes followed by staining or specific ligand binding.

Rothe G. Electrophoresis of enzymes. Springer Verlag, Berlin, (1994).

Paper and thin-layer electrophoresis: These methods have mostly been abandoned in profit of gel electrophoresis, because of improved separation and the higher loading capacity of agarose and polyacrylamide gels. Electrophoretic separations on thin-layer silica gel plates linked to buffer tanks are only carried out for the analysis of polysaccharides of high molecular weight and lipopolysaccharides, which would obstruct the pores of the gels (Scherz, 1990).

Scherz H. Electrophoresis 11 (1990) 18–22.

Cellulose acetate membrane electrophoresis: cellulose acetate membranes have large pores and therefore hardly exert any sieving effect on proteins (Kohn, 1957). This means that these electrophoretic separations are entirely based on charge density.

Kohn J. Nature 180 (1957) 986-988.

The matrix exerts little effect on diffusion so that the separated zones are relatively wide while the resolution and limit of detection are low. On the other hand they are easy to handle and separation and staining are rapid. The cellulose acetate strips are suspended in the tank of a horizontal apparatus, so that both ends dip in the buffer; no cooling is necessary during separation. This technique is widely used for routine clinical analysis and related applications for the analysis of serum or isoenzymes.

Because the resolution and reproducibility of separations in agarose and polyacrylamide gels are better, cellulose acetate membranes are more and more often replaced by gel electrophoresis.

Gel electrophoresis

The gel: The gel matrix should have adjustable and regular pore sizes, be chemically inert and not exhibit electroendosmosis. Vertical cylindrical gel rods or plates as well as horizontal gel slabs are employed, the latter being usually cast on to stable support film to facilitate handling (Fig. 6).

The use of gel rods has become very rare, because of the complicated handling and difficult pattern comparison. Slab gels for vertical and flatbed systems can be easily polymerized in the laboratory, but prefabricated gels of many different types are available from various suppliers. An overview over the features, benefits and drawbacks of vertical and horizontal slab gel systems is given in Table 1.

The most instructions in the second part are describing horizontal gels on support films since these can be used for all applications and with universally applicable equipment.

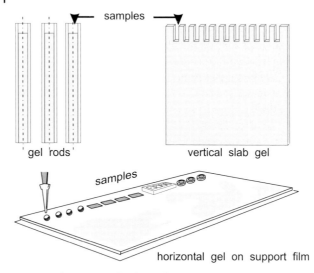

Fig. 6: Gel geometries for electrophoretic separations.

Tab. 1: Comparison of flatbed and vertical gel systems

	Flatbed Systems:	Vertical Systems:
Up to 3 mm thick gels can be used in vertical systems.	Gel thickness is limited, because cooling is only possible from one side	Higher protein loading capacity, because thicker gels can be used, which are cooled from both sides
Also, on flatbed systems mostly gels on film supports are employed. But films can be removed.	One gel per instrument is run	Blotting is easier because of higher gel thickness Multiple gel runs possible
In IEF, samples very often have to be loaded inside the pH gradient.	Very versatile for different methods, ideal for isoelectric focusing	Limited technical possibilities, not optimal for isoelectric focusing
Very thin gels show higher sensitivity of detection, and are easier and quicker to stain.	Thin layers can easily be used, easy sample application	The thinner the gel, the more complicated is sample application
The buffer strip concept reduces chemical and radioactive liquid waste considerably (Kleine et al. 1992).	Buffer strips (polyacrylamide or filter paper) can be used instead of large volumes of liquid buffers	
On flatbed systems mostly gels on film supports are employed.	Easy to handle and to clean, no glass plates necessary, thus ideal for routine applications	Many pieces to set up and to clean
In a flatbed system the buffers can not leak into each other.	Higher electric safety	

Starch gels were introduced by Smithies (1955) and are prepared from hydrolyzed potato starch which is dissolved by heating and poured to a thickness of 5 to 10 mm. The pore size can be adjusted by the starch concentration of the solution. Because of the low reproducibility and the impractical handling these gels have been largely replaced by polyacrylamide gels.

Smithies O. Biochem J. 61 (1955) 629–641.

Starch is a natural product whose properties can vary greatly.

Agarose gels are mostly used when large pores for the analysis of molecules over 10 nm in diameter are needed. Agarose is a polysaccharide obtained from red seaweed.

By removal of the agaropectin, gels of varying electroendosmosis and degrees of purity can be obtained. They are characterized by their melting point (35 °C to 95 °C) and the degree of electroendosmosis (m_r).

m_r is dependent on the number of polar groups left. The definition is the same like for relative electrophoretic mobility.

The pore size depends on the concentration of agarose: one usually refers to the weight of agarose and the volume of water. The unavoidable losses of water which occur during heating can vary from batch to batch, so in practice, this value cannot be absolutely exact. In general gels with a pore size from 150 nm at 1% (w/v) to 500 nm at 0.16% are used.

For pore diameters up to 800 nm (0.075% agarose): Serwer P. Biochemistry 19 (1980) 3001–3005.

Agarose is dissolved in boiling water and then forms a gel upon cooling. During this process double helixes form which are joined laterally to form relatively thick filaments (Fig. 7).

For DNA separations 1 to 10 mm thick gels are cast on UV-transparent trays, because the bands are usually stained with fluorescent dyes: Ethidium bromide or SYBR Green.

The gels are run under buffer in order to prevent drying out due to electroendosmosis.

For protein electrophoresis the gels are made by coating horizontal glass plates or support films with a solution of agarose. The thickness of the gel – usually 1 – 2 mm – is determined by the volume of the solution and the surface it covers.

Very even gel thicknesses are obtained by pouring the solution in prewarmed molds.

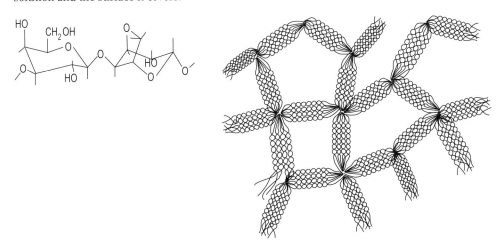

Fig. 7: Chemical structure of agarose and structure of the polymers during gel formation.

Kerenyi L, Gallyas F. Clin Chim Acta 38 (1972) 465–467.

The separated protein bands are mostly detected by Amido Black or Coomassie Brilliant blue staining of the gels after drying them. In order to improve the protein detection limit, the first silver staining technique had been developed for agarose gels to detect oligoclonal IgGs in cerebrospinal fluid (Kerenyi and Gallyas, 1972)

Raymond S. Weintraub L. Science. 130 (1959) 711–711.
The reaction is started with ammonium persulphate as catalyst, TEMED provides the tertiary amino groups to release the radicals.

Polyacrylamide gels were first used for electrophoresis by Raymond and Weintraub (1959). They are chemically inert and mechanically stable. By chemical co-polymerization of acrylamide monomers with a cross-linking reagent – usually N,N'-methylenebisacrylamide (Fig. 8) – a clear transparent gel which exhibits very little electroendosmosis is obtained.

Hjertén S. Arch Biochem Biophys Suppl 1 (1962) 147.

The pore size can be exactly and reproducibly controlled by the total acrylamide concentration T and the degree of cross-linking C (Hjertén, 1962):

$$T = \frac{a+b}{V} \times 100[\%], \ C = \frac{b}{a+b} \times 100[\%]$$

a is the mass of acrylamide in g,
b the mass of methylenebisacrylamide in g, and
V the volume in mL.

Gels with C > 5 % are brittle and relatively hydrophobic. They are only used in special cases.

When C remains constant and T increases, the pore size decreases. When T remains constant and C increases, the pore size follows a parabolic function: at high and low values of C the pores are large, the minimum being at $C = 5\%$.

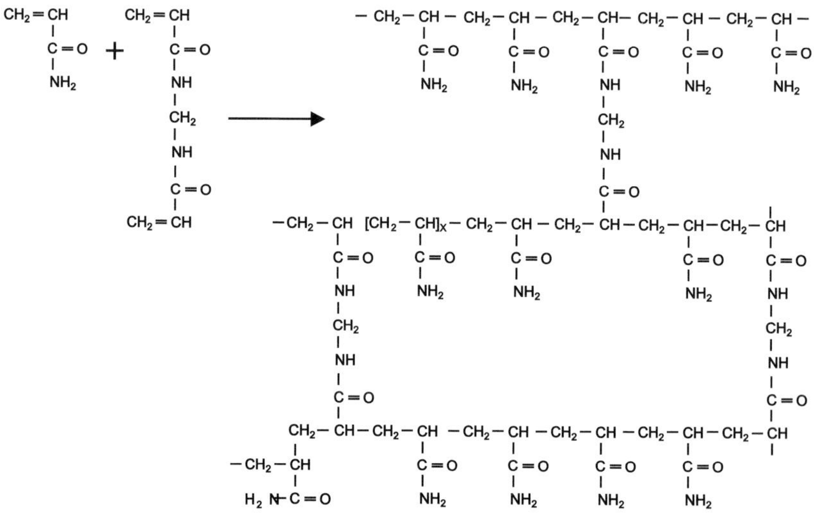

Fig. 8: The polymerization reaction of acrylamide and methylenebisacrylamide.

Besides methylenebisacrylamide a number of other cross-linking reagents exist, they have been listed and compared by Righetti (1983). N,N′-Bisacryloylcystamine is mentioned here, it possesses a disulfide bond which can be cleaved by thiol reagents. Because of this, it is possible to solubilize the gel matrix after electrophoresis.

Righetti PG. Isoelectric focusing: theory, methodology and applications. Elsevier Biomedical Press, Amsterdam (1983).

Polymerization should take place under an inert atmosphere since oxygen can act as a free radical trap. The polymerization is temperature dependent: to prevent incomplete polymerization the temperature should be maintained above 20 °C.

The monomers are toxic and should be handled with precaution.

To minimize oxygen absorption gels are usually polymerized in vertical casting chambers: cylindrical gels in glass tubes and flat gels in moulds formed by two glass plates sealed together around the edges.

With horizontal casting oxygen intake is increased. That must be compensated by a higher amount of catalyst, often leading to problems during separation.

For electrophoresis in vertical systems the gel in glass rods or cassettes are placed into the buffer tanks, and are in direct contact with the electrode buffers. Gels for flatbed systems are polymerized on a support and removed from the mould before use.

For sample application wells are formed at the upper edge of the gel during polymerization (see Fig. 6). These are made by insertion of a sample comb between the glass plates. In horizontal gels, sample wells are not always necessary; the samples can be applied directly on the surface with strips of filter paper or silicone rubber.

In homogeneous buffer systems, narrow sample slots on the surface of horizontal gels are also important to obtain good results.

The various gel electrophoresis methods can be divided into those in restrictive and non-restrictive media. Restrictive gels work against diffusion so the zones are more distinctly separated and better resolved than in non-restrictive gels. The limit of detection is thus increased.

In restrictive gels, the molecule size has a major influence on the result of the separation.

1.1
Electrophoresis in non-restrictive gels

For these techniques the frictional resistance of the gel is kept negligibly low so that the electrophoretic mobility depends only on the net charge of the sample molecule. Horizontal agarose gels are used for high molecular weight samples such as proteins or enzymes and polyacrylamide gels for low molecular weight peptides or polyphenols.

1.1.1
Agarose gel electrophoresis

Zone electrophoresis
Agarose gels with concentrations of 0.7 to 1% are often used in clinical laboratories for the analysis of serum proteins. The separation

times are exceedingly low: about 30 min. Agarose gels are also used for the analysis of isoenzymes of diagnostic importance such as lactate dehydrogenase (Fig. 9) and creatine kinase.

Because of their large pore size, agarose gels are especially suited to specific protein detection by immunofixation: after electrophoresis the specific antibody is allowed to diffuse through the gel. The insoluble immunocomplexes formed with the respective antigen result in insoluble precipitates and the non-precipitated proteins can be washed out. In this way only the desired fractions are detected during development.

Immunoprinting functions in a similar way: after the electrophoretic separation, an agarose gel containing antibodies or a cellulose acetate membrane impregnated with antibodies is placed on the gel. The antigens then diffuse towards the antibodies and the identification of the zone is done in the antibody-containing medium. Immunoprinting is mainly used for gels with small pores.

Besides immunofixing and immunoprinting, immunoblotting also exists for protein identification: immobilizing membranes, for example nitrocellulose, are used on the surface of which the proteins are adsorbed, see "blotting" on page 67 ff.

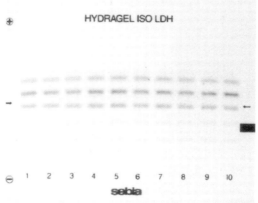

Fig. 9: Agarose electrophoresis of lactate dehydrogenase isoenzymes. Specific staining with the zymogram technique.

Immunoelectrophoresis

The principle of immunoelectrophoresis is the formation of precipitate lines at the equivalence point of the antigen and its corresponding antibody. In this method it is important that the ratio between the quantities of antigen and antibody be correct (antibody titer).

When the antibody is in excess, statistically at most one antigen binds to each antibody while when the antigen is in excess at most one antibody binds to each antigen. Yet at a specific antigen/antibody ratio (equivalence point) huge macromolecules are formed.

They consist of an antigen-antibody-antigen-antibody-… sequence and are immobilized in the gel matrix as an immunoprecipitate. The white precipitate lines are visible in the gel and can be revealed with protein stains. The method is specific and the sensitivity very high because distinct zones are formed. Immunoelectrophoresis can be divided into three principles (Fig. 10):

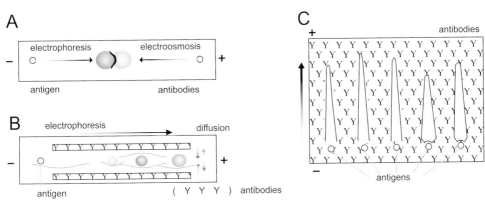

Fig. 10: The three principles of immunoelectrophoresis A, B and C, see text for details.

A. *Counter immunoelectrophoresis* according to Bussard (1959): in an agarose gel exhibiting high electroendosmosis, the buffer is set at a pH about 8.6 so that the antibody does not carry any net charge. The sample and the antibody are placed in their respective wells and move towards each other: the charged antigens migrate electrophoretically and the antibodies are carried by the electro-osmotic flow.
Bussard A. Biochim. Biophys Acta. 34 (1959) 258–260.

B. *Zone electrophoresis/immunodiffusion* according to Grabar and Williams (1953): first a zone electrophoresis is run in an agarose gel, followed by the diffusion of the antigen fraction towards the antibody which is pipetted into troughs cut in the side parallel to the electrophoretic run.
Grabar P, Williams CA. Biochim Biophys Acta. 10 (1953) 193.

C. The *"rocket"* technique according to Laurell (1966) and the related methods: antigens migrate in an agarose gel which contains a definite concentration of antibody. As in method A the antibodies are not charged because of the choice of the buffer. As the sample migrates one antibody will bind to one antigen until the ratio of concentrations corresponds to the equivalence point of the immunocomplex.
Laurell CB. Anal Biochem. 15 (1966) 45–52.

The result is that rocket shaped precipitation lines are formed, the enclosed areas are proportional to the concentration of antigen ion in the sample. A series of modifications to this technique exist, including two-dimensional ones.

Affinity electrophoresis

Bøg-Hansen TC, Hau J. J
Chrom Library. 18 B (1981)
219–252.

This is a method related to immunoelectrophoresis which is based on the interactions between various macromolecules for example lectin-glycoprotein, enzyme-substrate and enzyme-inhibitor complexes (Bøg-Hansen and Hau, 1981)

All the techniques known from immunoelectrophoresis can be employed. For example, specific binding lectin collected worldwide from plant seeds are examined with line affinity electrophoresis. In this way carbohydrate changes in glycoproteins during different biological processes can be identified. In Fig. 11 an application of affinity electrophoresis to differentiate between alkaline phosphatase of liver and bone is shown.

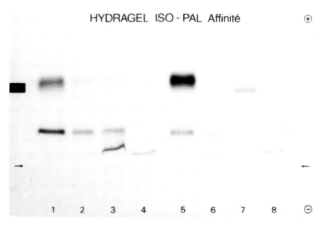

Fig. 11: Affinity electrophoresis of isoenzymes of alkaline phosphatase from the liver and the bones. The wheat germ agglutinin specifically binds the bone fraction which is recognizable as a characteristic band close to the application point. Alkaline phosphatase staining.

1.1.2
Polyacrylamide gel electrophoresis of low molecular weight substances

See method 1
*According to the guide-lines
of the SI, the use of the term
Dalton for 1.6601 × 10⁻²⁷ kg
is no longer recommended.
However it is still a current
unit in biochemistry.

Since low molecular weight fractions cannot be chemically fixed in the matrix, horizontal ultra-thin layer polyacrylamide gels on film supports are used. Those are dried at 100 °C immediately after electrophoresis and then sprayed with specific reagents. With this method for example, dyes with molecular weights of approximately 500 Da* can be separated.

1.2
Electrophoresis in restrictive gels

1.2.1
The Ferguson plot

Although during electrophoresis in restrictive gels, electrophoretic mobility depends both on net charge and on molecular radius this method can also be used for the physico-chemical analysis of proteins. The principle was formulated by Ferguson (1964): the samples are separated under identical buffer, time and temperature conditions but with different gel concentrations (g/100 mL for agarose, %T for polyacrylamide). The distances traveled will vary: m_r is the relative mobility. A plot of log 10 m_r versus the gel concentration yields a straight line.

Ferguson KA. Metabolism. 13 (1964) 985–995.

The slope (see Fig. 12) is a measure of the molecular size and is called the retardation coefficient K_R.

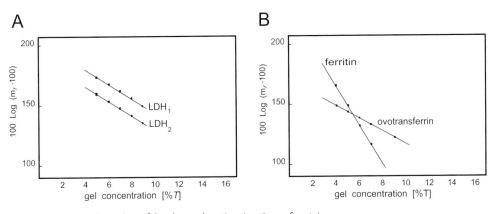

Fig. 12: Ferguson plots: plots of the electrophoretic migrations of proteins versus gel concentrations. (A) Lactate dehydrogenase isoenzymes; (B) Different proteins. See text for further details.

For globular proteins there is a linear relationship between K_R and the molecular radius r (Stokes radius), so the molecular size can be calculated from the slope of the plot. Once the free mobility and the molecular radius are known the net charge can also be calculated (Hedrick and Smith, 1968). For protein mixtures the following deductions can be made according to the appearance of the plots:

Hedrick JL, Smith AJ. Arch Biochem Biophys. 126 (1968) 155–163.

- The lines are parallel: The proteins have the same size but different mobilities e.g. iosenzymes. *Fig. 12A!*
- The slopes are different but the lines do not cross: the protein corresponding to the upper curve is smaller and has a higher net charge.

Fig. 12B!

Same net charge, different molecular sizes.

- The lines cross beyond *T*=2%: the larger protein has the higher charge density and intercepts the y-axis at a higher value.
- Several lines cross at a point where *T*< 2%: these are obviously the various polymers of one protein.

1.2.2
Agarose gel electrophoresis

Proteins

Jovin TM, Dante ML, Chrambach A. Multiphasic buffer systems output. Natl Techn Inf Serv. Springfield VA USA PB (1970) 196 085–196 091.

Since highly concentrated agarose gels above 1% (1g/100 mL agarose in water) are cloudy and the electro-osmotic flow is high, agarose gels are only used for the separation of very high molecular weight proteins or protein aggregates. Since agarose gels do not contain catalysts which can influence the buffer system, they have also been used to develop a series of multiphasic discontinuous buffer systems (Jovin *et al.* 1970).

Nucleic acids

Maniatis T, Fritsch EF, Sambrook J. Molecular cloning a laboratory manual. Cold Spring Laboratory (1982). Rickwood D, Hames BD. Gel electrophoresis of nucleic acids. IRL Press Ltd. (1982).

Perlman D, Chikarmane H, Halvorson HO. Anal Biochem. 163 (1987) 247–254.

These dyes have to be handled with care, because they are mutagens.

Agarose electrophoresis is the standard method for separation, DNA restriction fragment-analysis and purification of DNA and RNA fragments (Maniatis *et al.* 1982; Rickwood and Hames, 1982). The fragment sizes analysed are in the range between 1,000 and 23,000 bp. Horizontal "submarine" gels are used for these nucleic acid separations: the agarose gel lies directly in the buffer (Fig. 13). This prevents the gel from drying out.

When a narrow pore size gel is required, agarose can be partially substituted by polysaccharides (Perlman *et al.* 1987).

The gels are stained with fluorescent dyes like Ethidium bromide or SYBR Green, and the bands are visible under UV light. Their sensitivites ranges are between 100 pg and 1 ng / band. Because they are intercalating in the helix, the sensitivity is dependent on the size of the DNA fragment and is lower for RNA detection.

For RFLP (restriction fragment length polymorphism) analysis, the separated DNA fragments are transferred onto an immobilizing membrane followed by hybridization with radiolabelled probes (s. 4 Blotting).

For a permanent record, mostly instant photos had been taken from the gels in a darkroom. Video documentation systems take the images inside a box, print the results on thermopaper, or feed them to a computer.

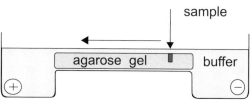

Fig. 13: The "submarine" technique for the separation of nucleic acids.

Pulsed field gel electrophoresis

For chromosome separation, pulsed field electrophoresis (PFG) according to Schwartz and Cantor (1984) is used; it is a modified submarine technique.

Schwartz DC, Cantor CR. Cell. 37 (1984) 67–75.

High molecular weight DNA molecules over 20 kb align themselves lengthwise during conventional electrophoresis and migrate with the same mobility so that no separation is achieved.

kb kilobases

In PFG the molecules must change their orientation with changes in the electric field, their helical structure is first stretched and then compressed. The "viscoelastic relaxation time" is dependent on the molecular weight. In addition, small molecules need less time to reorient themselves than large ones. This means that after renewed stretching and reorientation, larger molecules have – for a defined pulse – less time left for actual electrophoretic migration. The resulting electrophoretic mobility thus depends on the pulse time or on the duration of the electric field: a separation according to the molecular weight up to the magnitude of 10 megabases is obtained.

For shorter DNA fragments the resolution with PFG is also better than with conventional submarine electrophoresis.

For the analysis of chromosomes, the sample preparation including cell disintegration, is done in agarose blocks which are placed in

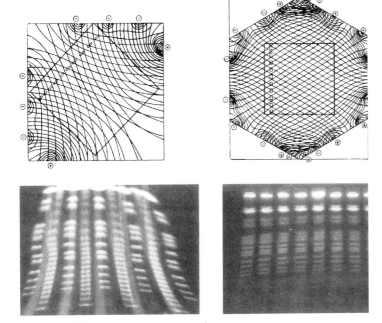

Fig. 14: Field lines and separation results for two types of PFG electrophoresis: *left* orthogonal doubly inhomogeneous fields and *right* homogeneous fields for hexagonally arranged point electrodes.

the pre-formed sample pockets. These molecules would be broken by the shear forces. 1.0 to 1.5% agarose gels are used for the separation.

The electric fields should have an angle of at least 110° relative to the sample. This is obtained for example by an inhomogeneous field with point electrodes mounted on orthogonal rails or in hexagonal configuration. The pulse time is of 1 s to 90 min for these techniques, depending on the length of the DNA molecules to be separated. Large molecules are better separated when the pulse time is long, small molecules needs short pulse times. The separations can last several days.

Fig. 14 shows the field lines for an orthogonal configuration with an inhomogeneous field and for an hexagonal configuration with a homogeneous field as well as the corresponding separations.

Pulsed field gel electrophoresis is mainly employed for research, but it has also found its place in routine analysis for bacterial taxonomy.

There are in addition other field geometries:
Field Inversion *(FI) electrophoresis: the electric field is pulsed back and forth in one direction.*
Transverse Alternating Field *electrophoresis (TAFE): The gel is mounted vertically in an aquarium-like tank and the field is pulsed back and forth between electrode pairs mounted on the top and the bottom of both sides of the gel.*

1.2.3
Polyacrylamide gel electrophoresis of nucleic acids

DNA sequencing

In the DNA sequencing methods according to Sanger and Coulson (1975) or Maxam and Gilbert (1977), the last step is electrophoresis in a polyacrylamide gel under denaturing conditions. The four reactions – containing variously long fragments of the DNA strand to be analyzed, each terminating with a specific base – are separated one beside the other. Determination of the order of the bands in these four lanes from the bottom to the top of the gel yields the base sequence, that is, the genetic information.

Tris-borate EDTA (TBE) buffer is used. To completely denature the molecule, the process is usually carried out at a temperature over 50 °C and in the presence of urea. Irregular heat distribution results in the *"smiling"* effect, when the bands are turned up at the ends. For this reason, it has proved effective to prewarm the gels with thermoplates independent from the electric field.

Manual sequencing: in the manual technique the bands are mostly revealed by autoradiography. Nucleotides or primers labelled with ^{32}P or ^{35}S are used. The gels are usually thinner than 0.4 mm since they must be dried for autoradiography.

Alternative nonradioactive detection methods have been developed:

Sanger F, Coulson AR. J Mol Biol. 94 (1975) 441–448. Maxam AM, Gilbert W. Proc Natl AcadSci USA. 74 (1977) 560–564.

Smiling effect: When the temperature in the middle of the gel is higher than at the edges the DNA fragments migrate faster.

*In practice, **vertical** gel slabs are used, which are – in most cases – heated by the electric field. An aluminum plate behind one of the glass plates distributes the heat evenly.*

This requires biotinylated or fluorescent primers, nucleotides or probes.

- Chromogenic or chemiluminescent detection on a membrane after the separated DNA fragments have been transferred from the gel.

- Silver staining of the gel.

This requires cycle sequencing.

The use of wedge shaped gels has proved useful: they generate a field strength gradient which induces a compression of the band pattern in the low molecular weight area and enables the analysis of substantially more bases in one gel.

The samples are introduced in sample wells (formed in the gel by a sample comb during polymerization) with microcapillaries or syringes with an extra thin needle. A typical sequencing autoradiogram is shown in Fig. 15.

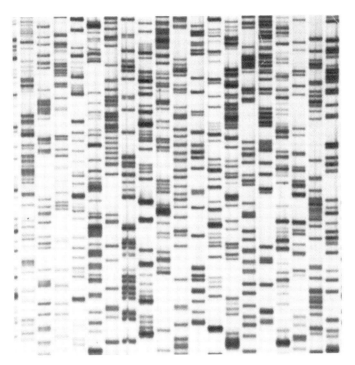

Fig. 15: Autoradiogram of DNA sequencing.

Automated sequencing: samples with fluorescent tags are used. There are two principles:

Also here almost exclusively vertical slab gels are employed.

1. *Single track system:* for the four necessary reactions – with the base endings A, C, G, T – four different fluorescent markers are used. For separation, the four reagents are applied on the gel and the zones which migrate in one track are measured with selective photodetectors.

*Ansorge W, Sproat BS,
Stegemann J, Schwager C.
J Biochem Biophys Methods.
13 (1986) 315–323.*

*Since the introduction of the
Cy5 label, a red laser can be
employed.*

2. *Four track system:* This principle is based on the traditional Sanger method (Sanger and Coulson 1975). Only one dye is used, for example fluoresceine, which is used to mark the primer. The samples are separated in four tracks per clone. A fixed laser beam constantly scans the whole width of the gel in the lower fifth of the separation distance. At this height, a photovoltaic cell is fixed to the glass plate behind each band. When the migrating bands reach that spot, the fluorescent DNA fragments will be excited and emit a light signal (Ansorge *et al.* 1986). Since a single photo cell corresponds to each band, the migrating bands will be registered one after the other by the computer, giving the sequence. In one track systems, the raw data must be processed so that the mobility shifts due to the different markers are compensated. In four track systems, the sequence can be recognized directly from the raw data (see Fig. 16).

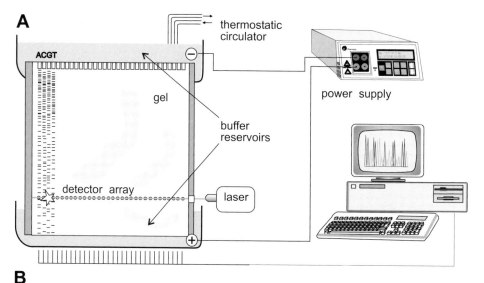

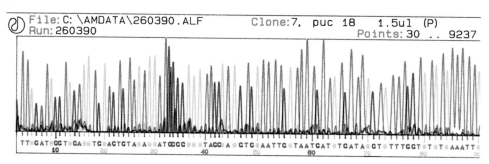

Fig. 16: (A) Instrumentation for automated DNA sequencing with a four track system; (B) Typical trace after treatment of the crude data by a computer.

Automated sequencing has many advantages over the manual technique:

- Since fluorescent markers are used, the use of radioactivity in the laboratory can be avoided. *No need for isotope laboratory.*
- Neither extensive treatment of the gel after separation nor time-consuming autoradiography are necessary.
- The laborious reading of the bands becomes unnecessary.
- The sequences are directly fed into the computer.
- The reactions labelled with the fluorescent label can easily be kept for a long time, so that the separation can be repeated later in case of doubt.
- The high sensitivity of fluorescent labelling also allows the sequencing of cosmids and lambda DNA as well as the products of the polymerase chain reaction PCR®*). In addition restriction analyses can be carried out.

*) The PCR process is covered by U.S. patents 4,683,195 and 4,683,302 owned by Hoffman-La Roche Inc. Use of the PCR process requires a license.

This "on-line electrophoresis" setup can also be employed for various DNA typing methods.

For high throughput genome sequencing, multi-capillary instruments have replaced the slabgel technique. The capillaries are usually filled with linear – non-crosslinked – polyacrylamide. The entire procedure, including sample application is automated. *Those are the biggest and most expensive electrophoresis instruments existing.*

DNA typing

Many new techniques and applications have recently been developed in this field. Because those are almost exclusively based on PCR® technology, the size range of the DNA fragments to be analysed lies between 50 and 1,500 bp. In this range the sensitivity and resolution of agarose electrophoresis with Ethidium bromide staining is coming to its limits, because the gel pores are too large for proper sieving and the intercalating fluorescent dyes are much less sensitive than for larger fragments. *Amplification of fragments larger than 1,500 bp is possible, however, with a lot of problems with reproduciblity.*

PAGE and silver staining:

The use polyacrylamide gels leads to much sharper bands and higher resolution; with subsequent silver staining a sensitivity of 15 pg per band can be achieved (Bassam *et al.* 1991). Vertical and horizontal slab gels can be used. Whereas in agarose electrophoresis the mobilities of DNA fragments are solely proportional to their sizes, the band positions in polyacrylamide gels are partly influenced by the base sequence as well. A and T rich fragments migrate slower than others. *Bassam BJ, Caetano-Annollés G, Gresshoff PM. Anal Biochem. 196 (1991) 80–83. Silver staining of DNA is much easier than of proteins, because fixation is very easy.*

Silver stained DNA bands can be directly reamplified after scratching them out of the gel without intermediate purification. About 20 % of the DNA molecules of a band remain undestroyed by the silver staining procedure. They are locked inside the stained band, thus DNA fragments do not contaminate the gel surface during staining. *Reamplification of DNA works only, when silver staining techniques specially designed for DNA detection are employed.*

Görg A, Postel W, Wester-meier R, Gianazza E, Righetti PG, J Biochem Biophys Methods. 3 (1980) 273–284.

Horizontal (flatbed) electrophoresis

Flatbed polyacrylamide systems have a number of advantages over the vertical ones when ultrathin gels polymerized on support films are used (Görg *et al.* 1980): simple handling, easy use of ready-made gels and buffer strips instead of large buffer volumes; good cooling efficiency and temperature control; possibility of washing, drying and rehydrating the gels; possibility of automation.

Amplified ribosomal DNA restriction analysis (ARDRA)

This method is derived from ribotyping and is mainly employed fro the identification of bacteria species.

Fragments of ribosomal DNA with polymorphic restriction sites of an organism are amplified with a primer pair and subsequently digested with a restriction enzyme. After gel electrophoresis and silver staining, species specific pattern are obtained.

Random amplified polymorphic DNA (RAPD)

Welsh J, McClelland M. Nucleic Acids Res. 18 (1990) 7213–7218.

Williams JGK, Kubelik AR, Livak KJ, Rafalski JA, Tingey SV. Nucleic Acids Res. 18 (1990) 6531–6535.

Berg DE, Akopyants NS, Kersulyte D. Meth Molec Cell Biol. 5 (1994) 13–24.

Caetano-Annollés G, Bassam BJ, Gresshoff PM. Bio/Technology 9 (1991) 553–557.

This method is applied for rapid detections of DNA polymorphisms of a wide variety of organisms: bacteria, fungi, plants, and animals. One single short oligonucleotide primer (10mer) of arbitrary sequence is used to amplify fragments of the genomic DNA (Welsh and McClelland, 1990; Williams *et al.* 1990). The low stringency annealing conditions lead to an amplification of a set of multiple DNA fragments of different sizes. Berg *et al.* (1984) have found a series of primers, which allow a very good differentiation of microorganisms. When optimized and uniform PCR conditions are employed, specific and reproducible band patterns are achieved.

A modification using 5mer primers is called DNA amplification fingerprinting (DAF) and has been introduced by Caetano-Annollés *et al.* (1991).

Even one additional band detected can make a big difference in the evaluation.
With optimized separation and detection strain-specific pattern are achieved.
See Method 12.

RAPD samples can be run on agarose gels with Ethidium bromide staining or on polyacrylamide gels with subsequent silver staining. As the resolution and sensitivity of the latter method is much higher, more variety differences can be detected. Figure 17 shows the RAPD patterns of different fungus varieties separated in a horizontal polyacrylamide gel and silver stained. The primers are based on those published by Berg *et al.* (1984).

Mutation detection methods

Landegren U, Ed. Laboratory protocols for mutation Detection. Oxford University Press (1996).

A comprehensive description of mutation detection methods can be found in the book "Laboratory Protocols for Mutation Detection", edited by Ulf Landegren (1996).

The most certain and sensitive method for the detection of mutations is the DNA sequence analysis. However, this method is too costly and time-consuming for screening purposes.

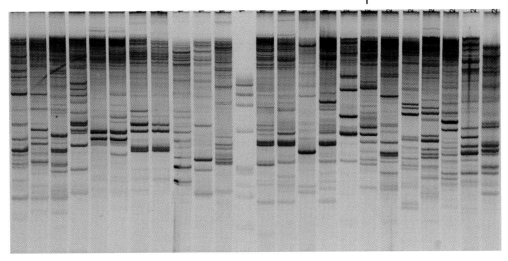

Fig. 17: RAPD electrophoresis of fungi varieties in a horizontal polyacrylamide gel. Silver staining. By kind permission of Birgit Jäger and Dr. Hans-Volker Tichy, TÜV Südwest GmbH – Biological Safety Division, Freiburg im Breisgau.

Single strand conformation polymorphism (SSCP)

The principle: Variations in the sequence as small as one base exchange alter the secondary structure of ssDNA, e.g. by different intramolecular base pairing. The changes in the sequence cause differences in the electrophoretic mobility, which are observed as band shifts (Orita *et al.* 1989).

Orita M, Iwahana H, Kanazewa H, Hayashi K, Sekiya T. Proc Natl Acad Sci USA. 86 (1989) 2766–2770.

The mechanism of SSCP is described as: Differential transient interactions of the bent and curved molecules with the gel fibers during electrophoresis, causing the various sequence isomers to migrate with different mobilities.

Single strands migrate much slower than the corresponding double strands.

Before screening, the mutants have to be defined by direct sequencing. The sequences fo the appropriate primer pair have to be found. The PCR products are denatured by heating with formamide or sodium hydroxid, and loaded onto a non-denaturing polyacrylamide gel for electrophoresis. Silver staining has to be employed for detection of the DNA fragments.

SSCP analysis is not a replacement but an addition to sequencing, when 100 % of defined mutations have to be detected. Intercalating dyes do not work here.

A high number of samples can be screened with a considerably lower effort than direct sequencing in a relatively short time, namely within a few hours.

However, the band shifts do not show up automatically for all mutations and under all conditions. Unfortunately, there is not a single and unique separation condition, which can be applied to the separations of all exons. The parameters influencing the result have been reviewed by Hayashi and Yandell (1993) and will be further discussed in Method 13 in part II.

Hayashi K, Yandell DW. Hum Mutat. 2 (1993) 338–346.

For this method, good cooling and temperature control system is very important.

Rehbein H, Mackie IM, Pryde S, Gonzales-Sotelo C, Perez-Martin R, Quintero J, Rey-Mendez M. Inf. Fischwirtsch. 42 (1995) 209–212.

SSCP of the mitochondrial cytochrome b gene is also employed for differentiation of animal species. Rehbein *et al.* (1995) have used the method for the identification of the species in canned tuna.

Heteroduplex and DSCP

Keen JD, Lester D, Inglehearn C, Curtis A, Bhattacharya. Trends Genet. 7 (1991) 5.

White MB, Carvalho M, Derse D, O'Brien SJ, Dean M. Genomics 12 (1992) 301–306.

Barros F, Carracedo A, Victoria ML, Rodriguez-Calvo MS. Electrophoresis 12 (1991) 1041–1045.

Dockhorn-Dworniczak B, Aulekla-Acholz C, Dworniczak B. Pharmacia LKB Offprint A37 (1990).

Single base substitutions can also be detected by heating the mixtures amplified wild type and mutant DNA and run the resulting heteroduplexes on a native polyacrylamide gel electrophoresis (Keen *et al.* 1991; White *et al.* 1992). The mobilities of heteroduplexes lie between the mobilities of the corresponding homoduplexes and single strands. Different mutations cause different mobility shifts of heteroduplexes. The bands can be detected with Ethidiumbromide or with silver staining.

Sometimes the technique is also called DSCP (double strand conformation polymorphism) (Barros *et al.* 1992). But it should not be forgotten, that also homoduplexes can show band shifts in native gels due to the influence of the contents of A and T.

For DNA diagnosis, DNA point mutations can quickly be revealed with the Primer Mismatch process in combination with electrophoresis of the amplification products in horizontal polyacrylamide gels (Dockhorn-Dworniczak *et al.*, 1990).

Denaturing gradient gel electrophoresis (DGGE) and constant denaturing gel electrophoresis (CDGE)

Fischer SG, Lerman LS. Proc Natl Acad Sci. 60 (1983) 1579-1583.

With DGGE single base exchanges in segments of DNA can be detected with almost 100 % efficiency. The principle of DGGE is based on the different electrophoretic mobilities of partially denatured molecules caused by differences in DNA melting (Fischer and Lerman, 1983).

Typically the 100 % denaturant solution contains 6 to 7 mol/L urea and 20 to 40 % formamide. The gels are run at temperatures between 40°C and 60°C.

With a denaturant gradient perpendicular to the electrophoresis direction, the region of a point mutation can be identified. Denaturant gradients parallel to the electrophoresis runs are better for screening applications.

The practical aspects and the gradient casting technique are described in method 14.

Constant denaturing gel electrophoresis (CDGE) is employed for screening, when the denaturant concentration of differential melting of a DNA segment has been detected with DGGE. Figure 18 is a schematic representation of perpendicular and parallel DGGE. As DGGE is not very easy to perform, it is only employed, when the techniques other than sequence analysis fail in detecting a mutation.

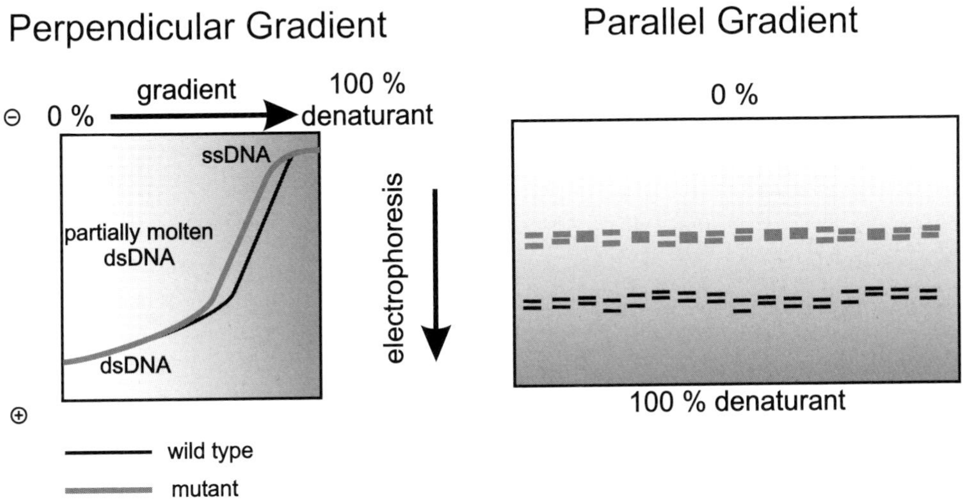

Perpendicular Gradient

Parallel Gradient

Fig. 18: Schematic representation of typical results of a perpendicular and a parallel DGGE.

Temperature gradient gel electrophoresis (TGGE)

Temperature gradient gel electrophoresis resolves homo- and hetero-duplexes according to their thermal stabilities (Riesner *et al.* 1989). In this technique, denaturing gels are run on a plate with a cold (15 °C) side at the cathode and a hot side (60 °C) at the anode. The method is well suitable for screening purposes. Suttorp *et al.* (1996) have described how to change a standard horizontal electrophoretic chamber into a TGGE device.

Riesner D, Steger G, Wiese U, Wulfert M, Heibey M, Henco K. Electrophoresis 10 (1989) 377–389.
Suttorp M, von Neuhoff N, Tiemann M, Dreger P, Schaub J, Löffer H, Parwaresch R, Schmitz N. Electrophoresis 17 (1996) 672–677.

Single Nucleotide Polymorphism (SNP) analysis

According to estimations, one single nucleotide (SNP) occurs approximately in every 100–300 bases of the human genome. Single nucleotide polymorphisms are present in both the coding and non-coding regions. The SNPs found in the coding regions of the genome are interesting for clinical research, because they may be indicators for the different responses of different patients to drug treatment and other factors. High-throughput systems are preferably employed, like the multi-capillary electrophoresis instruments used for DNA sequencing.

Denaturing PAGE of microsatellites

Denaturing gels provide a very high resolving power, thus they are very useful for separating of microsatellites with very short repeats down to 2 bp. Because the Taq-polymerase used in PCR adds an addi-

It is not always necessary to apply completely denaturing conditons on the gel: 7 mol/L urea in the gel and 25 °C separation temperature are often sufficient.

tional A to the 3'-end of a part of the single strands, double bands are frequently seen after silver staining of denaturing gels.

When labelling techniques like radioactivity or fluorescence are employed, only one of the primer pair is marked to avoid visualization of the duplets.

Instructions for denaturing electrophoresis are found in method 14.

Native PAGE of mini- and microsatellites

Möller A, Wiegand P, Grüschow C, Seuchter SA, Baur MP, Brinkmann B. Int J Leg Med 106 (1994) 183–189.

Puers C, Hammond HA, Jin L, Caskey CT, Schumm JW. Am J Hum Genet 53 (1993) 953–958.

Variable number of tandem repeats (VNTR) and short tandem repeats (STR) analysis are used in forensic laboratories: They are performed in denaturing and in non-denaturing gels. In both cases, the assignment of alleles with well-defined (sequenced) allelic ladders of the respective VNTR or STR locus, which are run in the same gel, proved to be the most reliable method (Puers *et al.* 1993; Möller *et al.* 1994).

Schickle HP. GIT Labormedizin. 19 (1996) 228–231.

With new special gel media fragments differing by 2 bp (in the size range of 100 to 200 bp) can be resolved in a 10 cm long gel under native conditions.

However, additionally to the regular types with length variations, there are sequence variants existing in some STR loci, which can only be identified by sequencing the fragments or running them on non-denaturing polyacrylamide gels. In order to achieve adequate resolution in native gels, long separation distances or special gel media and buffer systems have to be used (Schickle, 1996).

Differential Display PCR Electrophoresis

Not only cell regulation and differentiation can be monitored, but also miscontrolled cells can be visualized in cancer research.

Liang, P, Pardee AB. Science 257 (1992) 967–971.

Bauer D, Müller H, Reich J, Riedel H, Ahrenkiel V, Warthoe P, Strauss M. Nucleic Acid Res. 21 (1993) 4272–4280.

The additional bands are identified on the developed X-ray film; after cutting a hole in this position the film is matched with the gel again, the band is scratched out.

This is a method to screen the total amount of cDNAs coming from the messenger RNA-pool of specific cell lines. The purpose is to display only the active genes besides the total amount of ca. 1 Million genes of a cell.

The method "DDRT" (Differential display reverse transcription) has been introduced by Liang and Pardee (1992) and improved by Bauer *et al.* (1993). The mRNA from the original cell and the stimulated cell are processed in parallel. Extracted mRNA is reverse transcribed with oligo-dT-NN anchor primers. The resulting 12 cDNA pools are amplified with the respective oligo-dT primer and a set of arbitrary 10mer primers. After high resolution electrophoresis of the amplification products, those additional bands, which have been expressed by the stimulated cell, are cut out and reamplified for cloning and sequencing (see Fig. 19). The original technique employs autoradiography for the detection of the bands.

Lohmann *et al.* (1995) have taken a big step forward with their "REN" (rapid, efficient, nonradioactive) technique: they use horizontal film-supported gels and cut out the silver stained DNA bands for reamplification. In this way, the method can be performed much faster, cheaper, and with a higher success of finding a gene, which has been expressed as a response of the cell.

Lohmann J, Schickle HP, Bosch TCG. BioTechniques 18 (1995) 200–202.
Urea and native gels can be employed. Sometimes it is necessary to use long gels, because the bands are spread over a wide range basepair-lengths.

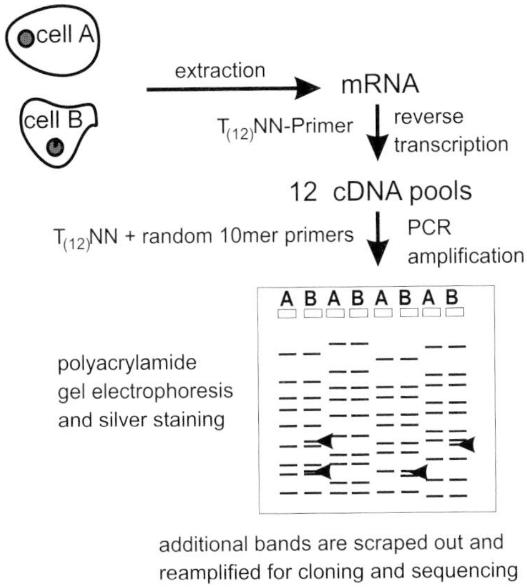

Fig. 19: The steps in a DDRT experiment.

Practical hints for improved reamplification can be found in the paper by Böckelmann *et al.* (1999).

Böckelmann R, Bonnekoh B, Gollnick H. Skin Pharmacol Appl Skin Physiol. 12 (1999) 54–63.

Two-dimensional DNA electrophoresis
Complex DNA samples can also be displayed with a two-dimensional electrophoresis: First the DNA is digested with a rare cutting restriction enzyme and the fragments are separated in an agarose gel; then the agarose gel is soaked in a mixture of selected restriction enzyme and applied on a polyacrylamide gel. Schickle *et al.* (1999) have converted the time consuming "handcraft" procedure into a faster technique, which is based on ready-made gels. The resulting tiny spots are visualized with autoradiography.

Schickle HP, Lamb B, Hanash SM. Electrophoresis 20 (1999) 1233–1238.

RNA and viroids
Bi-directional electrophoresis (Schumacher *et al.* 1986) is used for viroid tests: the plant extract (RNA fragment + viroid) is first separated under native conditions at 15 °C. After a certain separation time, the gel is cut behind a zone marked with a dye such as Bromophenol Blue or xylenecyanol.

Schumacher J, Meyer N, Riesner D, Weidemann HL. J Phytophathol. 115 (1986) 332–343.

The gel contains 4 mol/L urea. The molecules are denatured, that is unfolded by the combination of urea and elevated temperature. For practical reasons this method is only carried out in horizontal systems.

An electrophoretic separation under denaturing conditions is carried out. The viroid forms a ring which cannot migrate. The RNA fragments which migrate more slowly during the first native separation, do not lose their mobility at 50 °C and migrate out of the gel. If a viroid is present, only one band is found when the gel is stained. The position of the viroid in the gel depends on its kind. Several new viroids have been discovered in this way.

1.2.4
Polyacrylamide gel electrophoresis of proteins

For analytical PAGE of proteins, the trend is to go from cylindrical gels to flat and thinner ones. Because of the development of more sensitive staining methods such as silver staining for example, very small quantities of concentrated sample solution can be applied for the detection of trace amounts of proteins.

The advantages of thinner gels are:

- faster separation
- better defined bands
- faster staining
- better staining efficiency, higher sensitivity

Disc electrophoresis

Ornstein L. Ann NY Acad Sci. 121 (1964) 321–349.
Davis BJ. Ann NY Acad Sci. 121 (1964) 404–427.
See also page 45: Isotachophoresis

Discontinuous electrophoresis according to Ornstein (1964) and Davis (1964) solves two problems of protein electrophoresis: it prevents aggregation and precipitation of proteins during the entry from liquid sample into the gel matrix, and promotes the formation of well defined bands. The discontinuity is based on four parameters (see Fig. 20):

- the gel structure
- the pH value of the buffer
- the ionic strength of the buffer
- the nature of the ions in the gel and in the electrode buffer

The gel is divided into two areas: resolving and stacking gel. The resolving gel with small pores contains 0.375 mol/L Tris-HCL buffer pH 8.8, the stacking gel with large pores contains 0.125 mol/L Tris-HCL pH 6.8.

Glycine is used because it is very hydrophilic and does not bind to proteins.

The electrode buffer contains only glycine, the gel only Cl^- ions. Glycine has a pI 6.7, it has almost no net charge at pH 6.8: the pH of the stacking gel. Thus glycine has a low mobility.

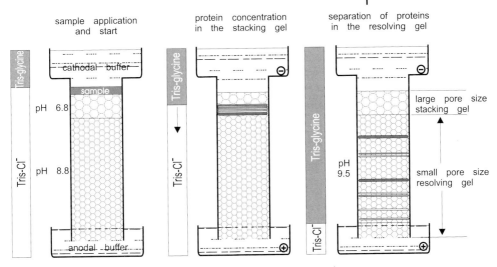

Fig. 20: Schematic diagram of the principles of disc electrophoresis according to Ornstein (1964). The buffer system shown is also employed for discontinuous SDS electrophoresis.

At first, the proteins are separated according to the *principle of iso-tachophoresis* and form stacks in the order of their mobility ("stacking effect"). The individual zones become concentrated. Because of the large pores in the stacking gel, the mobilities are dependent on the net charge, not on the size of the molecule.

Because of the relatively slow migration velocity of glycine, the samples enter the gel slowly without sudden concentrating. The stacking effect is described in chapter 2 Isotachophoresis.

The protein stack migrates – slowly and at constant speed – towards the anode, till it reaches the border to the resolving gel. The frictional resistance suddenly increases for the proteins, they migrate slower, and the zones become higher concentrated. The low molecular weight glycine is not affected by this, passes the proteins, and becomes higher charged in the resolving zone; the new Cl⁻ / glycine⁻ front moves ahead of the proteins.

2nd zone sharpening effect!

Several events now occur simultaneously:

- The proteins are in a homogeneous buffer medium, destack and start to separate according to the principles of zone electrophoresis.

A discontinuity now only exist at the front.

- Their mobility now depends on their charge as well as on their size. The ranking of the protein ions changes.
- The pH value rises to 9.5 and because of this, the net charge of the proteins and increases.

The separation becomes faster.

Maurer RH. Disk-Electro-phorese – Theorie und Praxis der diskontinuierlichen Poly-acrylamid-Electrophorese. W de Gruyter, Berlin (1968).
For SDS electrophoresis see page 35 ff.

Disc electrophoresis affords high resolution and good band definition. In the example cited above, proteins with pIs higher than pH 6.8 migrate in the direction of the cathode and are lost. Another buffer system must be chosen to separate these proteins. A selection can be found in the works of Maurer (1968) and Jovin (1970). Alternatively SDS can be added to the gel and running buffer to in order to have all proteins negatively charged.

The stacking gels is only cast onto the resolving gel just before electrophoresis because, when the complete gel is left standing for a long time the ions diffuse towards one another.

Gradient gel electrophoresis

Rothe GM, Purkhanbaba M. Electrophoresis 3 (1982) 33–42.

By continuously changing the acrylamide concentration in the polymerization solution, a pore gradient gel is obtained. Gradient gels have a zone sharpening effect and can be used to determine the molecular diameter of proteins in their native state (Rothe and Purkhanbaba, 1982).

The determination of molecular weights in this manner can be problematic, since different proteins have different tertiary structures. Structural proteins cannot be compared with globular proteins.

When the acrylamide concentration and cross-linking are high enough in the small pore area, the protein molecules can be driven to an end point, where they are trapped in the tight gel matrix. Since the speed of migration of the individual protein molecules depends on their charge, the electrophoresis must be carried out long enough so that the molecule with the lowest net charge also reaches its end point.

When several gels are cast simultaneously, the solutions are injected from the bottom. In this case, the solutions in the mixing chamber and the reservoir are interchanged (see page 238).

There are various ways of making gels with linear or exponential porosity gradients. All are based on the same principle: two monomer solutions with different acrylamide concentrations are prepared. During casting, the concentrated solution is continuously mixed with the diluted solution, so that the concentration in the casting mold decreases from bottom to top (Fig. 21). For single gels the solution is poured into the top of the cassette.

The density of the highly concentrated solution is increased with glycerol or sucrose so that the layers in the molds do not mix. In principle a concentration gradient is poured. The mixing of the less dense dilute solution with the high densitiy solution takes place in the mixing chamber using a magnetic stirrer bar.

Exponential gradients are formed when the mixing chamber is sealed. The volume in the mixing chamber stays constant, the same quantity of dilute solution flows in as solution out of the mixing chamber (see Fig. 21).

If the mixing chamber is left open at the top, the principle of communicating vases is valid: so that the height of both fluids stays equal, half as much of the dilute solution flows in as of solution flowing out of the mixing chamber. A linear gradient is thus formed. A compensating stick in the reservoir compensates the volume of the stirrer bar and the difference in the densities of both solutions (see page 219 and 236 for porosity gradients, pages 265 and 281 for pH gradients, and pages 269 and 321 ff for additive gradients).

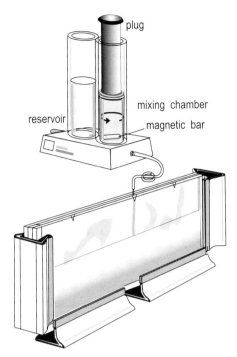

plug

mixing chamber

reservoir

magnetic bar

Fig. 21: Casting of an exponential gradient gel with a gradient maker. The stirrer bar is rotated with a magnetic stirrer (not shown).

SDS electrophoresis

SDS electrophoresis – SDS being the abbreviation for *sodium dodecyl sulphate* – which was introduced by Shapiro *et al.* (1967) separates exclusively according to molecular weight. By loading with the anionic detergent SDS, the charge of the proteins is so well masked that anionic micelles with a constant net charge per mass unit result: 1.4 g SDS per g protein.

Shapiro AL, Viñuela E, Maizel JV. Biochem Biophys Res Commun. 28 (1967) 815–822.

In addition, the tertiary and secondary structures are cancelled because of the disruption of the hydrogen bonds and unfolding of the molecules.

Thus there is no influence of the shape of the protein on the running condition.

Disulfide bonds between cysteine residues can only be cleaved by a reducing thiol agent such as 2-mercaptoethanol or dithiothreitol. The SH groups are often protected by a subsequent alkylation with iodoacetamide, iodoacetic acid or vinylpyridine (Lane, 1978).

Lane LC. Anal Biochem. 86 (1978) 655–664.

The unfolded amino acid chains, bound to SDS, form ellipsoids with identical central axes. During electrophoresis in restrictive polyacrylamide gels containing 0.1% SDS there is a linear relationship between the logarithm of the molecular weight and the relative distance of migration of the SDS-polypeptide micelle.

This linear relationship is only valid for a certain interval, which is determined by the ratio of the molecular size to the pore diameter.

Marker proteins for various molecular weight intervals are available.

Gels with a pore gradient show a wider separation range and a larger linear relationship than gels with a constant pore size. In addition, sharper bands result since a gradient gel minimizes diffusion (Fig. 22). The molecular weight of the proteins can be estimated with a calibration curve using marker proteins (Fig. 23).

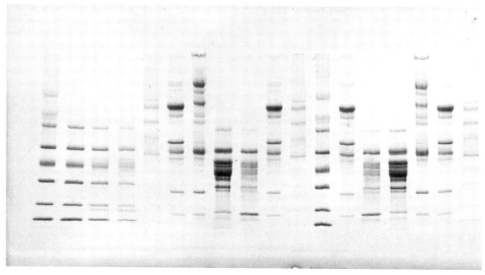

Fig. 22: Separation of proteins in a linear pore gradient gel T = 8% to 18% by SDS electrophoresis. Staining with Coomassie Brilliant Blue (Cathode on top).

For example when it is not reduced, albumin shows a molecular weight of 54 kDa instead of 68 kDa since the polypeptide chain is only partially unfolded.

For separation of physiological fluids or analysis of urine proteins for example, the reduction step is left out to prevent the breakdown of the immunoglobulins into subunits. In these cases the incomplete unfolding of certain proteins must be taken into account and therefore the molecular weight cannot be determined exactly.

There are a number of practical advantages to SDS electrophoresis:

Even very hydrophobic and denatured proteins.

- SDS solubilizes almost all proteins.

This ensures rapid separations.

- Since SDS-protein complexes are highly charged, they possess a high electrophoretic mobility.

Towards the anode.

- Since the fractions are uniformly negatively charged, they all migrate in one direction.

This limits diffusion.

- The polypeptides are unfolded and stretched by the treatment with SDS and the separation is carried out in strongly restrictive gels.

Sharp zones.

- This affords high resolution.

No strong acids are necessary.

- The bands are easy to fix.

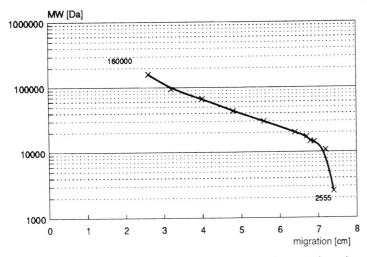

Fig. 23: Semi-logarithmic representation of a molecular weight curve. The molecular weights of the marker proteins are represented as a function of their migration. (SDS linear pore gradient gel according to Fig. 22).

- The separation is based on one physico-chemical parameter, the molecular weight.

  It is an easy method for molecular weight determination.

- Charge microheterogeneities of isoenzymes are cancelled out.

  There is one band for one enzyme.

- Proteins separated with SDS bind dyes better.

  The detection limit increases ten-fold compared to native PAGE.

- After electrophoretic transfer on an immobilizing membrane, the SDS can be removed from the proteins without eluting the proteins themselves.

  See chapter 4: Blotting.

SDS electrophoresis can be carried out in a continuous phosphate buffer system (Weber and Osborn, 1968) or in a discontinuous system (see page 34):

Weber K, Osborn M. J Biol Chem. 244 (1968) 4406–4412.

Lämmli (1970) has directly adopted the disc electrophoresis method according to Ornstein (1964) and Davis (1964), for proteins charged with SDS, though the discontinuities in pH value and ionic strength are in most cases not necessary.

Lämmli UK. Nature 227 (1970) 680–685.

- Because the protein-SDS micelles have very high negative charges, the mobility of glycine is lower than that of the proteins in the stacking gel at the beginning of electrophoresis, even at pH 8.8; it does not bind SDS.

  However, the discontinuity of the anions and the different gel porosities are very important.

- During stacking no field strength gradient results, since there are no charge differences within the sample: so no low ionic strength is necessary.

The overlayering of the resolving gel with butanol for example can thus be avoided and especially the laborious removal of the overlay before pouring the stacking gel.

This means that SDS disc electrophoresis gels can be cast in one step: Glycerol is added to the resolving gel and then the stacking gel, which contains the same buffer but no glycerol, is directly cast on top of it. In addition, the run time is shorter since the separation starts more quickly.

For ready-made gels with longer shelf lives, another buffer system with pH values around 7 should be chosen.

Since there are no diffusion problems between the stacking and the resolving gel buffers with these gels, they can be stored longer than conventional disc gels. Yet their shelflife is limited by the high pH value of the gel buffer, since, after about 10 days, the polyacrylamide matrix starts to hydrolyse.

Since Tricine is much more expensive than glycine, it is only used at the cathode, the anode contains Tris-acetate.
Kleine B, Löffler G, Kaufmann H, Scheipers P, Schickle HP, Westermeier R, Bessler WG. Electrophoresis 13 (1992) 73–75.

Long shelflife gels: Tris-acetate buffer with a pH of 6.7 has proven to have the best storage stability and separation capacity. Tricine is used instead of glycine as the terminating ion. The principle of this buffer system with polyacrylamide electrode buffer strips in a ready-made SDS gel can be seen in Fig. 24. These buffer strips simplify electrophoresis considerably and reduced chemical and radioactive liquid waste Kleine *et al.* (1992).

Schägger H, von Jagow G. Anal Biochem. 166 (1987) 368–379. The major effect is caused by the use of tricine instead of glycine. Also the "long shelflife gels" show a markedly better separation of small peptides.

Low molecular weight peptides: the resolution of peptides below 14 kDa is not sufficient in conventional Tris-glycine-HCl systems. This problem has been solved by the development of a new gel and buffer system by Schägger and von Jagow (1987). In this method an additional spacer gel is introduced, the molarity of the buffer is increased and tricine used as terminating ion instead of glycine. This method yields linear resolution from 100 to 1 kDa.

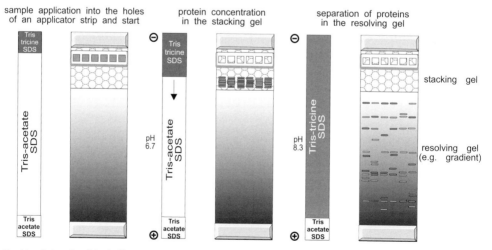

Fig. 24: Principle of the buffer systems of ready-made gels for discontinuous SDS-electrophoresis. Horizontal gels with Tris-Tricine buffer strips.

Glycoproteins migrate too slowly in SDS electrophoresis, since the sugar moiety does not bind SDS. When a Tris-borate-EDTA buffer is used, the sugar moieties are also negatively charged, so the speed of migration increases (Poduslo, 1981).

Poduslo JF. Anal Biochem. 114 (1981) 131–139.
The use of gradient gels is also beneficial for better MW estimations.

Blue Native Electrophoresis

Membrane proteins: When membrane proteins are solubilized with nonionic detergents, these detergents would interfere with the SDS. Schägger and von Jagow (1991) have developed "Blue Native electrophoresis" of membrane protein complexes to solve this problem:

Schägger H, von Jagow G. Anal Biochem. 199 (1991) 223–231.

In a vertical chamber Coomassie Blue G-250 is added to the cathodal buffer in a native polyacrylamide gel electrophoresis.

During the run the dye competes with the nonionic detergent and binds to the membrane proteins and complexes and charges them negatively analogous to SDS. All these protein-dye complexes migrate towards the anode, also basic proteins. They are soluble in detergent-free solution, and – as the negatively charged protein surfaces repel each other – aggregation between proteins is minimized.

The membrane proteins and complexes can be isolated in enzymatically active form. The gels do not need to be stained, because the proteins and complexes migrate as blue bands.

The separation lanes of individual samples can be cut out from the polyacrylamide slab and applied directly on SDS polyacrylamide gel for a separation into a second dimension. In presence of SDS the complexes are dissolved and the partners of the respective complexes are displayed in the gel. The procedure has been described in detail (H. Schägger, 1994).

Schägger H. Chapter 4: Native electrophoresis. In: A Practical guide to membrane protein Purification (Von Jagow G, Schägger H, eds.) Academic Press, New York (1994) 81–103.

Cationic detergent electrophoresis

Strongly acidic proteins do not bind SDS and very basic nucleoproteins behave abnormally in SDS gels. The alternative is to use cationic detergents, for instance cetyltrimethylammonium bromide (CTAB), in an acidic medium at pH 3 to 5 (Eley *et al.* 1979). This allows a separation according to the molecular weight in the direction of the cathode. This cationic detergent also causes less damage to the protein than SDS, so CTAB electrophoresis can be used as a form of native electrophoresis (Atin *et al.* 1985). See also next page.

Eley MH, Burns PC, Kannapell CC, Campbell PS. Anal Biochem. 92 (1979) 411–419.

Atin DT, Shapira R, Kinkade JM. Anal Biochem. 145 (1985) 170–176.

Rehydrated polyacrylamide gels:

In washed gels the SDS Tris-HCl / Tris-glycine buffer system shows poor results. However good results are obtained with the Tris-acetate / Tris-tricine system.

In this method, the gel is rehydrated in Tris-acetate pH 8.0 using a horizontal tray. If, for highly concentrated protein samples, a discontinuity in pH and molarity between stacking and resolving gel is required, the stacking zone can be selectively equilibrated in a higher diluted Tris-acetate buffer pH 5.6 using a vertical chamber (see page 228).

The performance of SDS buffer systems are obviously highly influenced by catalysts and / or monomers of acrylamide.

This procedure of washing, drying, rehydration and equilibration can only be performed with gels polymerized on carrier films, which are used in horizontal systems.

The ionic catalysts APS and TEMED would destabilize these buffer systems, see method 4, pages 175 and following.
Hsam SLK, Schickle HP, Westermeier R, Zeller FJ. Brau-wissenschaft 3 (1993) 86–94.
Rehbein H. Electrophoresis 16 (1995) 820–822.

Native electrophoresis in amphoteric buffers: the polymerization catalysts can be washed out of the polyacrylamide gels on support films used in horizontal systems with deionized water. By equilibration with amphoteric buffers such as HEPES, MES or MOPS for example, there is a wide spectrum for electrophoresis under native conditions. This method proved to be particularly useful for acidic electrophoresis of basic hydrophobic barley hordeins (Hsam *et al.* 1993) and basic fish sarcoplasmic proteins (Rehbein, 1995).

Two-dimensional electrophoresis techniques

Several aims are pursued by the combination of two different electrophoretic methods:

- Proteins separated by electrophoresis are then identified by crossed immunoelectrophoresis.

Altland K, Hackler R. In: Neuhoff V, Ed. Electrophoresis' 84. Verlag Chemie, Weinheim (1984) 362–378.

- A complex protein mixture is first separated by zone electrophoresis and then further purified by IEF, or vice versa (Altland and Hackler, 1984).

Langen H, Takács B, Evers S, Berndt P, Lahm H-W, Wipf B, Gray C, Fountoulakis M. Electrophoresis 21 (2000) 411–429.

- Hydrophobic proteins, such as membrane-bound proteins, are separated first in an acidic gel at pH 2.1 in presence of the cationic detergent 16-BAC, followed by an SDS electrophoresis (Langen *et al.* 2000). As the separation patterns in 16-BAC and in SDS differ substantially, a decent resolution is obtained.
- Membrane protein complexes are first separated by Blue native electrophoresis (see above) and then separated with SDS PAGE for the display and identification of complex partners.

The highest resolution is obtained by first separating according to the isoelectric points, the second dimension according to the molecule mass.
A flat-bed gel can also be cut into strips after the first separation and transferred onto the second gel.

- Highly heterogeneous mixtures of proteins such as cell lysates or tissue extracts should be completely fractionated into individual proteins so as to obtain an overall picture of the protein composition and to enable location of individual proteins.

O'Farrell PH. J Biol Chem. 250 (1975) 4007–4021.

For these techniques, the first-dimensional runs are carried out in individual gel rods or strips and loaded onto the second-dimensional gels.

High resolution 2-D electrophoresis: This method had been introduced by O'Farrell (1975). The sample is denatured with a lysis buffer, the first dimension is isoelectric focusing in presence of 8 or 9 molar urea and a non-ionic detergent. SDS electrophoresis is run as the second dimension. In Fig. 25 the principle of the traditional 2-D electrophoresis methodology is shown.

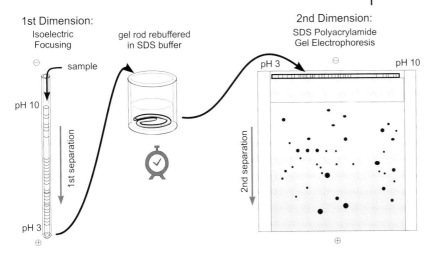

Fig. 25: The principle of the classical high-resolution 2-D electrophoresis according to O'Farrell (1975).

The method has recently become highly interesting, because the protein spots obtained can be further analysed with new methods of mass spectrometry. Protein spots are then identified with the help of genomic databases. This approach is used for "Proteome analysis" (Wasinger *et al.* 1995), which will be described in more detail in the dedicated chapter 6 (page 91ff) and in the book "Proteomics in Practice" (Westermeier and Naven, 2002).

Formerly, the identification of protein spots was very complicated and time consuming: genomic data were not available.

Westermeier R, Naven T. Proteomics in Practice. A laboratory manual of proteome analysis. WILEY-VCH, Weinheim (2002).

2
Isotachophoresis

As already mentioned earlier, "Isotachophoresis" means "migration with the same speed". To understand the effects and the features of the technique, one should imagine *the four facts of isotachophoresis*, which are equally important and which happen at the same time:

Many "Non physisists" continuously complain, that they do not understand isotachophoresis as easily as other biophysical methods. The reason: several things occur simultaneously.

- Migration of all ions with the same speed
- Separation of components as an "Ion train"
- Zone sharpening effect
- Concentration regulating effect

In Fig. 26 all these effects are displayed together.

The main prerequisite for an isotachophoretic separation is a *discontinuous buffer system* with a leading and a terminating electrolyte. If the anions in a sample are to be determined, the leading anions must have higher mobility, and the terminating anions must have a lower mobility than the sample components.

Either anions or cations can be separated at one time, but not both simultaneously. The terminating ions are sometimes called "trailing ions".

In an anionic separation, the *leading* electrolyte will be at the anodal, and the terminating electrolyte at the cathodal side. The sample is applied between the two. The system also contains a common cationic counter-ion.

A practical example: Chloride is the leading ion, and Glycine is the terminating ion, Tris is the counter ion, see page 34.

Isotachophoresis is carried out at *constant current* so as to maintain a constant field strength within the zones. The speed of migration then also stays the same during the separation.

2.1
Migration with the same speed

When an electric field is applied, the ions start to migrate with the speed of the terminating ion, the ion with the lowest mobility. All ions conduct the current. The sample and the leading ions can not move faster than the terminating ion, because this would cause an ion gap: no current would be transported (see Fig. 26 A).

When applied in gels, this effect has the advantage of slow sample entry, thus avoiding sample precipitation.

Electrophoresis in Practice, Fourth Edition. Reiner Westermeier
Copyright © 2005 WILEY-VCH Verlag GmbH & Co. KGaA, Weinheim
ISBN: 3-527-31181-5

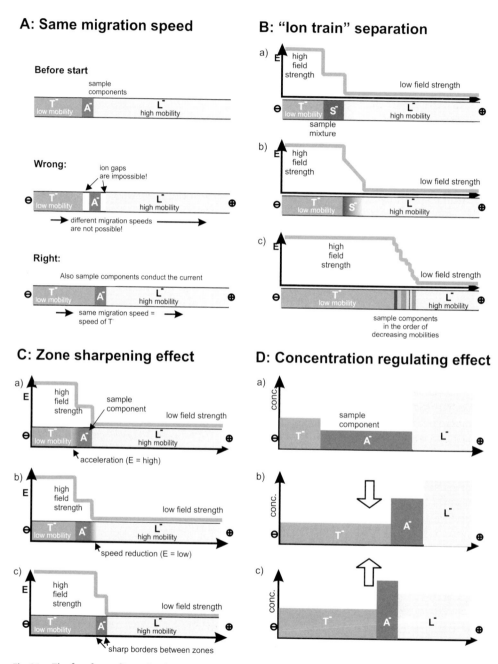

Fig. 26: The four facts of isotachophoresis. A: In a discontinuous buffer system the ions are forced to migrate with the same speed. B: The ions are separated, but each zone travels in immediately after the other one. C: Because of the slowing down and acceleration of ions in the different field strength areas, zones are sharpened. D: The concentration regulating effect is the basis of quantification in isotachophoresis, see text for further explanations.

2.2
"Ion train" separation

Because all the ions are forced to migrate at the same velocity, the field strength is higher in the area of the ions of lower mobility, and is lower in the area containing the more mobile ions. During this migration pure contiguous zones containing the individual substances are formed within the sample.

At equilibrium, the ion with the highest mobility migrates in the front, the others migrate behind in order of decreasing mobility:

The sample ions form stacks.

$$m_{L^-} > m_{A^-} > m_{B^-} > m_{T^-}$$

m: mobility, L^- leading ion, T^-, terminating ion, A^- and B^- sample ions

The zone with the highest mobility has the lowest field strength, the one with the lowest mobility has the highest field strength. The product of the field strength and the mobility of each zone is constant (see Fig. 26 B).

2.3
Zone sharpening effect

Automatically sharp uones are formed: Would an ion diffuse into a zone with a higher mobility, it would be slowed down because of the lower electric field. It will migrate back to its own zone. Should an ion fall behind, it will be accelerated out of the neighbouring zone by the higher field strength (see Fig. 26 C)

The system works against diffusion and results in a distinct separation of the individual substances. In contrast to other separation techniques, the fractions are contiguous.

2.4
Concentration regulation effect

The basis of quantitative analysis with isotachophoresis is the "regulating function" (beharrliche Funktion) of Kohlrausch (1897). It defines the conditions at the boundary between two different ions L^- and A^- with the same counter-ion R^+ during the migration of this boundary in the electrical field. The ratio of the concentrations CL^- and CA^- of the ions L^-, A^- and R^+ is the following:

Kohlrausch F. Ann Phys. 62 (1897) 209–220.

$$\frac{C_{L^-}}{C_{A^-}} = \frac{m_{L^-}}{m_{L^-} + m_{R^+}} \times \frac{m_{A^-} + m_{R^+}}{m_{A^-}}$$

m, the mobility is expressed in $cm^2/V \times s$ and is constant for each ion under defined conditions.

This leads to the concentrating effect: the higher the concentration of the leading ion, the more concentrated the zones.

At a given concentration of the leading electrolyte L^-, the concentration of A^- is fixed since all the other parameters are constant. This can be applied to the next zone: since the concentration of A^- is defined, the concentration of B^- is determined, and so on...

Figure 26 D shows how this regulating effect converts situations (a) or (b) into the stable situation (c).

Quantitative analysis

Kohlrausch's equation can be more simply expressed as :

$$C_{A^-} = C_{L^-} \times \text{constant}$$

The bands are not "peaks" (Gaussian distribution) as in conventional electrophoresis or chromatography but "spikes" (concentration dependent bands). For this reason standard interpretation programs cannot be used.

At equilibrium the concentration of the sample ions C_{A^-} is proportional to the concentration of the leading ions C_{L^-}. This means that the ionic concentration is constant in each zone. The number of ions in each zone is proportional to the *length* of the zone. A characteristic of isotachophoresis is that the quantification of the individual components is done by measuring the *length* of the zone. Fig. 26 D shows how, during the isotachophoretic run, states a and c automatically determine state b.

To determine the concentration of a substance at least *two runs* must be performed: first the unmodified sample is separated, and then during the second run, a known amount of pure substance is added. The original quantity of the substance to be analyzed can be deduced from the lengthening of the zone.

*Everaerts FM, Becker JM, Verheggen TPEM. Isotachophoresis, theory, instrumentation and applications.
J Chromatogr Library. Vol. 6. Elsevier, Amsterdam (1976).Hjalmarsson S-G, Baldesten A. In: CRC Critical review in anal. chem. (1981) 261–352.*

Isotachophoresis is usually performed in Teflon *capillaries* (Everaerts *et al.* 1976; Hjalmarsson and Baldesten, 1981) new developments being quartz capillaries (Jorgenson and Lukacs, 1981; Hjertén, 1983). Voltages up to 30 kV and currents of the order of several µA are used. For effective differentiation between directly contiguous zones, current and thermometric conductivity detectors are also used. Fig. 27 shows the isotachophoretic separation of penicillins.

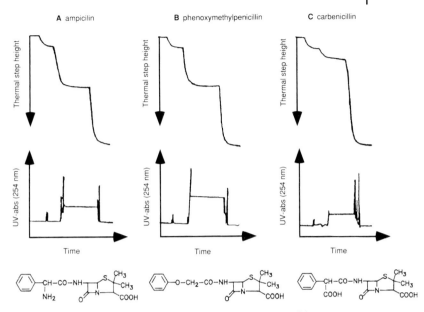

Fig. 27: Isotachophoresis of penicillins. Simultaneous detection of the zones with a thermocouple detector and an UV detector. From the application laboratory of Pharmacia LKB, Sweden.

3
Isoelectric focusing

abbreviation: IEF

3.1
Principles

The use of isoelectric focusing is limited to molecules which can be either positively or negatively charged. Proteins, enzymes and peptides are such amphoteric molecules. The net charge of a protein is the sum of all negative and positive charges of the amino acid side chains, but the three-dimensional configuration of the protein also plays a role (Fig. 28).

The substances to be separated must have an isoelectric point at which they are not charged.

At low pH values, the carboxylic side groups of amino acids are neutral:

$$R–COO^- + H^+ \rightarrow R–COOH$$

At high pH values, they are negatively charged:

$$R–COOH + OH^- \rightarrow R–COO– + H_2O$$

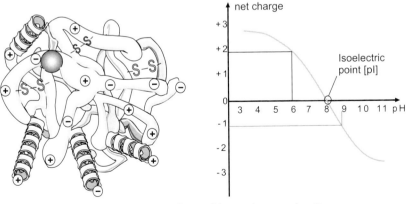

Fig. 28: Protein molecule and the dependence of the net charge on the pH value. A protein with this net charge has two positive charges at pH 6 and one negative charge at pH 9.

Electrophoresis in Practice, Fourth Edition. Reiner Westermeier
Copyright © 2005 WILEY-VCH Verlag GmbH & Co. KGaA, Weinheim
ISBN: 3-527-31181-5

The amino, imidazole and guanidine side-chains of amino acids are positively charged at low pH values:

$$R\text{-}NH_2 + H^+ \rightarrow R\text{--}NH_3^+$$

at high pH values, they are neutral:

$$R\text{--}NH_3^+ + OH^- \rightarrow R\text{--}NH_2 + H_2O$$

Many of the microheterogeneities in IEF patterns are due to these modifications in the molecules.

For composite proteins such as glyco- or nucleoproteins, the net charge is also influenced by the sugar or the nucleic acid moieties. The degree of phosphorylation also has an influence on the net charge.

The net charge curve is characteristic of a protein. With the titration curve method explained in chapter 3.8 it can easily be reproduced in a gel.

If the net charge of a protein is plotted versus the pH (Fig. 28), a continuous curve which intersects the x-axis at the isoelectric point pI will result. The protein with the lowest known pI is the acidic glycoprotein of the chimpanzee: pI = 1.8. Lysozyme from the human placenta has the highest known pI: pI = 11.7.

It is important to find the optimum place in the gradient at which the proteins enter the gel without any trouble, do not aggregate and at which no protein is unstable.

When a mixture of proteins is applied at a point in a pH gradient, the different proteins have a different net charge at this pH value (see Fig. 1). The positively charged proteins migrate towards the cathode, the negatively charged towards the anode, until they reach the pH value, where they are isoelectric.

The fact of the time stability of the pattern is not always true: carrier ampholytes pH gradients drift after some time, some proteins are not – or not very long – stable at their pI.

In contrast to zone electrophoresis, isoelectric focusing is an end point method. This means, that the pattern – once the proteins have reached their pIs – is stable without time limit. Because of the focusing effect sharp protein zones and a high resolution are obtained.

The book by Righetti (1983) is recommended for further information.

Isoelectric focusing is employed with great success for protein isolation on a preparative scale. It is, however, mainly used for the identification of genetic variations and to investigate chemical, physical and biological influences on proteins, enzymes and hormones. In the beginning, sucrose concentration gradient columns in liquid phase were used, whereas gel media are almost exclusively employed nowadays.

Svensson H. Acta Chem Scand. 15 (1961) 325–341.

The definition of the resolving power of isoelectric focusing was derived by Svensson (1961):

$$\Delta pI = \sqrt{\frac{D[d(pH)/dx]}{E[-du/d(pH)]}}$$

ΔpI is the minimum pI difference needed to resolve two neighboring bands.
See also: titration curve analysis.

ΔpI: resolution capacity
D: diffusion coefficient of the protein
E: field strength (V/cm)
d(pH)/dx: pH gradient
du/d(pH): mobility slope at pI

This equation shows how resolution can be increased:

- When the diffusion coefficient is high, a gel with small pores must be chosen so that diffusion is limited.
- A very flat pH gradient can be used.

But it also illustrates the limits of isoelectric focusing:

- Though the field strength can be raised by high voltages, it cannot be increased indefinitely.
- It is not possible to influence the mobility at the pI.

Nowadays isoelectric focusing can also be performed in capillary electrophoresis equipment, but in the following only gel IEF methods are desribed.

3.2
Gels for IEF

Analytical focusing is carried out in polyacrylamide or agarose gels. It is advantageous to use very thin gels with large pore sizes cast onto support films (Görg et al. 1978).

Görg A, Postel W, Westermeier R. Anal Biochem. 89 (1978) 60–70.

Polyacrylamide gels

Ready polymerized carrier ampholyte polyacrylamide gels and rehydratable polyacrylamide gels with or without immobilized pH gradients (explained below) are commercially available.

These ready-made gels are all polymerized on support films.

The use of washed, dried and rehydrated gels has been published soon after introduction of polyacrylamide for IEF by Robinson (1972), the methodology has been considerably improved by Allen and Budowle (1986). The benefits are listed at page 197.

Robinson HK. Anal Biochem. 49 (1972) 353–366.
Allen RC, Budowle B, Lack PM, Graves G. In Dunn M, Ed. Electrophoresis '86. VCH, Weinheim (1986) 462–473.

Hydrophobic proteins need the presence of 8 to 9 molar urea to stay in solution. Because of the buffering capacity of urea, there is a light increase in the pH in the acid part of the gel. High urea contents in the gel lead to configurational changes in many proteins and disruption of the quarternary structure. The solubility of very hydrophobic proteins, such as membrane proteins for example, can be increased by the addition of non-ionic detergents (e.g. Nonidet NP-40, Triton X-100) or zwitterionic detergents (e.g. CHAPS, Zwittergent).

Because the gels do not co-polymerize with the support films in the presence of non-ionic detergents, it is recommended to rehydrate a prepolymerized, washed and dried gel in the relevant solution.

Agarose gels

Carboxylic and sulfate groups which can be charged always remain.

Agarose gels for isoelectric focusing have only been available since 1975, when it became possible to eliminate the charges of agarose by removing or masking the agaropectin residues in the raw material. Agarose IEF exhibits stronger electroendosmosis than polyacrylamide gel electrophoresis IEF.

Disadvantages: Silver staining does not work as well for agarose gels as for polyacryl-amide gels. In the basic area, electroendosmosis is particu-larly strong.

Separations in agarose gels, usually containing 0.8 to 1.0 % agar-ose, are more rapid. In addition macromolecules larger than 500 kDa can be separated since agarose pores are substantially larger than those of polyacrylamide gels. Another reason to use agarose for IEF is: Its components are not toxic and do not contain catalysts which could interfere with the separation.

Incorporation of linear polyacrylamide: Hoffman WL, Jump AA, Kelly PJ, Elanogo-van N. Electrophoresis 10 (1989) 741–747.

It is difficult to prepare stable agarose gels with high urea concen-trations because urea disrupts the configuration of the helicoidal structure of the polysaccharide chains. Rehydratable agarose gels are advantageous in this case (Hoffman *et al.* 1989).

3.3
Temperature

Righetti PG. J. Chromatogr. 138 (1977) 213–215.

Perella M, Heyda A, Mosca A, Rossi-Bernardi L. Anal Biochem 88 (1978) 212–224.

Cryoproteins are precipitating at low temperatures.

Since the pK values of the Immobilines, the carrier ampholytes and the substances to be analyzed are temperature dependent, IEF must be carried out at a constant controlled temperature, usually 10 °C. For the analysis of the configuration of subunits of specific proteins, li-gand bindings or enzyme-substrate complexes, cryo-IEF methods at temperatures below 0 °C are used (Righetti, 1977, and Perella *et al.* 1978). In order to increase the solubility of cryoproteins (like IgM), agarose IEF is performed at + 37 °C.

3.4
Controlling the pH gradient

Marker proteins for various pH ranges exist. These proteins are chosen so that they can focus independently of the point of application.

Note: Standard marker proteins cannot be used in urea gels, because their conforma-tions are changed, and thus their pIs.

Measurement of the pH gradient with electrodes is a problem since these react very slowly at low temperatures. In addition additives influence the measurement. CO_2 diffusing into the gel from the air reacts with water to form carbonate ions. Those form the anhydrid of carbonic acid and lowers the pH of the alkaline part. To prevent errors which can occur during the measurement of pH gradients, it is recommended to use marker proteins of known pIs. The pIs of the sample can then be measured with the help of a pH calibration curve.

3.5
The kinds of pH gradients

The prerequisite for highly resolved and reproducible separations is a stable and continuous pH gradient with regular and constant conductivity and buffer capacity.

There are two different concepts which meet these demands: pH gradients which are formed in the electric field by amphoteric buffers, the carrier ampholytes, or immobilized pH gradients in which the buffering groups are part of the gel medium.

3.5.1
Free carrier ampholytes

The theoretical basis for the realization of *"natural"* pH gradients was derived by Svensson (1961) while the practical realization is the work of Vesterberg (1969): the synthesis of a heterogeneous mixture of isomers of aliphatic oligoamino-oligocarboxylic acids. These buffers are a spectrum of low molecular weight ampholytes with closely related isoelectric points.

Vesterberg, O. Acta Chem. Scand. 23 (1969) 2653–2666.

The general chemical formula is the following:

$$- CH_2 - N - (CH_2)_x - N - CH_2 -$$

$$(CH_2)_x \qquad (CH_2)_x$$

$$NR_2 \qquad COOH$$

Where R = H or
−(CH₂)ₓ–COOH,
x = 2 or 3

These carrier ampholytes possess the following properties:

- a high buffering capacity and solubility at the pI,
- good and regular conductivity at the pI,
- absence of biological effects,
- a low molecular weight.

Naturally occurring ampholytes such as amino acids and peptides do not have their highest buffering capacity at their isoelectric point. They can therefore not be employed.

Most of the commercially available solutions contain 40 % (w/v) carrier ampholytes. The product "Pharmalytes^{TM}" are produced with a different chemistry, the concentration can therefore not be specified. However, they are used with the same volumes like a 40 % solution.

The pH gradient is produced by the electric field. For example, in a focusing gel with the usual concentration of 2 to 2.5 % (w/v) carrier ampholyte (e.g. for gradients from pH 3 to 10) the gel has a uniform average pH value. Almost all the carrier ampholytes are charged: those with the higher pI positively, those with the lower pI negatively (Fig. 29).

By controlling the synthesis and the use of a suitable mixture the composition can be monitored so that regular and linear gradient result.

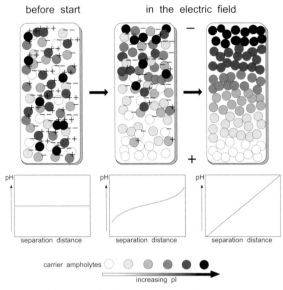

Fig. 29: Diagram of the formation of a carrier ampholyte pH gradient in the electric field.

The anodal end of the gel becomes more acidic and the cathodal side more basic.

When an electric field is applied, the negatively charged carrier ampholytes migrate towards the anode, the positively charged ones to the cathode and their velocity depends on the magnitude of their net charge.

The carrier ampholytes lose part of their charge so the conductivity of the gel decreases.

The carrier ampholyte molecules with the lowest pI migrate towards the anode and those with the highest pI towards the cathode. The other carrier ampholytes align themselves in between according to their pI and will determine the pH of their environment. A stable, gradually increasing pH gradient from pH 3 to 10 results (Fig. 29).

The proteins are considerably larger than the carrier ampholytes – their diffusion coefficient is considerably smaller – they focus in sharper zones.

Since carrier ampholytes have low molecular weights they have a high rate of diffusion in the gel. This means that they diffuse away from their pI constantly and rapidly and migrate back to it electrophoretically: because of this, even when there are only a limited number of isomers a "smooth" pH gradient results. This is particularly important when very flat pH gradients, for example between pH 4.0 and 5.0, are used for high resolution.

Electrode solutions

These electrode solutions are particularly important for long lasting separations in gels containing urea, for basic and for flat gradients. They are not necessary for short gels.

To maintain a gradient as stable as possible, strips of filter paper soaked in the electrode solutions are applied between the gel and the electrodes, an acid solution is used at the anode and a basic one at the cathode. Should, for example, an acid carrier ampholyte reach the anode, its basic moiety would acquire a positive charge from the medium and it would be attracted back by the cathode.

The native IEF in Fig. 30 could be carried out without electrode solutions because a washed and rehydrated gel with a wide pH gradient was used.

Electrode solutions for agarose IEF are listed in chapter 5, for polyacrylamide gels – if required – in chapter 6 of part II.

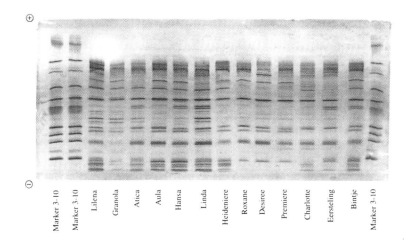

Fig. 30: Isoelectric focusing in a washed and rehydrated polyacrylamide gel. Press sap of potatoes of different varieties. Coomassie Brilliant Blue staining. (Anode on top). From Pharmacia, Freiburg.

Urea IEF

For a number of samples it is necessary to avoid protein-protein interactions and / or to increase the solubility by adding 8 molar urea to the sample solution and to the gel. This causes denaturation of the proteins. The separation is slower than for native IEF, because of the higher viscosity in the urea solution. Urea IEF gels require electrode solutions, partly because of the extended separation times. Very pure urea must be used: When it is partly degraded to isocyanate, proteins become carbamylated, resulting in artifactual additional bands.

It is recommended to prepare urea solutions fresh, and remove isocyanate from the urea solution with mixed bed ion exchanger short before use.

Sometimes a nonionic or zwitterionic detergent is added in order to increase the solubility further, and to avoid crystallysation of the urea. In these cases the copolymerization of the gel and the film support does not work any more.

Urea gels are not available as ready-made gels, because of the limited stability of the urea in solution. But prepolymerized and dried gels on film supports can be soaked in a fresh urea- carrier ampholyte solution short before use. This procedure was, for instance, applied on the differentiation of fish varieties by Rehbein *et al.*

Rehbein H, Kündiger R, Pineiro C, Perez-Martin RI. Electrophoresis 21 (2000) 1458–1463.

Separator IEF

Brown RK, Caspers ML, Lull JM, Vinogradov SN, Felgenhauer K, Nekic M. J Chromatogr. 131 (1977) 223–232.

Ever since the introduction of IEF, modifications of the pH gradients have been investigated. If the resolution is not satisfactory, it is often possible to add *separators* (Brown *et al.* 1977):

These are amino acids or amphoteric buffer substances which flatten the pH gradient in the area of their pI. Their position in the gradient can be changed by adapting the temperature conditions and separator concentration, so that complete separation of neighboring protein bands can be achieved.

Jeppson JO, Franzen B, Nilsson VO. Sci Tools. 25 (1978) 69–73.

One example is the separation of glycosylated HbA from the neighbouring main hemoglobin band in the pH gradient 6 to 8 by the addition of 0.33 mol/L β-alanine at 15 °C (Jeppson *et al.* 1978).

Plateau phenomenon

Problems with carrier ampholytes can arise when long focusing times are necessary. For example, when narrow pH intervals are are used, or in the presence of highly viscous additives such as urea or non-ionic detergents, the gradient begins to drift in both directions but specially towards the cathode.

Righetti PG, Drysdale JW. Ann NY Acad Sci. 209 (1973) 163–187.

This leads to a plateau in the middle with gaps in the conductivity. Part of the proteins migrate out of the gel (Righetti and Drysdale, 1973) and are not included. Because of the limited number of different homologues, the gradients cannot be flattened and the resolution capacity not increased at will.

A gel can "burn" through at the conductivity gaps.

The procedure of a carrier ampholyte IEF run

The IEF running conditions should always be described in a protocol or a publication.

As isoelectric focusing is in principle a nondenaturing method, the optimization of the running conditions is very important to prevent precipitation and aggregation of proteins, and to achieve good reproducibility.

pIs are highly dependent on the temperature.

• Temperature setting

To establish the gradient.

• Prefocusing

On the optimized location with the optimized mode.

• Sample loading

At low field strength to prevent aggregation.

• Sample entry

Volthour integration is often used.

• Separation time is a compromise between letting all proteins reach their pIs and keeping the gradient drift to a minimum.

Proteins have to be fixed during the carrier ampholytes are washed out.

• Fixing (with TCA or by immunofixation) and staining – or alternatively – application of zymogram detection.

3.5.2
Immobilized pH gradients

Because of some limitations of the carrier ampholytes method, an alternative technique was developed: immobilized pH gradients or IPG (Bjellqvist *et al.* 1982). This gradient is built with acrylamide derivatives with buffering groups, the Immobilines, by co-polymerization of the acrylamide monomers in a polyacrylamide gel.

Bjellqvist B, Ek K, Righetti PG, Gianazza E, Görg A, Westermeier R, Postel W. J Biochem Biophys Methods. 6 (1982) 317–339.

The general structure is the following:

$$CH_2 = CH - \underset{\underset{O}{\|}}{C} - \underset{\underset{H}{|}}{N} - R$$

R contains either a carboxylic or an amino group.

An *Immobiline* is a weak acid or base defined by its pK value.
At the moment the commercially available ones are:

- two acids (carboxylic groups) with pK 3.6 and pK 4.6.
- four bases (tertiary amino groups) with pK 6.2, pK 7.0, pK 8.5 and pK 9.3.

To be able to buffer at a precise pH value, at least two different Immobilines are necessary, an acid and a base. Fig. 31 shows a diagram of a polyacrylamide gel with polymerized Immobilines, the pH value is set by the ratio of the Immobilines in the mixture.

The wider the pH gradient desired, the more Immobiline homologues are needed.

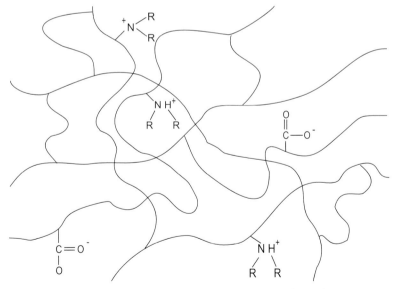

Fig. 31: Diagram of a polyacrylamide network with co-polymerized Immobilines.

Here the pH gradient is absolutely continuous.

A pH gradient is obtained by the continuous change in the ratio of Immobilines. The principle is that of an acid base titration and the pH value at each stage is defined by the *Henderson-Hasselbalch* equation:

C_A and C_B are the molar concentrations of the acid, and basic Immobiline, respectively.

$$pH = pK_B + \log \frac{C_B - C_A}{C_A}$$

when the buffering Immobiline is a base.

If the buffering Immobiline is an acid, the equation becomes:

$$pH = pK_A + \log \frac{C_B}{C_A - C_B}$$

Preparation of immobilized pH gradients

0.5 mm thick Immobiline gels, polymerized on a support film have proved most convenient.

In practice immobilized pH gradients are prepared by linear mixing of two different polymerization solutions with a gradient maker (see Fig. 21), as for pore gradients. In principle a concentration gradient is poured. Both solutions contain acrylamide monomers and catalysts for the polymerization of the gel matrix.

The catalysts must be washed out of the gel because they interfere with IEF. This is more rapid if the gel is thin.

Immobiline stock solutions with concentrations of 0.2 mol/L are used. The solution which is made denser with glycerol is at the acid end of the desired pH gradient, the other solution is at the basic end. During polymerization, the buffering carboxylic and amino groups covalently bind to the gel matrix.

Applications of immobilized pH gradients

It has proved very practical to dry the gels after washing them and to let them soak in the additive solution afterwards.

Immobilized pH gradients can be exactly calculated in advance and adapted to the separation problem. Very high resolution can be achieved by the preparation of very flat gradients with up to 0.01 pH units per cm.

These features are particularly useful for the first dimension in high resolution two-dimensional electrophoresis. The gels are cut in narrow strips for individual sample runs (see chapter 6 and method 10).

Since the gradient is fixed in the gel it stays unchanged during the long separation times which are necessary for flat gradients, but also when viscous additives such as urea and non-ionic detergents are used. In addition there are no wavy iso-pH lines: the gradient is not influenced by proteins and salts in the solution.

The broadest pH gradient which can, at present, be prepared with commercially available Immobilines encompasses 6 pH units: from 4.0 to 10.0; the broadest commercially available gradient spans over 8 pH units: 3–11.

Recipes for the preparation of narrow and wide immobilized pH gradients are given in this book in the section on methods for immobilized pH gradients (part II, method 10). The quantities necessary for the 0.2 molar Immobiline stock solutions for the acid and basic starter solutions are given in mL for the standard gel volume.

A very comprehensive source of informations on immobilized pH gradients is the book by Righetti (1990).

Righetti PG. Immobilized pH gradients: theory and methodology. Elsevier, Amsterdam (1990).

Further developments

Altland (1990) and Giaffreda et al. (1993) have published software for personal computers which permit the calculation the desired pH gradients with optimization of the distribution of buffer concentration and ionic strength.

Altland K. Electrophoresis 11 (1990) 140–147.
Giaffreda E, Tonani C, Righetti PG. J Chromatogr. 630 (1993) 313–327.

In the meantime it has been possible to expand the pH range mentioned before in both directions by using additional types of Immobilines and also to prepare very acidic (Chiari M et al. 1989a) and basic narrow pH gradients (Chiari M et al. 1989b). These are an additional acid with pK 0.8 and a base with pK 10.4. At these pH extremities the buffering capacity of the water ions H^+ and OH^- must be taken into consideration. Furthermore dramatic differences occur in the voltage gradient, which must be compensated by the gradual addition of additives to the gradient.

Acidic Immobiline:
Chiari M, Casale E, Santaniello E, Righetti PG. Theor Applied Electr. 1 (1989a) 99–102.

Basic Immobiline:
Chiari M, Casale E, Santaniello E, Righetti PG. Theor Applied Electr. 1 (1989b) 103–107.

For adequate and reproducible analysis of very basic proteins like lysozyme, histones, and ribosomal proteins in an immobilized pH gradient 9–12 several methodical modifications are necessary (Görg et al. 1997).

Görg A, Obermaier C, Boguth G, Csordas A, Diaz J-J, Madjar J-J. Electrophoresis 18 (1997) 328–337.

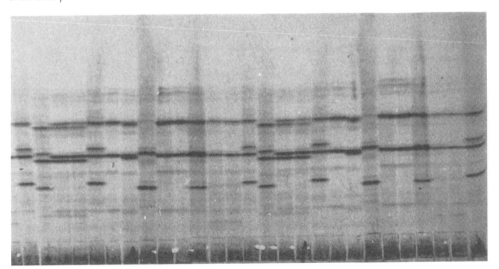

Fig. 32: IEF in immobilized pH gradient pH 4.0 to 5.0. Isoforms of α_1-antitrypsin (protease inhibitors) in human serum. By kind permission of Prof. Dr. Pollack and Ms. Pack, Institut für Rechtsmedizin der Universität Freiburg im Breisgau. (Anode at the top).

Jenne DE, Denzel K, Blätzinger P, Winter P, Obermaier B, Linke RP, Altland K. Proc Natl Acad Sci USA. 93 (1996) 6302–6307.

This means that the pore size is limited towards the top.

The use of immobilized pH gradients is at present restricted to polyacrylamide gels only.

Denaturing gradient gel IEF: Immobilized pH gradient gels with a perpendicular urea gradient from 0 to 8 mol/L have been employed by Altland and Hackler (1984) and Jenne *et al.* (1996) for the separation of human plasma proteins to detect various mutations, which cause diseases. The technique is similar to DGGE described on page 30, but is easier to perform, and – as it is done on the protein level – provides more informations on the danger of a mutation.

Fig. 32 shows an isoelectric focusing result of α_1-antitrypsin isoforms in IPG pH 4.0 to 5.0.

3.6
Protein detection in IEF gels

Because the proteins are present in native form, and large pore size gels are used for isoelectric focusing, the proteins need to be fixed much more intensively before staining than after zone elctrophoresis separations. Usually 20 % (w/v) TCA is used. Ammoniacal silver staining shows a much better sensitivity for protein detection in IEF gels than all other modifications of silver staining. In native gels zymogram techniques for the functional detection of enzymes can be employed. The zymogram techniques are also applied on titration curve gels (see 3.8) and work best, when no acrylamide monomers are present, like in agarose and washed and rehydrated gels.

3.7
Preparative isoelectric focusing

Radola BJ. Biochim Biophys Acta. 295 (1973) 412–428. The procedure is thoroughly described in: Westermeier R: In Cutler P, Ed. Protein purification protocols. Second edition. Methods in molecular biology Volume 244. Humana Press, Totowa, NJ (2004) 225–232.

Preparative carrier ampholyte IEF is mainly carried out in horizontal troughs in granular gels (Radola, 1973). A highly purified dextran gel is mixed with the carrier ampholyte and poured in the trough. Here focusing is done over a long separation distance: about 25 cm. After prefocusing to establish the pH gradient a section of the gel is removed from a specific part of the gradient, mixed with the sample and poured back into place.

For the elution, small columns with nylon sieves are used.

Carrier ampholyte IEF

After IEF the protein or enzyme zones can be detected by staining a paper replica. To recover them, the gel is fractionated with a lattice and the fractions eluted out of the gel with a buffer. Proteins quantities of the order of 100 mg can thus be isolated.

The method has lately experienced a renaissance as a very useful tool for prefractionation of highly heterogeneous protein mixtures under denaturing conditions for high resolution 2-D electrophoresis in narrow pH intervals (Görg *et al.* 2002).

Görg A, Boguth G, Köpf A, Reil G, Parlar H, Weiss W. Proteomics 2 (2002) 1652–1657.

Immobilized pH gradients

Immobilized pH gradients are also very useful for preparative separations:

- They offer a high loading capacity.
- The buffering groups are fixed in the gel.
- The conductivity is low, so even gels which are 5 mm thick hardly heat up.

Righetti PG, Gelfi C. J Biochem Biophys Methods. 9 (1984) 103–119.

Polyacrylamide gels with IPG bind proteins more strongly than other media, so electrophoretic elution methods must be used (Righetti and Gelfi, 1984).

This technique is especially useful for low molecular peptides since the buffering groups of the gradient stay in the gel (Gianazza *et al.* 1983). Peptides are the same size and – after IEF – possess the same charge as the carrier ampholytes so they cannot be separated.

Gianazza E, Chillemi F, Duranti M, Righetti PG, J Biochem Biophys Methods. 8 (1983) 339–351.

Isoelectric membranes: An important approach is the application of the principles and chemistry of immobilized pH gradients on high resolution separation of proteins in a gel-free liquid. Righetti *et al.* (1989) have designed a multicompartment apparatus, whose segments are divided by isoelectric Immobiline membranes. The electrodes are located in the two outer segments. The separation happens between the isoelectric membranes in gel-free liquid, which is constantly recirculated.

Righetti PG, Wenisch E, Faupel M. J Chromatogr. 475 (1989) 293–309.

The highlight of the system are the membranes with defined pH values (Wenger *et al.* 1987): glass microfiber filters are soaked in acrylamide polymerization solutions, which are titrated exactly to the desired pH values with Immobilines. Thus "crystal grade" proteins are obtained without further contamination. Speicher and Zhou (2000) have successfully used such an instrument for prefractionation of complex protein mixtures prior to 2-D electrophoresis.

Wenger P, de Zuanni M, Javet P, Righetti PG. J Biochem Biophys Methods 14 (1987) 29–43.
Speicher DW, Zuo X. Anal Biochem. 284 (2000) 266–278.

The Fig. 33 shows the principle of the purification of a protein in a three-chamber setup, where the chambers are divided by two membranes with pH values closely below and above the pI of the protein to be purified.

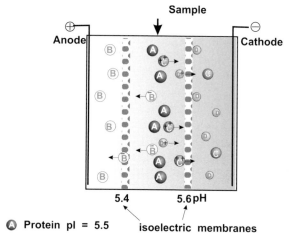

Ⓐ Protein pI = 5.5   isoelectric membranes

Fig. 33: Purification of a protein (A) between two isoelectric membranes. All contaminating charged substances and proteins with pIs lower than pH 5.4 (B) or higher than pH 5.6 (C, D) migrate out of the central chamber, which is enclosed by the two membranes.

3.8
Titration curve analysis

Rosengren A, Bjellqvist B, Gasparic V. In: Radola BJ, Graesslin D. Ed. Electrofocusing and isotachophoresis. W. de Gruyter, Berlin (1977) 165– 171.

Carrier ampholyte gels can also be used to determine the charge intensity curve of proteins. This method is very useful for several reasons: it yields extensive information about the characteristics of a protein or enzyme, for example the increase in mobility around the pI, conformational changes or ligand bindings properties depending on the pH. The pH optimum for separation of proteins with ion-exchange chromatography and for preparative electrophoresis can also be established (Rosengren *et al.* 1977).

A gel with large pore sizes (4 to 5 % T) is used, in order to avoid influence s of the molecule sizes on the mobilities.

A pre-run is performed in the square gel without any sample until the pH gradient is established. The gel, placed on the cooling plate, is rotated by 90° and the sample is applied in a long trough previously polymerized into the gel (see Fig. 34 A).

In practice a series of native electrophoresis runs under various pH conditions are carried out.

When an electric field is applied perpendicular to the pH gradient, the carrier ampholytes will stay in place since their net charges are zero at their pIs.

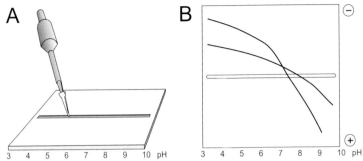

Fig. 34: Titration curves. A) Application of the sample in the sample trench *after* the pH gradient has been established. B) Titration curves.

The sample proteins will migrate with different mobilities according to the pH value at each point and will form curves similar to the classical acid-base titration curves (Fig. 34 B). The pI of a protein is the point at which the curve intersects the sample trough.

There is a representation standard for titration curves for purposes of comparison: the gel is oriented so that the pH values increase from left to right and the cathode is on top (Fig. 35).

As can be seen, no buffer reservoirs are necessary for native electrophoresis in amphoteric buffers. This forms the basis of an electrophoretic method, which is described in method 4.

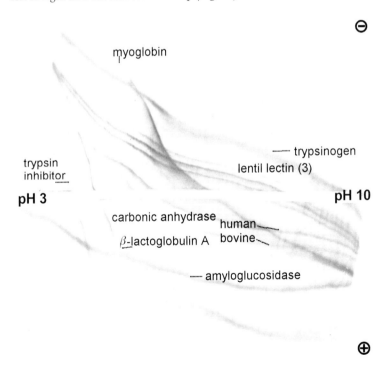

Fig. 35: Titration curves of a pI marker protein mixture pH 3 to 10. Cathode at the top.

4
Blotting

4.1
Principle

Blotting is the transfer of large molecules on to the surface of an immobilizing membrane. This method broadens the possibilities of detection for electrophoretically separated fractions because the molecules adsorbed on the membrane surface are freely available for macromolecular ligands, for example antigens, antibodies, lectins or nucleic acids. Before the specific detection the free binding sites must be blocked with substrates, which do not take part in the ensuing reaction (Fig. 36).

In addition blotting is an intermediate step in protein sequencing and an elution method for subsequent analyses.

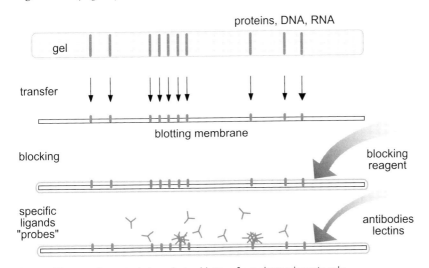

Fig. 36: The most important steps during blotting from electrophoresis gels.

Electrophoresis in Practice, Fourth Edition. Reiner Westermeier
Copyright © 2005 WILEY-VCH Verlag GmbH & Co. KGaA, Weinheim
ISBN: 3-527-31181-5

4.2
Transfer methods

4.2.1
Diffusion blotting

Quantitative transfers cannot be achieved with this method, especially not with larger molecules.

The blotting membrane is applied on to the gel surface as when making a replica. The molecules are transferred by diffusion. Since the molecules diffuse regularly in every direction, the gel can be placed between two blotting membranes thus yielding two mirror-image transfers (Fig. 37). The diffusion can be accelerated by increasing the temperature, the technique in then known as *thermoblotting*. It is mostly used after electrophoresis in gels with large pores.

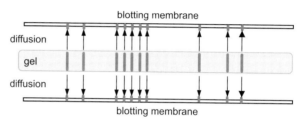

Fig. 37: Bi-directional transfer of proteins by diffusion blotting from a gel with large pores.

4.2.2
Capillary blotting

Southern EM. J Mol Biol. 98 (1975) 503–517.
Alwine JC, Kemp DJ, Stark JR. Proc Natl Acad Sci USA, 74 (1977) 5350–5354.

Olsson BG, Weström BR, Karlsson BW. Electrophoresis 8 (1987) 377–464.

This technique is a standard one for subsequent hybridization according to Southern (1975) (Southern blot) during DNA separations. The transfer of RNA on to a covalently binding film or nylon membrane which is now known under the name Northern blot also uses this technique (Alwine *et al.* 1977).

This kind of transfer can also be used for proteins, which were separated in a gel with large pores (Olsson *et al.* 1987). Buffer is drawn from a reservoir through the gel and the blotting membrane to a stack of dry paper tissues by capillary force. The molecules are carried to the blotting membrane on which they are adsorbed. The transfer occurs overnight (Fig. 38).

4.2.3
Pressure blotting

Desvaux FX, David B, Peltre G. Electrophoresis 11 (1990) 37–41.

Pressure blots from agarose gels cast on GelBond film are obtained very easily: a wet blotting membrane is laid on the gel, covered by one or several dry filter paper sheets, a glass plate and a 1 kg weight for

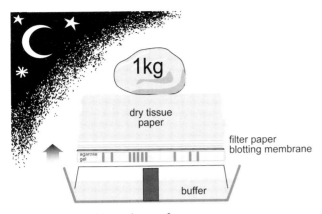

Fig. 38: Capillary blotting, the transfer occurs overnight.

100 cm^2. The transfer is very fast, only a few seconds! Even multiple successive and identical blots can be obtained from one single gel (Desvaux *et al.* 1990).

The most effective transfer from isoelectric focusing gels is pressure blotting according to Towbin *et al.*

Towbin H, Özbey Ö, Zingel O. Electrophoresis 22 (2001) 1887–1893.

4.2.4
Vacuum blotting

This technique is mostly used instead of capillary blotting (Olszewska and Jones, 1988). It is important to have a controlled low vacuum with, depending on the case, a 20 to 40 cm high water column to prevent the gel matrix from collapsing. An adjustable pump is used since a water pump yields a vacuum that is too high and irregu-

Olszewska E, Jones K. Trends Gen. 4 (1988) 92–94.

Also pressure can be employed; this is also called pressure blotting.

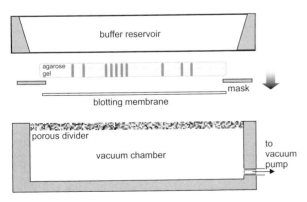

Fig. 39: Transfer of nucleic acids with vacuum blotting in 30 to 40 min.

lar. The surface of the gel is accessible to reagents during the entire procedure. A diagram of a vacuum blotting chamber is represented in Fig. 39.

Vacuum blotting possesses some advantages over capillary blotting, it:

in the tank
Gel remains in the tank.

- is faster: 30 to 40 min, instead of overnight;
- is quantitative, there are no back transfers; leads to sharper zones and better resolution;
- allows faster depurination, denaturation and neutralization;
- reduces the mechanical stress on the gel;
- saves expenses for solutions and paper.

4.2.5
Electrophoretic blotting

Towbin H, Staehelin T, Gordon J. Proc Natl Acad Sci USA. 76 (1979) 4350–4354.
Burnette WN. Anal Biochem. 112 (1981) 195–203.

Electrophoretic transfers are mainly used for proteins SDS electrophoresis (Towbin *et al.* 1979; Burnette, 1981). Only in some cases also nucleic acids are transferred with the help of an electric field. Either different samples are applied on a gel and analysed together on the membrane, or the antigen solution is separated across the entire gel width and the membrane is cut into narrow strips for probing in different antibody solutions (e.g. patients sera).

Burgess R, Arthur TM, Pietz BC. Meth Enzymol. 328 (2000) 141–157.

Electrophoretic protein blotting followed by immunodetection is frequently called *"Western Blotting"* following the methodical evolution from *"Southern"* and *"Northern Blotting"*. The modification of blotting proteins and probing the membrane with a non-antibody protein to detect specific protein-protein interactions is called *"Far-Western blotting"* (Burgess *et al.* 2000). The probing protein is then tagged with a labeled antibody for detection.

Tank blotting

The buffer should be cooled so that the blotting sandwich does not warm up too much.

Originally, vertical buffer tanks with coiled platinum wire electrodes fixed on two sides were used. For this technique, the gel and blotting membrane are clamped in grids between filter papers and sponge pads and suspended in the tank filled with buffer. The transfers usually occur overnight.

Semi-dry blotting

Kyhse-Andersen J. J Biochem Biophys Methods. 10 (1984) 203–209.
Tovey ER, Baldo BA. Electrophoresis 8 (1987) 384–387.

Semi dry blotting between two horizontal graphite plates has gained more and more acceptance in the last few years. Only a limited volume of buffer, in which a couple of sheets of filter paper are soaked, is necessary. This technique is simpler, cheaper and faster and a discontinuous buffer system can be used (Kyhse-Andersen, 1984; Tovey and Baldo, 1987).

The *isotachophoresis* effect occurs here: the anions migrate at the same speed, so that a regular transfer takes place. A system of graphite plates does not need to be cooled. A current not higher than 0.8 to 1 mA per cm^2 of blotting surface is recommended. The gel can overheat if higher currents are used and proteins can precipitate.

Graphite is the best material for electrodes in semi dry blotting because it conducts well, does not overheat and does not catalyze oxidation products.

The transfer time is approximately one hour and depends on the thickness and the concentration of the gel. When longer transfer times are required as for thick (1 mm) or highly concentrated gels, a weight is placed on the upper plate so that the electrolyte gas is expelled out of the sides. Fig. 40 shows a diagram of a semi dry-blotting setup.

In semi dry blotting as well, several blots can be made simultaneously. They are piled in layers called "trans units".

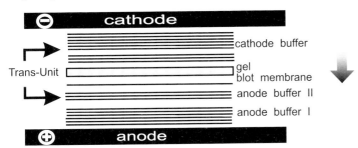

Fig. 40:   Diagram of a horizontal graphite blotter for semi dry blotting. Up to six trans units can be blotted at one time (Kyhse-Andersen, 1984).

It is also possible to perform electrophoretic transfers on two membranes simultaneously: *Double Replica Blotting* (Johansson, 1987). An alternating electric field is applied on a blotting sandwich with a membrane on each side of the gel with increasing pulse time, so that two symmetrical blots result.

Johansson K-E. Electrophoresis 8 (1987) 379–383.

Blotting of gels supported by films

Ready-made or self-made gels backed by support films are used more and more often for electrophoresis of proteins and electrofocusing. These films which are inpermeable to current and buffer must be separated from the gels so that electrophoretic transfers or capillary blotting can be carried out. To separate the gel and the film without damage an apparatus exists with a taut thin steel wire which is pulled between them (see Fig. 41).

Fig. 41: Instrument for complete and trouble-free separation of gels backed by films.

4.3
Blotting membranes

Disadvantages: limited binding capacity and poor mechanical stability.

Nitrocellulose is the most commonly used membrane. It is available in pore sizes from 0.05 μm to 0.45 μm. The pore size is a measure of the specific surface: the smaller the pores, the higher the binding capacity.

Salinovich O, Montelaro RC. Anal Biochem. 156 (1986) 341–347.

Proteins adsorbed on nitrocellulose can be reversibly stained so that the total protein can be estimated before specific detection (Salinovich and Montelaro, 1986).

see the instructions in the second part.

It is also possible to render the blots totally transparent by an imbedding technique.

Montelaro RC. Electrophoresis 8 (1987) 432–438.

Nitrocellulose is occasionally also used for preparative methods and the proteins can be eluted out again (Montelaro, 1987).

Handmann E, Jarvis HM. J Immunol Methods. 83 (1985) 113–123.

A better adsorption of glycoproteins, lipids and carbohydrates is obtained by ligand precoating of nitrocellulose (Handmann and Jarvis, 1985).

Disadvantages: staining is not reversible.

Matsudaira P. J Biol Chem. 262 (1987) 10035–10038.

Polyvinylidenedifluoride (PVDF) membranes on a Teflon base possess a high binding capacity and a high mechanical stability like nylon membranes. PVDF membranes can also be used for direct protein sequencing (Matsudaira, 1987).

DBM and DPT are more and more often replaced by nylon membranes.

Diazobenzyloxymethyl and *diazophenylthioether* papers (DBM, DPT), which must be chemically activated before use, enable a two step binding with molecules: electrostatic and covalent.

Positively and negatively charged but also neutral nylon membranes exist.

Nylon membranes possess a high mechanical stability and a high binding capacity, usually due to electrostatic interactions. This means that staining can be a problem because small molecules are also strongly bound.

Karey KP, Sirbasku DA. Anal Biochem. 178 (1989) 255–259.

Fixing with glutardialdehyde after transfer is recommended to increase the binding of low molecular weight peptides to nylon membranes (Karey and Sirbasku, 1989).

Ion-exchange membranes, diethylaminoethyl (DEAE) or carboxymethyl (CM) are used for preparative purposes because of the reversibility of the ionic bonds.

Activated glass fiber membranes are used when blotted proteins are directly sequenced. Several methods to activate the surface exist: for example, bromocyanide treatment, derivatization with positively charged silanes (Aebersold *et al.* 1986) or hydrophobation by siliconation (Eckerskorn *et al.* 1988).

Unfortunately a blotting membrane which binds 100% of the molecules does not exist yet. For example, during electroblotting small proteins often migrate through the film while larger proteins have not yet completely left the gel. It is therefore necessary to try and obtain as regular transfers as possible.

Disadvantage: These membranes are often very brittle.

Aebersold RH, Teplow D, Hood LE, Kent SBH. J Biol Chem. 261 (1986) 4229–4238.

Eckerskorn C, Mewes W, Goretzki H, Lottspeich F. Eur J Biochem. 176 (1988) 509–519.

4.4
Buffers for electrophoretic transfers

A. Proteins

Tank blotting

A Tris-glycine buffer of pH 8.3 is usually used for tank blotting of SDS gels (Towbin *et al.* 1979). This moderate pH should cause little damage to the proteins. 20% methanol is often added to the buffer to increase the binding capacity of the blotting membrane and to prevent the gel from swelling. At this relatively low pH value the proteins are not highly charged so the transfers are time-consuming and the buffer heats up gradually.

Sometimes to speed up the transfer, ready-stained proteins for blotting and 0.02 to 0.1% SDS to keep hydrophobic proteins in solution are added to the buffer. But this can lead to problems with the binding capacity of the membrane.

Since SDS, methanol, heat and long transfer times are also harmful to proteins, it can be an advantage to use buffers with higher pH values, this saves a considerable amount of time. For example titrate 50 mmol/L CAPS to pH 9.9 with sodium hydroxide.

This transfer takes about 50 min so the buffer does not have time to warm up.

Acid gels with basic proteins and isoelectric focusing are blotted in a tank with 0.7% acetic acid (Towbin *et al.* 1979).

The proteins then migrate towards the cathode.

For glycoproteins, polysaccharides and also lipopolysaccharides 10 mmol/L sodium borate pH 9.2 (Reiser and Stark, 1983) is recommended since boric acid binds to the sugar moieties and thus, in a basic medium, the molecules acquire a negative charge.

Reiser J, Stark GR. Methods Enzymol. 96 (1983) 205–215.

Semi dry blotting

A continuous buffer must of course be used when it is necessary to carry out a double replica electroblotting with the semi dry technique.

Up to now a continuous Tris-glycine-SDS buffer was often recommended for horizontal blotting (semidry or graphite plate blotting). Yet experience has shown that in general a discontinuous buffer system is preferable since it yields sharper bands and more regular and efficient transfers.

Continuous buffer:

48 mmol/L Tris, 39 mmol/L glycine, 0.0375% (w/v) SDS, 20% methanol.

Discontinuous buffer system (acc. to Kyhse-Andersen, 1984):

This buffer system can be used for SDS as well as for native and IEF gels.

Anode I:	0.3 mol/L Tris, 20% methanol
Anode II:	25 mmol/L Tris, 20% methanol
Cathode:	40 mmol/L 6-aminohexanoic acid
	same as ε-aminocaproic acid),
	20% methanol, 0.01% SDS

This is mainly applicable to SDS gels.

If the transfer efficiency of high molecular weight proteins (80 kDa) is not satisfactory, the gel can be equilibrated in the cathode buffer for 5 to 10 min before blotting.

For enzyme detection the buffer must not contain any methanol, otherwise biological activity is lost. Brief contact with a small amount of SDS does not denature the proteins.

Caution: proteins can diffuse rapidly out of IEF gels.

For transfers from urea IEF gels the urea should first be allowed to elute out of the gel by soaking it in cathode buffer. Otherwise the proteins, which possess no charge after IEF cannot bind to SDS as required for electrophoretic transfer.

B. Nucleic acids

Tank blotting

Smith MR, Devine CS, Cohn SM, Lieberman MW. Anal Biochem. 137 (1984) 120–124.

Acid buffers are often used for DNA blotting: 19 mmol/L sodium phosphate, 54 mmol/L sodium citrate pH 3.0 (Smith *et al.* 1984).

Semi-dry blotting

Fujimura RK, Valdivia RP, Allison MA. DNA Prot Eng Technol. 1 (1988) 45–60.

For neutral blotting 10 mmol/L Tris-HCl, 5 mmol/L sodium acetate, 0.5 mmol/L EDTA, pH 8.7 are used. For alkaline blotting 0.4 mol/L NaOH is used (Fujimura *et al.* 1988). Both these techniques can be carried out in tanks with equal efficiency. However, alkaline blotting damages the plastic material of the tank while graphite plates are resistant to sodium hydroxide.

4.5
General staining

It is often desirable to check the electrophoresis and/or the results of the transfer overall before carrying out a precise evaluation.

Nucleic acids are in general visualized with ethidium bromide, which is often added to the gel before separation. The nucleic acids can then be seen under UV light.

For *proteins*, besides staining with Amido Black or Coomassie Brilliant Blue, mild staining methods such as the very sensitive Indian Ink method (Hancock and Tsang, 1983) exist as well as reversible ones with Ponceau S (Salinovich and Montelaro, 1986) or Fast Green FCF (see part II, method 9). The sensitivity of Indian Ink staining and the antibody reactivity of the proteins can be enhanced by alkaline treatment of the blotting membrane (Sutherland and Skerritt, 1986).

Hancock K, Tsang VCW. Anal Biochem. 133 (1983) 157–162. Sutherland MW, Skerritt JH. Electrophoresis 7 (1986) 401–406. Kittler JM, Meisler NT, Viceps-Madore D. Anal Biochem. 137 (1984) 210–216. Moeremans M, Daneels G, De Mey J. Anal Biochem. 145 (1985) 315–321.

In many cases one can also use:

- a general immunostain (Kittler *et al.* 1984),
- colloidal gold (Moeremans *et al.* 1985),
- autoradiography,
- fluorography (Burnette, 1981).

Nylon membranes bind anionic dyes very strongly so normal staining is not possible but nylon membranes can be stained with caccodylate iron colloid (FerriDye) (Moeremans *et al.* 1986).

Moeremans M, De Raeymaeker M, Daneels G, De Mey. J. Anal Biochem. 153 (1986) 18–22.

4.6
Blocking

Macromolecular substances which do not take part in the visualization reaction are used to block the free binding sites on the membrane.

Denhardts buffer is used for *nucleic acids* (Denhardt, 1966): 0.02% BSA, 0.02% Ficoll, 0.02% polyvinylpyrollidene, 1 mmol/L EDTA, 50 mmol/L NaCl, 10 mmol/L NaCl, 10 mmol/L Tris-HCl pH 7.0, 10 to 50 mg heterologous DNA per mL.

Denhardt D. Biochem Biophys Res Commun. 20 (1966) 641–646.

A number of possibilities exist for proteins, 2% to 10% bovine albumin is used most often (Burnette, 1981). The cheapest blocking substances and the ones that cross-react the least are: skim milk or 5% skim milk powder (Johnson *et al.* 1984), 3% fish gelatin, 0.05% Tween 20. The blocking step is quickest and most effective at 37 °C.

Johnson DA, Gautsch JW. Sportsman JR. Gene Anal Technol. 1 (1984) 3–8.

4.7
Specific detection

Hybridization

- *Radioactive probes*

Radioactive labeling is mainly used for the evaluation of RFLP analysis.

A higher detection sensitivity can be obtained for the analysis of DNA fragments with radioactive DNA or RNA probes which bind to complementary DNA or RNA on the blotting membrane (Southern, 1975; Alwine *et al.* 1977).

- *Non-radioactive probes*

as for immunoblotting

There is now a trend to avoid radioactivity in the laboratory, so accordingly the samples can be marked with biotin-streptavidin or dioxigenin.

Characterization of individuals in forensic medicine.

This method is also used for DNA fingerprinting.

Enzyme blotting

The transfer of native separated enzymes on to blotting membranes has the advantage that the proteins are fixed without denaturation and thus do not diffuse during slow enzyme-substrate reactions and the coupled staining reactions (Olsson *et al.* 1987).

Immunoblotting

Specific binding of immunoglobulins (IgG) or monoclonal antibodies are used to probe for individual protein zones after blocking. An additional marked protein is then used to visualize the zones. Once again several possibilities exist:

- *Radioiodinated protein A*

Renart J, Reiser J, Stark GR. Proc Natl Acad Sci USA. 76 (1979) 3116–3120.

The use of radioactive protein A which attaches itself to specific binding antibodies enables high detection sensitivities (Renart *et al.* 1979). But, ^{125}I-protein A only binds to particular IgG subclasses; in addition radioactive isotopes are now avoided as much as possible in the laboratory.

- *Enzyme coupled secondary antibodies*

Taketa K. Electrophoresis 8 (1987) 409–414.
Blake MS, Johnston KH, Russell-Jones GJ. Anal Biochem. 136 (1984) 175–179.

An antibody to the specific binding antibody is used and it is conjugated to an enzyme. Peroxidase (Taketa, 1987) or alkaline phosphatase (Blake *et al.* 1984) are usually employed as the conjugated reagent.

The ensuing enzyme-substrate reactions have a high sensitivity. The tetrazolium method has the highest sensitivity in the peroxidase method (Taketa, 1987).

- *Gold coupled secondary antibody*

Detection by coupling the antibody to *colloidal gold* is very sensitive (Brada and Roth, 1984): in addition the sensitivity can be increased by subsequent *silver enhancement*: the lower limit of detection lies around 100 pg (Moeremans *et al.* 1984).

Brada D, Roth J. Anal Biochem. 142 (1984) 79–83.

Moeremans M, Daneels G, Van Dijck A, Langanger G, De Mey J. J Immunol Methods. 74 (1984) 353–360.

- *Avidin biotin system*

Another possibility is the use of an amplifying enzyme detection system. The detection results from enzymes which are part of a non-covalent network of polyvalent agents (antibodies, avidin): for example biotin-avidin-peroxidase complexes (Hsu *et al.* 1981) or complexes with alkaline phosphatase.

Hsu D-M, Raine L, Fanger H. J Histochem Cytochem. 29 (1981) 577–580.

- *Chemiluminescence*

The highest sensitivity without using radioactivity can be achieved with enhanced chemiluminescent detection methods (Laing, 1986). In Fig. 42 it is shown, how such a signal can be further increased by combining the biotin-streptavidin peroxidase system with enhanced chemiluminescence.

Laing P. J Immunol Methods. 92 (1986) 161–165.

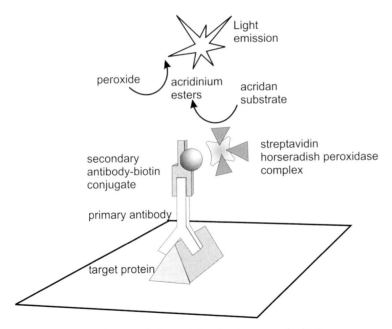

Fig. 42: Schematic diagram of enhanced chemiluminescence plus biotin-streptavidin complexes, which afford the highest detection sensitivity.

To detect the signal, the membrane is exposed to a film or to an appropriate CCD camera in an absolutely dark closet for a certain time period. New multipurpose scanners can also be used in a chemiluminescence detection mode.

Immunological detection on blots can be automated, for example with the staining unit of the PhastSystem® (Prieur and Russo-Marie, 1988).

Prieur B, Russo-Marie F. Anal Biochem. 172 (1988) 338–343.

Lectin blotting

Bayer EA, Ben-Hur H, Wilchek M. Anal Biochem. 161 (1987) 123–131.

The detection of glycoproteins and specific carbohydrate moieties can be performed with lectins. Visualization is carried out by aldehyde detection or, analogous to immunoblotting, with the avidin-biotin method (Bayer *et al.* 1987).

4.8
Protein sequencing

Vandekerckhove J, Bauw G, Puype M, Van Damme J, Van Montegu M. Eur J Biochem. 152 (1985) 9–19.
Eckerskorn C, Lottspeich F. Chromatographia. 28 (1989), 92–94.
Aebersold RH, Pipes G, Hood LH, Kent SBH. Electrophoresis 9 (1988) 520–530.
Eckerskorn C, Strupat K, Karas M, Hillenkamp F, Lottspeich F. Electrophoresis 13 (1992) 664–665.
Strupat K, Karas M, Hillenkamp F, Eckerskorn C, Lottspeich F. Anal Chem. 66 (1994) 464–470.

The use of blotting for direct protein sequencing has been a big step forward for protein chemistry and molecular biology (Vandekerckhove *et al.* 1985). Blotting is mostly performed out of one-dimensional SDS gels or 2D gels (Matsudaira *et al.* 1987; Aebersold *et al.* 1986; Eckerskorn *et al.* 1988; Eckerskorn and Lottspeich, 1989). If the proteins to be sequenced have to be separated by isoelectric focusing, an immobilized pH gradient should be used because carrier ampholytes would interfere with the sequencing signals (Aebersold *et al.* 1988).

Also matrix-assisted laser desorption ionization mass spectrometry (MALDI-MS) has been employed to measure the molecular mass of proteins from blots with high precision (Eckerskorn *et al.* 1992; Strupat *et al.* 1994).

Simpson RJ, Moritz RL, Begg GS, Rubira MR, Nice EC. Anal Biochem. 177 (1989) 221–236.

The reference by Simpson *et al.* (1989) presents a review of the different methods.

4.9
Transfer problems

- *Poor solubility*, especially of hydrophobic proteins, can prevent *The addition of 6 to 8 mol/L* a transfer. In such cases the blotting buffer should contain a *urea is only reasonable in semi* detergent, for example SDS. The addition of urea can also *dry blotting (small buffer* increase the solubility. *volume).*

- During *native immunoblotting* according to Bjerrum *et al. Bjerrum OJ, Selmer JC,* (1987), the electrophoresis gel contains non-ionic detergents. *Lihme A. Electrophoresis 8* Its binding capacity is hindered in case of direct contact with *(1987) 388–397.* the blotting membrane. This can be prevented by inserting a 2 to 3 mm thick agarose gel layer containing transfer buffer but no detergent between the gel and the blotting membrane. Nitrocellulose with increased binding capacity can also be used.

- High molecular weights cause a slower migration out of the gel. But when blotting is carried out for a long time and/or at high field strengths, low molecular weight proteins detach themselves again from the membrane and are lost. Several possibilities to obtain regular transfers over a wide molecular weight spectrum exist:

- The use of pore gradients in SDS-PAGE, *After the separation the proteins are distributed according to their molecular weight in areas with small or large pores.*

- The use of a discontinuous buffer system for semi dry blot- *The isotachophoresis effect* ting, *induces a regular velocity.*

- The treatment of high molecular weight proteins with protease *Limited proteolysis usually does* after electrophoresis, *not damage the antigenicity.*

- The use of a buffer with another pH, *This increases the mobility.*

- The use of a buffer without methanol, *The pores become larger when the gel swells.*

- The addition of SDS (0.01 to 0.1%) to the buffer, *Caution: too much SDS reduces the binding capacity.*

- The use of native immunoblotting (Bjerrum *et al.* 1987), *Gels with large pores can be used here,*

- Blotting for a longer time and placing a second blotting mem- *To trap low molecular weight* brane behind. *proteins.*

Applications of protein blotting

A series of review articles has been published:

Gershoni JM, Palade GE. Anal Biochem. 112 (1983) 1–15.

Beisiegel U. Electrophoresis 7 (1986) 1–18.

Bjerrum OJ. Ed. Paper symposium protein blotting. Electrophoresis 8 (1987) 377–464.

Baldo BA, Tovey ER. Ed. Protein blotting. Methodology, research and diagnostic applications. Karger, Basel (1989).

Baldo BA. In Chrambach A, Dunn M, Radola BJ. Eds. Advances in electrophoresis 7. VCH, Weinheim (1994) 409–478.

5
Interpretation of electropherograms

5.1
Introduction

5.1.1
Purity control

To control purity, electrophoretic methods are usually used in combination with chromatography. The physico-chemical properties of the substances to be investigated are generally known.

SDS electrophoresis is the most frequently used technique for protein analysis. During SDS electrophoresis, configurational differences and irrelevant charge heterogeneities of polypeptides or enzymes are eliminated so that only real impurities appear as extra bands of different molecular weights. For analysis of low molecular weight peptides special designed gels or buffer systems are usually applied. In some cases the load of the major fraction has to be very high, because low amounts of contaminating proteins must be detected and – sometimes – quantified.

When the kind of glycosylation or the charge properties of a protein are significant, isoelectric focusing is used. In some cases band heterogeneity caused by the various conformations of a molecule must be taken into account when interpreting results.

Agarose submarine gels are often used for nucleic acids. Ethidium bromide staining is used for detection, either with hybridization in the gel or after blotting on an immobilizing membrane.

Gel electrophoresis has the advantage that a large number of samples can be analyzed at one time.

Depending on the type and degree of impurity, staining methods with high or low sensitivity are employed; blotting techniques are used for many experiments.

In cases of very high protein loads, a complete discontinuous buffer system has to be applied (see page 34).

5.1.2
Quantification prerequisites

Quantification is either carried out with direct UV measurement of the zones in capillary systems or – when support materials such as gels or films are used – indirectly by autoradiography or staining of the zones followed by densitometric measurement.

Electrophoresis in Practice, Fourth Edition. Reiner Westermeier
Copyright © 2005 WILEY-VCH Verlag GmbH & Co. KGaA, Weinheim
ISBN: 3-527-31181-5

In *capillary* electrophoresis, the scans resemble chromatograms and the peaks can be integrated as in standard chromatography or HPLC.

In *immunoelectrophoresis*, the distance between the precipitation line and the origin is a measure of the concentration of the substance in the sample, regardless of whether the process involves electrophoretic or electroosmotic migration or diffusion.

The quality of the results depends on the quality of the antibodies: impurities with cross-reacting antibodies must be eliminated.

It is, however, important to know the quantity of antibody introduced as well as the titre of the antigen-antibody mixture. The clearest and most reproducible results are obtained with the rocket method of Laurell (1966). The areas enclosed by the precipitation lines are proportional to the amount of antibody in the sample. In many cases, it is accurate enough to simply measure the height of the precipitin arc.

The success of quantitative determinations using electrophoretic separations on gels or other supports depends on several factors:

For practical reasons, methods for sample preparation are described in connection with each separation method.

- During sample preparation, substance losses through adsorption on membranes or column material during desalting or concentration should be avoided as well as the formation of irreversible deposits during extraction or precipitation. Complexing or chelating agents must also be removed from the sample or else complex formation must be inhibited.
- The application method must be chosen in such a way that all substances completely penetrate the separation medium. This is especially critical during isoelectric focusing of heterogeneous protein mixtures since different proteins are unstable at different pH values or have a tendency to aggregate. In such cases it can be assumed that all proteins will not penetrate the medium at the same point. During sample application, the volumetric precision of the syringe or micropipette is also important.

Qualitative differences during separations can be compensated with the help of marker proteins.

- The quality of the separation is crucial for densitometric measurements. Wavy distorted zones, which result for example from the salt concentration being too high, lead to questionable densitograms. In addition, a zone can only be properly quantified when it is well separated from the neighboring bands.

Hot staining and colloidal staining methods are recommended.

- The prerequisite for a reliable quantification is an effective staining of the bands while avoiding destaining during background washing.

Nevertheless, it should always be presumed that differences in staining effectiveness exist. For this reason a protein mixture consisting of known pure substances (e.g. marker proteins) at various concentrations (serial dilutions) should always be applied and run in parallel as is done in qualitative comparison of bands for molecular weight or isoelectric point determination.

Since each protein possesses a specific affinity to the dye used which is different from other proteins, it should always be remembered that only relative quantitative values can be determined.

For example measurements relative to albumin.

Gradient gels almost always display an increasingly or decreasingly shaded background after staining. This should be accounted for during photometric measurement of the electrophoresis bands and the ensuing integration of the surfaces.

5.2
Image analysis

In many cases it is sufficient to compare the bands and spot patterns visually, to photograph the separations or keep the original gels in dry or humid form. However, measurement and further analysis of the electropherograms is necessary for a number of applications:

e.g. the identification of substances or simple comparison of electrophoresis patterns.

- It is difficult, even impossible to determine the intensity of individual fractions visually. For example:
  Homozygous and heterozygous genotypes must be differentiated during genetic investigations. Usually only the presence or absence of a band or spot can be detected visually.
  When protein metabolism kinetics is studied, the increase or decrease of certain fractions must be recognized.

  Homozygote means: intense band. Heterozygote means: partial intensity.

- Single fractions or groups of fractions should be quantified. This is only possible by scanning the separation traces and integrating the surfaces of the peak diagram with a densitometer. It is necessary to measure the zones as exactly as possible since they can present different forms and zone widths depending on the method used.

  A whole series of factors must be taken into account during quantification. They will be discussed later.

- The interpretation of 2-D electropherograms with several hundreds to a few thousand spots is complicated and time-consuming, which is why computers are essential. The patterns are normally recorded with scanners.

  see 2-D electrophoresis, pages 42 and 92 ff.

- Densitograms are the usual form of representation in many areas such as clinical chemistry for example.
- Electrophoresis patterns can often be compared more exactly by densitometry than by the simple visual comparison.
- Data processing is used more and more often because of the amount of data collected in laboratories. To be able to evaluate, save and process results of electrophoresis by computer, the lanes must be digitalized by a scanner.

  In many laboratories, scanning of gels and subsequent printing with a laser or ink jet printer has replaced photographing.

- Molecular weights or isoelectric points of samples can be assigned by computer.

- For microbiological taxonomy, plant variety determinations, population studies, and clinical genetics lane relationship dendrograms have to be developed.

Magnifying glass effect.

- For many uses, especially in routine analysis, the separation distances are shortened and the results more difficult to interpret visually. Bands which lie close to one another can be resolved by high resolution scanning and enlarged by the computer.

5.2.1
Hardware for image analysis

Prior to the analysis, the electropherograms have to be fed into a computer with a video camera, a desktop scanner or a densitometer.

The background of blotting membranes sometimes reaches 2.5 O.D.

Optical density: A prerequisite for correct and linear values is a sufficient light intensity. In high resolution electrophoresis the zones can have an optical densities (O.D.) over 3 O.D.

The unit O.D. for the optical density is mostly used in biology and biochemistry and is defined as follows: 1 O.D. is the amount of substance, which has an absorption of 1 when dissolved and measured in 1 mL in a cuvette with a thickness of 1 cm.

The intensity of the light absorption of a substance is called extinction.

According to Lambert-Beer's law the extinction of light of a specific wavelength shining through a dissolved substance is proportional to its concentration.

The quantification of results is only possible when the absorptions are linear. High performance desktop scanners show linear measurements up to 3.5 O.D.

CCD cameras provide digital signals. They can be coupled directly to a computer.

Cameras are mainly used for relatively small gels up to 15 cm separation distance. For high resolution gels, like those used for 2-D electrophoresis, scanners are recommended.

Video Cameras: The resolution and sensitivity of video cameras has been considerably improved in the last years. They can be used for visible and UV light. Video cameras are mostly employed for the evaluation of relatively small gels, blotting membranes, and Ethidium bromide stained agarose gels. With cooled CCD cameras very high sensitivity can be achieved. A camera has the advantage, that it can accumulate signals over a certain time period for detecting weak signals. Modern camera systems are built into light-shielded viewing chambers, no darkrooms are required for fluorescence and chemiluminescence detection.

Because they are relatively slow, they are more and more replaced by scanners.

Densitometers: Those are mobile photometers which measure electrophoresis or thin-layer chromatography lanes. The Rf values and the extinction (light absorption) of the individual zones are determined. The result is a peak diagram (densitogram). The surfaces under the peaks can be used for quantification. Such densitometers are still used in routine clinical analysis for the evaluation of cellulose acetate and agarose gels.

Desktop Scanners: High performance instruments are available, which can scan in both, reflectance and transmittance mode, and have a liquid leakage-free scanning bed (Fig. 43). They scan very fast, provide very high resolution, and they are less expensive than densitometers. For quantitative results calibration and scanning in transmission mode is a must.

Blotting membranes are mostly scanned in the reflectance mode.

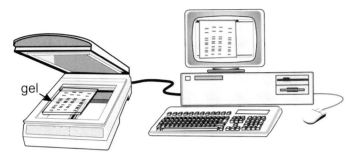

gel

Fig. 43: Modified desk top scanner with liquid leakage-free scanning bed for rapid high resolution scanning of one- and two-dimensional electrophoresis gels.

Storage Phosphor Screen Scanners: Autoradiography techniques are the most sensitive. The detection is much faster with phosphor screen scanners than with exposures on X-ray films, and they have a much higher linear dynamic range for quantification. After the exposure of a dried gel or a blotting membrane, the storage screen is scanned with a HeNe laser at 633 nm.

For reuse, the screen is exposed to extra-bright light to erase the image.

Differentially labelled samples can be analysed in the following way: with direct exposure both ^{35}S and ^{32}P signals are recorded; with a second exposure through a thin copper foil only ^{32}P labelled proteins are detected.

Fluorescence Scanners: Staining or labelling proteins with fluorescent dyes show less sensitivity than silver staining, but a much wider linear dynamic range. Modern instruments have a confocal scan head to cancel signals from scattered excitation light, and to reduce fluorescence background coming from glass plates and other supporting material. The laser light excites the fluorescent label or bound fluorescent dye, the emitted light of a different wavelength is bundled with a collection lens and transported to the detector via a fibre optic cable (see Fig. 44). Signals emitted from bands or spots, which are excited by stray light, are focused out, they will not hit the "peep hole", and will thus not be conducted by the fibre optic cable.

There is still a drawback for fluorescence detection: the high price for fluorescence dyes compared to other staining and labelling methods.

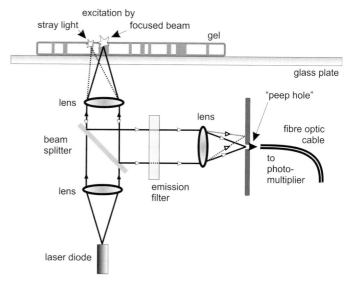

Fig. 44: Schematic diagram of the principle of fluorescence detection with confocal optics.

There is a trend to replace radioactivity in the laboratories wherever possible by fluorescence detection.

Multipurpose Scanners: The functions for storage phosphor imaging, multicolor fluorescence detection, and chemiluminescence can be combined in one instrument. Lasers with different wavelengths are combined with different filters for the various scanning modes. The detectors are usually very sensitive photomultipliers. In Fig. 45 such an instrument is displayed.

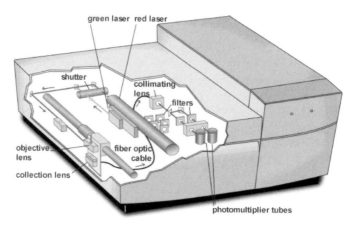

Fig. 45: Variable mode laser scanner. With kind permission from Amersham Pharmacia Biotech, Sunnyvale, USA.

Computer Hardware: Because today's personal computers have very fast processors, excellent graphic devices, and can be equipped with high capacity hard disks for reasonable prices, it is no longer necessary to perform the image analysis on a work station or a UNIX® based computer. The data can easily be further processed with office softwares.

Personal computers are much easier to operate than work stations. The new scanning and data processing softwares are very convenient and easy to handle.

However, there are still two standards existing:

- In numerous academic laboratories Apple Macintosh computers and software are preferred.
- In the industry – but also in many academic laboratories – Windows is used.

Some suppliers provide software for both computer standards.

5.2.2
Software for image analysis

Either the evaluation software is linked to the scanning software, or an image can be imported as a grey TIF file for further analysis.

Color scans are impossible to evaluate.

Here are a few general points for what a "good" electrophoresis imaging software should afford:

- Ease of use, as intuitive as possible.

  A manual should not be needed

- Mathematical correct algorithms.

  Reproducible calculations

- Fast and automated for high throughput applications.

  Evaluation is often a bottleneck

- Original data must not be changed.

  Except permissible adjustments

- Editing functions to correct errors.

  With reporting

One-dimensional gel software
Usually an automated 1-D evaluation follows this sequence:

Additional functions allow:
– lane matching,
– normalization for the density comparison between different gels, and
– data base methods like
– clustering for dendrogram constructions and
– sample identification.

1. lane detection
2. correction of distortions (like a smiling front)
3. lane correction (if necessary)
4. band detection
5. band matching
6. R_f, M_r, or pI calculation
7. background subtraction
8. band quantification
9. report tables

In Fig. 46 a typical screen of a SDS gel evaluation is shown. The molecular weights are assigned with the help of marker protein values.

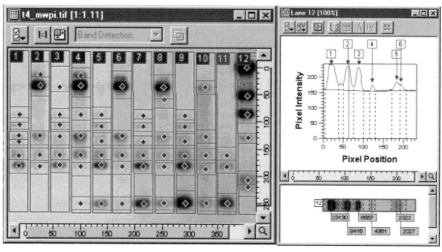

Fig. 46: Computer screen showing the detected lanes of an SDS gel and the densitogram of a selected lane with annotated molecular weights.

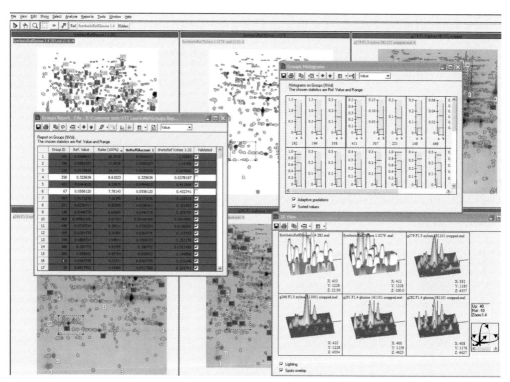

Fig. 47: Computer screen of 2-D electrophoresis image analysis software. In the background: 2-D images with automatically detected spots; in the foreground: volume table, histograms and three-dimensional representation of selected spots.

Two-dimensional gel software

For automated 2-D evaluation the following path is taken:

There is a major difference to 1-D evaluation: One gel corresponds to one sample.

1. spot detection and spot filtering
2. background correction
3. spot volume calculation
4. 2-D calibration
5. averaging multiple gels of a sample
6. choose reference gel
7. spot matching
8. normalization
9. pattern comparison
10. statistical evaluation

Fig. 47 displays the computer screen of digitized and processed images of 2-D electrophoresis gels. The samples were extracts from different bacterial cultures grown on different sugar media. The volume tables and histograms of spot intensities represent the quantitative contents of selected bacterial proteins grown on the different media. The three-dimensional representation is particularly useful for the detection of low expressed proteins next to highly abundant proteins.

Particularly for 2-D gels the use of quick and automated evaluation software is important, because it is impossible to find pattern differences with the eye. Sophisticated database functions are required to search across different images and experiments.

Critical issues in quantification

a) Ratio absorption : concentration

It should first be understood that the conventional laws of photometry do not apply to densitometry. While dilute solutions (mmol concentrations) are measured in the UV and visible ranges during photometry, the absorbing medium is a highly concentrated precipitate in

It is very complicated, if not impossible, to calculate a constant for a protein-dye aggregate.

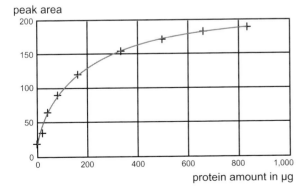

peak area

Fig. 48: Color intensity curve of carbonic anhydrase, separated with SDS-PAGE and stained with colloidal Coomassie Brilliant Blue G-250.

gel densitometry. It usually is a protein which is bound to a chromophore. A typical protein curve is shown in Fig. 48. Similar problems occur during densitometric evaluation of X-ray films. The Lambert-Beer law cannot be applied to densitometry because it is only valid for very dilute, "ideal" sample solutions.

b) External standard

An external standard should be used for a series of experiments. If a mixture of proteins is used, the amount of each standard protein applied is known. If the amount of a specific protein is to be calculated in mg, it should be remembered that every protein has a different affinity for a dye, e.g. Coomassie (Fig. 49).

The diagram is based on the calculation of the amount of marker

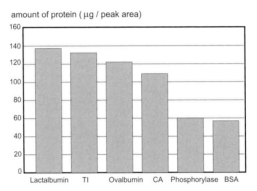

amount of protein (µg / peak area)

Fig 49: Diagram of the protein amount of individual markers per integrated peak area. TI: trypsin inhibitor from soybeans. Staining with Coomassie Brilliant Blue G-250.

protein in the mixture per measured and integrated peak area:

Lactalbumin	138 µg	Carbonic anhydrase (CA)	108 µg
Trypsin inhibitor	133 µg	Phosphorylase B	60 µg
Ovalbumin	122 µg	BSA	57 µg

Recalculated means: if the protein to be determined should later become available as pure substance, a comparison can then be carried out.

Therefore an unknown protein should not be calculated directly using albumin for calibration, the protein itself should be available in the pure form. But if the values are recorded compared to albumin, the correct values can be recalculated later.

The path to an absolute value lies through the identification of a protein with immuno or lectin blotting and the use of the procedure described above. If there is no antibody or specific ligand available, a subject for a thesis has been found!

6
Proteome Analysis

6.1
General

Meanwhile the data of a many fully sequenced genomes are available, including the human genome. Depending on the type of organism, only a part of the sequence can be directly correlated with a biological function. It had been found, that there is very low correlation between mRNA abundance and protein level (Anderson and Seilhamer, 1997; Haynes *et al.* 1998). Furthermore, since it became apparent, that the human genome surprisingly contains less than about 26,000 genes, it became obvious, that biological events cannot be explained from the genomic information directly. It has thus become necessary to analyse the protein complement of the genome, which has been defined as the "Proteome" (Wasinger *et al.* 1995).

A proteome consists of considerably more proteins than expected by direct transcription and translation due to alternative splicing and post-translational modifications. Not all proteins are expressed at the same time, and in a living cell continuously changes are happening. This makes it impossible to create a proteome map like a static gene map. Therefore, during an experiment, a series of samples is analysed, and the quantitative changes of expression levels are monitored. This approach is called "proteome analysis" or "proteomics". The new concept is mainly applied to drug discovery, diagnostics, therapy, and agricultural research.

In practice cell lysates or tissue extracts are searched for up or down regulated proteins. In an experiment, cells are stimulated by gene deletion or overexpression, pharmaceutical treatment, withdrawal of nutrients, or by physical or chemical stimulation. The great challenge of the Proteomics approach is finding target proteins with high significance. This is not an easy task, because there are a lot of variants in the analysed biological systems, and the separation and detection techniques have limited reproducibility.

Anderson L, Seilhamer J. Electrophoresis 18 (1997) 533–537.

Haynes PA, Gygi SP, Figeys D, Aebersold R. Electrophoresis 19 (1998) 1862–1871.

Wasinger VC, Cordwell SJ, Cerpa-Poljak A, Yan JX, Gooley AA, Wilkins MR, Duncan MW, Harris R, Williams KL, Humphery-Smith I. Electrophoresis 16 (1995) 1090–1094.

Hanno Langen has called the proteome a "moving target".

The key to a successful result is the proper planning of the Proteomics experiment.

Electrophoresis in Practice, Fourth Edition. Reiner Westermeier
Copyright © 2005 WILEY-VCH Verlag GmbH & Co. KGaA, Weinheim
ISBN: 3-527-31181-5

Washburn MP, Wolters D, Yates JR III. Nature Biotechnol 19 (2001) 242–247.
Koller A, Washburn MP, Lange BM, Andon NL, Deciu C, Haynes PA, Hays L, Schieltz D, Ulaszek R, Wei J, Wolters D, Yates III JR. Proc Natl Acad Sci USA 99 (2002) 11969–11974.

The "classical" Proteomics approach is based on the separation of proteins by high-resolution 2-D electrophoresis with subsequent identification and further analysis of selected protein spots with mass spectrometry. During the last few years a complementary "non-gel" approach has been established: Multidimensional protein identification technology (MuDPIT, Washburn *et al.* 2001). Here the mixture of sample proteins is first digested by proteolysis and then analysed by several chromatography separations with subsequent mass spectrometry. This approach is also called "Shot-gun" Proteomics. It can detect proteins, which do not appear in a 2-D gel. On the other hand, proteins included in a 2-D gel do partly not show up in this peptide/chromatography-based technique. This was also confirmed in a recently published paper (Koller *et al.* 2002). Both approaches are complementary.

The peptide/chromatography-based techniques have to separate an enormous complex mixture; the evaluation of results is still a challenge. Approaches to reduce the complexity of the sample are under development. However, the technique shows very high sensitivity of detection and offers a useful tool for identification of proteins. 2-D electrophoresis reduces the complexity of protein analysis considerably, because the separation is carried out on the protein level rather than the peptide level. The proteins of interest are selected by image analysis: only the proteins, which show changes between the biological stages, are digested and analysed.

2-D gel-based Proteome analysis includes the following major steps:

- Sample preparation under maintenance of the protein composition at the actual status of the cell
- Two-dimensional electrophoresis
- Detection of protein spots
- Search for protein changes with image analysis
- Spot excision
- Enzymatic digestion of the proteins in the gel pieces
- Identification and characterization of the proteins by mass spectrometry and genomic database search
- Bioinformatics for protein identification and database searching

See also page 42.

2-D electrophoresis for Proteomics has a number of important features:

Results can be used to confirm protein identification.

- Physico-chemical parameters of proteins measured

Several thousand proteins are resolved.

- Extremely high resolving power

- High throughput possible · *With parallel runs*

- Tolerance to crude protein mixtures · *To avoid further modifications*

- Tolerance to relatively high sample loads · *To detect minor components*

- Isoforms and post-translational modifications are displayed · *Indicated by spot changes.*

- The separation is non-destructive for the proteins.

- 2-D gels are very efficient fraction collectors · *Several thousand protein spots can be stored*

- Proteins are protected inside the gel matrix for analysis · *Proteases have no access*

- Multiple detection, like specific staining or blotting, is applicable

- Difference gel electrophoresis (DIGE) with fluorescence protein labelling allows sample multiplexing. · *This offers the use of an internal standard.*

- With DIGE quantitative results can be obtained at a high confidence level · *With the internal standard, created by pooled sample mixture.*

In principle, the physico-chemical properties of proteins, like pI and molecular weight, can be calculated from the open reading frames by "in-silico" translation. However, the theoretical 2-D maps and the 2-D electrophoresis protein pattern do not match in reality: Many of the gene products are modified by complex gene interactions, cellular events, and environmental influence. A number of post-translational modifications (PTM) can happen to a gene product, like truncation, phosphorylation, different kinds of glycosylations, acetylation, etc.

PTMs alter the isoelectric points and the molecular weights, and cause more complex patterns than a theoretical 2-D map constructed from the open reading frames. Not all proteins are expressed at the same time.

Highly sensitive and accurate mass spectrometry methods, new software, and the existence and accessibility of genomic and protein databases allow the quick identification even of small protein spots.

Another important prerequisite for this approach was the improved and now more reliable 2-D electrophoresis methodology.

From the genome sequence of a relatively simple organism, Mycoplasma pneumoniae, gene products with isoelectric points between pH 3 and 13 are expected (Himmelreich *et al.* 1996). "In silico" translations of other sequenced genomes show, that higher organisms have a similar distribution of their protein spectra. Although most of these "theoretical" isoelectric points are altered by various posttranslational modifications, the average distribution of the isoelectric points will cover a range of pH 3 to 13.

Himmelreich R, Hilbert H, Plagens H, Pirkl E, Li B-C, Herrmann R. Nucl Acids Res. 24 (1996) 4420–4449.

Is it possible to analyse all these proteins in a 2-D gel?

Banks R. Dunn MJ, Forbes MA, Stanly A, Pappin DJ, Naven T, Gough M, Harnden P, Selby PJ. Electrophoresis 20 (1999) 689–700.
Anderson NL, Anderson NG. Mol Cell Proteomics 1 (2002) 845–867.

New prefractionation techniques and methods complementary to 2-D electrophoresis need to be applied and further developed.

We have to face the fact, that at least 10 to 15 % proteins can not be included in a gel. Very basic proteins (pI >11) are not easy to analyse with 2-D electrophoresis. Also some hydrophobic proteins and high molecular weight proteins (> 200 kDa) resist entering a 2-D gel. Some gene products, mostly regulatory proteins, are expressed in such low copy numbers, that the applied detection methods do not have enough sensitivity to visualize them. Or, for instance, proteins from tissue material acquired with laser capture microdissection are so scarce, that their detection requires extremely sensitive staining or labelling methods. The most challenging sample is human plasma, where highly abundant proteins like albumin, immunoglobulins and transferrin exist in very high concentrations, but the most interesting components like interleukins, cytokines and other disease markers are present in extremely low amounts: the dynamic range of protein concentrations span 10 orders of magnitude (Anderson NL and Anderson NG, 2002).

Görg A, Obermaier C, Boguth G, Weiss W. Electrophoresis 20 (1999) 712–717.

Since the introduction of 2-D electrophoresis, the technique has been considerably improved: A. Görg *et al.* (1999) have published separations in immobilized pH gradients as wide as pH 3 to 12 (see Fig. 50). Gels with pH gradients up to 3 to 11 are commercially available.

A few introductory and review papers provide a good picture of the possible applications and expectations on the proteome analysis approach:

Pennington SR, Wilkins MR, Hochstrasser DR, Dunn MJ. Proteome analysis: from protein characterization to biological function. Trends cell biol. 7 (1997) 168–173.
Haynes PA, Gygi SP, Figeys D, Aebersold R. Proteome analysis: biological assay or data archive? Electrophoresis 19 (1998) 1862–1871.
Some specific journals like "Proteomics" and "Molecular and Cellular Proteomics" have been launched.
Several books on proteome analysis have been published:
Wilkins MR, Williams KL, Appel RD, Hochstrasser DF, Eds. Proteome research: New frontiers in functional genomics. Springer, Berlin (1997).
Link AJ, Ed. 2-D proteome analysis protocols. Methods in molecular biology 112. Humana Press, Totowa, NJ (1999).
Dunn MJ, Ed. From genome to proteome. Advances in the practice and application of proteomics. WILEY-VCH Weinheim (1999).
Kellner R, Lottspeich F, Meyer H, Eds. Microcharacterization of proteins. Second edition. WILEY-VCH Weinheim (1999).
Rabilloud T, Ed. Proteome research: Two-dimensional gel electrophoresis and identification methods. Springer Berlin Heidelberg New York (2000).
James P, Ed. Proteome Research: Mass spectrometry. Springer Berlin (2001).
Pennington SR, Dunn MJ, Eds. Proteomics from protein sequence to function. BIOS Scientific Publishers Limited, Oxford (2001).
Westermeier R, Naven T. Proteomics in Practice. A laboratory manual of proteome analysis. WILEY-VCH, Weinheim (2002).

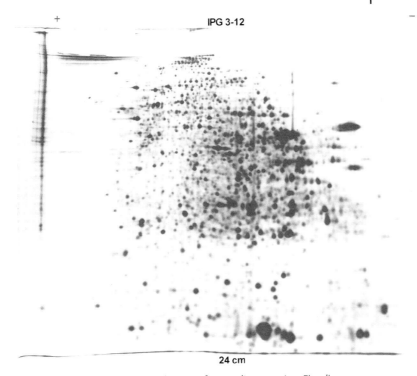

+   IPG 3-12   −

24 cm

Fig. 50: Two-dimensional electrophoresis of mouse liver proteins. *First dimension:* isoelectric focusing in an immobilized pH gradient pH 3–12 in a 24 cm long gel strip. *Second dimension:* SDS PAGE in a 13% gel. Silver stained. With kind permission of Professor A. Görg.

Simpson RJ. Proteins and proteomics. A laboratory manual. Cold Spring Harbor Laboratory Press, Cold Spring Harbor, New York (2003).
Simpson RJ, Ed. Purifying proteins for proteomics: A laboratory manual. Cold Spring Harbor Laboratory Press, Cold Spring Harbor, New York (2003).

A global as well as numerous local Proteomics societies have been founded, which are engaged in scientific and educational activities in proteomics, assist in the coordination of public proteome initiatives, facilitate opportunities for international cooperation, and have started initiatives for resources and technology, production of antibodies, model Proteomes like plasma, liver, brain, and define Proteomics standards. *More information is available under: http://www.HUPO.org.*

In the following chapter a short overview over the methodology in proteome analysis will be given.

6.2
Sample preparation

Hydrophobic proteins must be brought into solution.
Proteases must be removed or deactivated.
Nucleic acids, lipids, salts and solid material should be removed without losing proteins.

The sample treatment is the key to reasonable results. The protein composition of the cell lysate must be reflected in the pattern of the 2-D electrophoresis gel without any losses or modifications. "Co-analytical modifications" (CAM) of proteins must be avoided. This is not trivial, because various protein-protein interactions may happen in such a complex mixture, and prepurification of the sample can lead to uncontrolled losses of some of the proteins. Too much salt, like washing cells with PBS, and amphoteric buffers in cell cultures, like HEPES, have to be avoided. The chemicals used have to be of the highest purity.

Berkelman T, Stenstedt T. Amersham Biosciences Handbook 80–6429–60 (2002).

Valuable information on sample preparation – and the how to run the two-dimensional separations – can be found in a handbook by Berkelman and Stenstedt (2002).

The standard "lysis buffer" is composed of:
8 to 9 mol/L urea, 4 % CHAPS, 60 mmol/L DTT or DTE, 0.8 % carrier ampholytes, 0.007 % bromophenol blue.

Isocyanate impurities and heating must be avoided, because these would cause carbamylation of the proteins, resulting in artefactual spots.

The high *urea* concentration is needed to get proteins into a single conformation by cancelling the secondary and tertiary structures, to get hydrophobic proteins into solution, and to avoid protein-protein interactions.

Mass spectrometry is particularly sensitive to contaminations coming from the detergent.

CHAPS is a zwitterionic detergent, and preferred to non-ionic polyol mixtures like Triton X-100 and Nonidet NP-40, because of its higher purity. It increases the solubility of hydrophobic proteins.

Righetti PG, Tudor G, Gianazza E. J Biochem Biophys Methods 6 (1982) 219–227.

They are no longer used to establish the pH gradient, this is today done with immobilized pH gradients.

DTT or *DTE* prevent different oxidation steps of the proteins. 2-mercaptoethanol should not be used, because of its buffering effect above pH 8 (Righetti *et al.* 1982).

Carrier ampholytes improve the solubility of proteins by substituting ionic buffers. They do not disturb the IEF very much because they migrate to their pIs, where they become uncharged.

Bromophenol blue is very useful as a control dye.

Special cases

But be aware: SDS does not always completely separate from the proteins, even under high field strength.

Organisms with tough cell walls sometimes require boiling for 5 minutes in 1 to 2 % SDS before they are diluted with lysis buffer.

For special cases – particularly for improving the solubility of *membrane proteins* – the addition of 2 mol/L thiourea to 7 mol/L urea instead of using 8 mol/L urea has been proposed by Rabilloud (1998). This procedure results in a lot more spots in the gel, but artefactual spots and streaks have been reported as well.

Rabilloud T. Electrophoresis 19 (1998) 758–760.

Many attempts have been made, and many are tried at present to find more effective detergents, or substitutes for detergents (Vuillard *et al.* 1995), alternative reductants and other additives. Unfortunately most of the reported substances are highly toxic; some reductants are hydrophobic and decrease the solubility, which was gained with better detergents.

Vuillard L, Marret N, Rabilloud T. Electrophoresis 16 (1995) 295–297.

Recently it could be shown by Méchin *et al.* (2003), that a cocktail of several different chaotropes, detergents and reductants could improve the solubilization of particularly difficult plant proteins.

Méchin V, Consoli L, Le Guilloux M, Damerval C, Proteomics 3 (2003) 1299–1302.

Sample treatment

Nucleic acids are efficiently broken down by sonification.

Heating must be avoided.

Lipids are removed with an excess of detergent (> 2%) or with precipitation.

Salts can be dialysed away or removed by precipitation.

However, both procedures can lead to protein losses.

Protease inhibitors can inactivate the proteolysis activities, however in some cell lysates not completely. PMSF is frequently used, but it is a toxic compound and has to be added to the sample prior to the reductant. Pefabloc is less toxic, but might lead to charge modifications of some proteins.

Some proteases are already inhibited by the denaturing conditions, some by basic pH.

Precipitation: For plant samples or other very diluted samples the following procedure according to Damerval *et al.* is frequently and successfully applied:

The contents of the cell lysate are precipitated with 10 % TCA in acetone; the pellets are washed with acetone, dried under vacuum, and resuspended with lysis buffer. The acid also modifies most of the proteases; subsequently protease activity is inhibited irreversibly.

Damerval C, DeVienne D, Zivy M, Thiellement H. Electrophoresis 7 (1986) 53–54.

An alternative procedure employs sequential precipitation with methanol and chloroform (Wessel and Flügge, 1984).

Wessel D, Flügge UI. Anal. Biochem. 138 (1984) 141–143.

Novel clean-up kits based on acidic precipitation with detergent co-precipitants and organic washing solutions allow a sample clean-up within 1 ¹/₂ hours. Practice has shown that more protein spots are present in the 2-D gel subsequently to such a treatment, because it dissociates protein-lipid-polysaccharide complexes very effectively. A good example can be found in the paper by Stasyk *et al.* (2001).

Available from Amersham Biosciences.
Stasyk T, Hellman U, Souchelnytskyi S. Life Science News 9 (2001) 9–12.

In chapter 11 more detailed instructions for sample preparation are given.

At present the situation can be described in the following way: Optimized procedures for different sample types do exist. However, a general *"Prepares them all"* procedure is not available.

Sample preparation for DIGE

Ünlü M, Morgan ME, Minden JS. Electrophoresis 18 (1997) 2071–2077.

For Difference Gel electrophoresis the sample proteins are labelled with modified fluorescent cyanine dyes (Ünlü *et al.* 1997). Three dyes with different excitation and emission wavelengths are available: Cy2, Cy3 and Cy5. After labelling the samples are mixed and applied on the same gel. The patterns are scanned with a fluorescent laser scanner at different wavelengths. Two different approaches exist: minimum labelling and saturation labelling for scarce samples.

Minimum labelling

The labelling effectiveness is considerably increased, when the sample has been cleaned up by precipitation.

The Cy dyes bind to the ε-amino group of the lysine. Only limited amount of dye is added to the protein mixture: about 3 % of the proteins are labelled. In this way an increase in hydrophobicity of the proteins is avoided, multiply labelled proteins occur statistically below the range of sensitivity of detection. It is important to perform the labelling procedure under exactly controlled conditions:

By adding about 30 mmol/L Tris-base

- The pH value must be adjusted to pH 8.5.

These reagents are added after labelling.

- Absence of carrier ampholytes and reductants.
- Half hour labelling on ice-water.
- Stopping of reaction by adding a small amount of lysine after labelling.

Saturation labelling

This is a useful method for the detection of proteins acquired with laser capture microdissection.

Only Cy3 and Cy5 can be modified for saturation labelling. With this approach all available thiol groups of all present proteins are labelled. Also this procedure must occur under strictly controlled conditions:

- Reduction for one hour at 37 °C at pH 8.0 with a strong reductant: Tris-carboxyethyl phosphine (TCEP).

The concentrations of reductant and label must be optimized for the sample type.

- Labelling for half an hour at 37 °C at pH 8.0 with the Cy dye derivative.

Pre-fractionation

The reason for missing of proteins is very often the high complexity of the sample. There are various methods of pre-fractionation to reduce the complexity of the protein mixture:

Protein chromatography: There is a basic difference between protein purification and fractionation: the goal of purification is the removal of all contaminants, whereas in fractionation the losses of proteins have to be avoided. The principles of protein chromatography are comprehensively described in the book Protein Purification, edited by Janson and Rydén (1997).

Janson JC, Rydén L, Eds. Protein purification. Principles, high-resolution methods, and applications. WILEY-Liss, New York (1998).

Differential Solubility: Sequential protein extraction with increasingly strong solubilizing agents divides the sample mixture into smaller subsets.

But hydrophobicity of a protein is not correlated to its function.

Blue Native Electrophoresis   (see page 41): This method reveals also proteins with high hydrophobicity because of the reduction of complexity of the protein mixture. Werhan and Braun (2002) applied a "three-dimensional electrophoresis" with the first dimension separation of protein complexes by blue native polyacrylamide electrophoresis. The visible bands are eluted from the gel, destained, and separated by 2-D electrophoresis.

Information on which protein is complexing with which partners is obtained.
Werhahn W, Braun H-P. Electrophoresis 23 (2002) 640–646.

Selective Removal of Abundant Proteins: The most specific fractionation can be obtained with immuno affinity techniques. For human plasma analysis, there are affinity columns on the market, which remove a high amount of albumin and some other highly abundant proteins like the immunoglobulins, transferrin, and fibrinogen. Proteins sticking to albumin should be analyzed as well: the removed fraction should not be discarded.

Also plant samples may require such a procedure in order to remove Rubisco.

Subcellular Components: Different methodology can be employed:

The location of a protein in a cell is closely linked to its function.

- Ultracentrifugation (Pretlow and Pretlow, 1991).

Pretlow TG, Pretlow TP Methods: A Companion to Methods in Enzymology 2 (1991) 183–191.

- Free flow electrophoresis (Zischka *et al.* 2003), see also page 10.

Zischka H, Weber G, Weber PJA, Posch A, Braun RJ, Bühringer D, Schneider U, Nissum M, Meitinger T, Ueffing M, Eckerskorn C. Proteomics 3 (2003) 906–916.

- Detergent fractionation in cytosol, membrane-organelle, nuclear membrane and cytoskeletal-matrix proteins (Ramsby *et al.* 1994).

Ramsby ML, Makowski GS, Khairallah EA. Electrophoresis 15 (1994) 265–277.

According to pI: When isoelectric focusing should be carried out in narrow interval pH gradients.

see also page 10

- Free flow isoelectric focusing in a carrier ampholytes-containing liquid stream.

see page 63 f.
- Electrophoretic separation between isoelectric membranes.

see page 62 f.
- Carrier isoelectric focusing in Sephadex.

6.3
Two-dimensional electrophoresis

First dimension

O'Farrell PH. J Biol Chem. 250 (1975) 4007–4021.

Carrier ampholytes IEF: In the traditional method for high resolution 2-D electrophoresis according to O'Farrell (1975) the isoelectric focusing step was carried out with carrier ampholytes generated pH gradients in thin gel rods, sometimes called "tube gels".

This effect is called "cathodal drift" or "plateau phenomenon".

However, during the long focusing time, which is required for denaturing conditions and long separation distances, these gradients become instable and drift considerably. What has been a gradient pH 3 to 10 in the beginning becomes a final gradient of pH 4 to 7. Practically all the basic proteins have been lost.

NEPHGE: A remedy was the modification of the first dimension technique: non-equilibrium pH gradient electrophoresis: (O'Farrell *et al.* 1977). Here the sample is loaded onto the acidic end of the gradient. The proteins are separated while the basic part of the gradient drifts towards the cathode. The run is stopped after a defined time period.

In NEPHGE carrier ampholytes and proteins build stacks. The resolution is limited by the number of carrier ampholytes.

Due to the time factor, it is not easy to run this method with adequate reproducibility. Furthermore, the proteins are not focused like in IEF. This results in limited resolution.

See pages 59 ff.
Görg A, Postel W, Günther S. Electrophoresis 9 (1988a) 531–546.
Görg A, Obermaier C, Boguth G, Harder A, Scheibe B, Wildgruber R, Weiss W. Electrophoresis 21 (2000) 1037–1053.

Immobilized pH gradients in the first dimension have initiated a quantum leap in the 2-D electrophoresis technology. Here, the samples are separated in narrow film-supported gel strips containing fixed buffering groups. The strips are cut from a 0.5 mm thin slab gel. The technique has been introduced and further developed by Görg *et al.* (1988a). The latest state of the art can be found in a review by Görg *et al.* (2000).

The features of the immobilized pH gradients, listed below, have caused a shift from the traditional technique to the IPG strip method:

See Fig. 50.
This leads to much higher reproducibility.

- The film-supported gels are easy to handle.
- Immobilized pH gradients are very reproducible
- Wide and stable gradients are possible.
- The fixed gradients are not modified by the sample composition, and they do not drift.

- Very narrow pH intervals in long gel strips can be prepared, which allow high spatial resolution and high protein loading.

  This allows also the detection of low expressed proteins.

- Different ways of sample applications are feasible; in most cases the dried strips are rehydrated with the sample solutions themselves.

  Protein losses due to aggregation and precipitation are avoided with sample loading by rehydration.

- Various additives, like detergents and reductants, can be added to the rehydration solution.

  Many of them would inhibit gel polymerization.

With the considerably improved first dimension, 2-D electrophoresis has now become more reproducible and more useful for analytical as well as for micropreparative applications. An inter-laboratory comparison has demonstrated the high reproducibility of the method (Blomberg *et al.* 1995).

Blomberg A, Blomberg L, Norbeck J, Fey SJ, Larsen PM, Roepstorff P, Degand H, Boutry M, Posch A, Görg A. Electrophoresis 16 (1995) 1935–1945.

Probably the most powerful feature is the possibility to reach almost unlimited spatial resolution with very narrow pH intervals.

This is absolutely necessary for protein identification with mass spectrometry.

Horizontal streaking in basic pH gradients

In basic pH ranges horizontal streaking of spots was a frequently observed problem. The streaking was caused by deprotonation of DTT or DTT, which are weak acids. The thus negatively charged reductants migrate towards the anode, leaving the cysteins unprotected. This leads to back folding and aggregating of polypeptides, which is accompanied by gaining different isoelectric points leading to horizontal smearing. Use of alternative reductants like Tributyl phosphine (TBP) or TCEP was not solving the problem, it caused other negative effects. Alkylation of heterogeneous protein mixtures prior to IEF produces artefactual spots of incompletely alkylated polypeptides.

With oxidation of the cysteinyls to mixed disulfides during IEF the problem is completely abolished (Olsson *et al.* 2002). The samples, containing any reductant of choice, are loaded at the anodal end of the IPG strip, which has been rehydrated with the standard chaotropes, detergent, carrier ampholytes, and – instead of a reductant – with 100 mmol/L of a disulfide reagent. This is an equilibrium reaction with high specifity, without unwanted side reactions.

Olsson I, Larsson K, Palmgren R, Bjellqvist B. Proteomics 2 (2002) 1630–1632.

Second dimension

Nothing really revolutionary has happened for this part of the procedure: Usually SDS polyacrylamide gel electrophoresis in a Tris-chloride Tris-glycine buffer according to Laemmli (1970) is employed, mostly without a stacking gel. Vertical and horizontal flatbed systems can be used with similar results (Görg *et al.* 1995).

A stacking gel is not needed, because the proteins are pre-separated by IEF and migrate from a gel into another gel. Görg A, Boguth G, Obermaier C, Posch A, Weiss W. Electrophoresis 16 (1995) 1079–1086.

Small peptides are better resolved in Tris-tricine gels according to Schägger and Von Jagow (1987), or in Tris-tricine ready-made long-shelf life gels which are described on page 40 f.

Another problem is the fragility of the gels; they can easily get a crack or break into pieces.

Once a conventional gel leaves the glass cassette after the run, it behaves like a living animal: it swells and shrinks during staining. This makes evaluation with image analysis cumbersome and automated spot-cutting impossible.

Gels on covalently bound support films, which do not come off during staining, can be employed in horizontal flatbed and in vertical instruments.

Film-supported SDS polyacrylamide gels for the second dimension are much easier to use. The only little drawbacks are: The steps in silver staining require more time, because the liquid can diffuse into the gel only from one side. The films used show fluorescent background at certain wavelengths. It could however be shown, that spot detection with the sensitive fluorescent dye SYPRO Ruby® works well, when a fluorescence scanner with a confocal scanning head is employed (see also page 86).

With *dimension-stable gels* image analysis is much easier and faster. They can be used in an automated spot-picker, which uses the information from the image analysis system; the kind of the detection technique does not matter.

Shelf life stability of the gels: Due to the high pH value 8.8 of the Tris-Chloride gel buffer the polyacrylamide gel starts to hydrolyse and changes its sieving properties already after a week.

Here the cathodal buffer consists of Tris-glycine like in the Laemmli buffer: Thus the protein pattern is the same as with a conventional buffer system.

Recently a modification of Laemmli buffer system has been introduced: the Tris of the gel buffer has been replaced by PPA (piperidino propionamide) resulting in a pH value below 7. Thus ready-made gels can be used without stability limitations, and with an improved reproducibility of the spot positions also in the second dimension.

Still some papers can be found, where this convention is not respected. But this complicates the comparisons of the 2-D patterns even more.

There is a convention how to display a 2-D gel: Like in a Cartesian coordinate system, where the small figures are on the left hand side at the bottom, the acidic proteins (with low pI) are shown on the left side, the low MW proteins at the bottom.

6.4
Detection techniques

The ideal detection technique should have a wide linear dynamic range, should be very sensitive, quantitative, compatible with the further analysis using mass spectrometry, quick, non-toxic, reasonably priced, and should not require living cells for labelling. None of the existing techniques combines all these features.

Labeling of living cells

Radioactive labelling: The most sensitive and quantitatively reliable method is the labelling of the proteins with ^{35}S or ^{32}P isotopes, exposure of the gel on a storage phosphor screen, which is subsequently scanned with a laser (see page 85). Less than 1 pg protein can be detected.

However, as already mentioned, there is a trend to replace radioactivity in the laboratories wherever possible.

Stable isotope labelling with $^{14}N/^{15}N$ (ammonium), $^{12}C/^{13}C$ (glucose) is ideal for differential 2-D electrophoresis and quantification, but is expensive and requires high resolution MS for detection.

These methods require living cells, and can therefore not be applied on analysis of biopsy tissue, body fluids, plant material etc.

The following staining and labelling techniques do not require living cells:

Staining after 2-D electrophoresis

Coomassie Brilliant Blue stains almost all proteins and peptides with good quantitative linearity. It is compatible with mass spectrometry, but is not very sensitive (at least ca. 0.1 µg of protein per spot is required). Alcohol free procedures should be used to reach the highest sensitivity, see methods 7 and 10.

The sensitivity is not the same for all proteins. It is dependent on the binding properties of the proteins, see page 90.

Negative staining with imidazole zinc is more sensitive: down to 15 ng (Hardy et al. 1996). As only the background is stained, and not the proteins, the recovery yield is very good for further analysis with mass spectrometry (Matsui et al. 1997). However, it cannot be used for quantification.

Hardy E, Santana H, Sosa AE, Hernandez L, Fernandez-Patron C, Castellanos-Serra L. Anal Biochem. 240 (1996) 150–152.
Matsui NM, Smith DM, Clauser KR, Fichmann J, Andrews LE, Sullivan CM, Burlingame AL, Epstein LB. Electrophoresis 18 (1997) 409–417.

Silver staining, introduced for 2-D polyacrylamide gels by Merril et al. (1979), picks up protein amounts down to 0.2 ng. Although ca. 50 modifications have been published, there are two main types employed for 2-D gels: silver nitrate and silver diamine procedures. The latter shows a better sensitivity for basic proteins, but it contains some caustic solutions, does not work with tricine buffer in the gel, and a silver mirror development on the gel surface can easily happen. In the silver nitrate method the silver is more weakly bound to the proteins, it can be modified for mass spectrometry compatibility. Both methods require multiple steps; automation of the procedure is therefore very helpful. It is very important, that staining is performed in closed trays to prevent keratin contamination.

Merril CR, Switzer RC, Van Keuren ML. Proc Nat Acad Sci 76 (1979) 4335–4339.
A silver diamine procedure (ammoniacalic) is described in method 6, because it is the most sensitive technique for isoelectric focusing gels.

A very sensitive silver nitrate technique is described in methods 7 and 11, in the latter also the mass spectrometry compatible version.

Mackintosh JA, Choi H-Y, Bae S-H, Veal DA, Bell PJ, Ferrari BC, Van Dyk DD, Verrills NM, Paik Y-K, Karuso P. Proteomics 3 (2003) 2273–2288.
Berggren KN, Schulenberg B, Lopez MF, Steinberg TH, Bogdanova A, Smejkal G, Wang A, Patton WF. Proteomics 2 (2002) 486–498.
Alba FJ, Bermudez A, Daban J-R. BioTechniques 21 (1996) 625–626. Steinberg TH, Hangland RP, Singer VI. Anal Biochem. 239 (1996) 238–245.

See also spot identification on page 109.

These dyes are available from Molecular Probes.

Note, that glass plates or film-support must be non-fluorescent. For non-backed gels, the cellophane trick can be applied, which is described below.

The Cyanine dyes are specially modified for binding to proteins. The standard CyDyes used for DNA labelling cannot be used here.

Fluorescence staining methods are less sensitive than silver staining: down to 2 – 8 ng protein. But they have very wide linear dynamic ranges of ca. 10^4, and they are compatible with subsequent mass spectrometry analysis. Deep Purple^{TM} is the most sensitive dye (Mackintosh *et al.* 2003), followed by SYPRO® Ruby (Berggren *et al.* 2002); others are similar sensitive like Coomassie Brilliant Blue: Nile Red (Alba *et al.* 1996), SYPRO® Red and Orange (Steinberg *et al.* 1996). These dyes are relatively expensive, and a fluorescence scanner or a CCD camera is required.

Blotting of 2-D gels is nowadays mainly used for immuno detection. The transfer efficiency of electro blotting is not sufficient for the general detection of all the proteins.

New fluorescent dyes for *specific detection* of post-translational modifications have lately been introduced: Pro-Q-Diamond dye for phosphorylated proteins, and Pro-Q-Emerald for glycosylated proteins. It is possible to stain a gel sequentially with the different dyes and finally with a total stain.

Fluorescence stained or labelled spots are not visible. Accurate picking of selected spots for mass spectrometry analysis works best, when the gels are fixed on a glass plate or film-support. Two small self-adhesive fluorescent labels, internal reference markers, are placed on the gel support before or after casting. Because those are visible, the camera of the spot picker can find their positions. The spot picker software will – with the help of the pixel coordinates of the reference markers – convert the pixel coordinates of the selected spots into the machine coordinates.

DIGE (Difference gel electrophoresis)

Difference gel electrophoresis using fluorescence protein pre-labelling is not solely a detection technique; it offers a complete concept for accurate quantitative Proteomics. With the CyDye technology it is possible to label the proteins from different samples with dyes of different excitation and emission wavelengths. The samples are mixed and run together in one gel (multiplexing). The gels are cast in non-fluorescent glass cassettes and do not need to be removed from the cassettes during scanning. The two labelling concepts, lysine minimum labelling and cysteins saturation labelling have been briefly described under "6.2 Sample preparation" on page 98.

Minimum labelling

The different samples, for instance healthy and diseased tissue extracts, as well as a pooled standard are labelled with the different dyes, mixed and applied on one gel as shown schematically in figure 51 A. The different patterns are acquired with a fluorescence scanner by excitation and measurement at three different wavelengths (see Fig. 51). Identical proteins of the different samples migrate to the same positions, because the dyes are matched for charge and size:

- pI: Each of the CyDye™ DIGE fluors contains a buffering group with the same pK value like the ε-amino group of lysine. *In this way the pI is not altered.*
- M_r: The molecular weight added to the polypeptides is very similar for each of the fluors (from 434 to 464 Da). *The differences are far below the resolution of SDS electrophoresis.*

Therefore spot matching within a gel is not needed. Changes of protein expression levels are easily detected by overlaying the image channels with false color display as shown in figure 51 B.

The 2-D patterns are qualitatively the same as those obtained from non-labelled proteins; because some multiply labelled proteins will exist in such low concentrations, that they are not detected. In general, the sensitivity of detection is similar like for post-staining with fluorescence dyes. The scanned images are evaluated with dedicated software, which can co-detect the identical protein spots of the different samples and the standard. In order to make quantitative changes measurable, the software has to normalize the images to compensate the different emission intensities of the dyes. In mass spectrometry analysis the signals of the minimum labelled proteins (about 3 % only) are below the detection level. For spot picking the gels are post-stained, because in the molecular weight range below 30 kDa SDS electrophoresis is able to resolve 450 Da mass differences, the non-labelled proteins have migrated further than the labelled proteins. The non-labelled proteins are thus detected, matched with the labelled, and picked for mass spectrometry analysis.

The non-labelled proteins make up 97 % of a spot in the high and medium molecular weight area, in the lower molecular weight area the non-labelled proteins are set off the labelled ones.

Saturation labelling

This labelling concept offers the possibility of labelling the total amount of cystein-containing proteins of a sample, which leads to very high sensitivity of detection. Proteins with more than one cystein are multiply labelled, resulting in an additional increase in sensitivity of detection. The spot pattern looks therefore different from those obtained with minimum-labelling or non-labelled proteins. But the modified Cy3 and Cy5 fluors are size-matched, in order to assure co-migration of the differently labelled proteins. A mass of 672.85 Da is added to each cystein labelled with Cy3, 684.86 to each cystein labelled with Cy5. These masses have to be taken in account for the identification with peptide mass fingerprinting.

The method is particularly useful for labelling scarce protein samples like those acquired with laser capture microdisection.

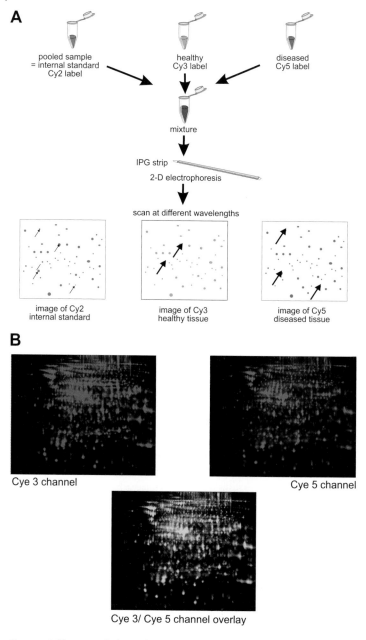

Fig. 51: Difference Gel Electrophoresis (DIGE): A: Schematic drawing of the concept of labelling two samples and a mixture of the samples, a pooled standard, with three different fluorescent dyes and running them together in 2-D electrophoresis. B: False color representation of the Cy 3 and Cye 5 image channels after scanning and overlay of the images. Channel overlay: Yellow spots: no changes, red spots: up-regulated, green spots: down-regulated proteins.

The internal standard

By employing the 2-D DIGE technique the number of gels to be run is considerably reduced. However, the particular strength of this multiplexing technique is the use of an internal standard for every protein. In this way gel-to-gel variations are eliminated. Each protein in the population must appear also in the standard; otherwise it will not be used in the evaluation. For multi-sample experiments the standard is run in each gel, this allows normalization between the gels according to the spot positions and the spot volumes of the standard. This allows routine detection of less than ten percent differences in protein expression levels with over ninety-five percent confidence within short time.

The "internal standard" converts 2-D electrophoresis from a semi-quantitative separation technique into a truly analytical method.

Alban *et al.* (2003) have published a paper which confirms the theoretical benefits of the internal standard in a spiking experiment. Friedman *et al.* (2004) have recently demonstrated the enormous practical progress achieved with DIGE using the internal standard for proteome analysis of human colon cancer.

Alban A, David S, Bjorkesten L, Andersson C, Sloge E, Lewis S, Currie I. Proteomics 3 (2003) 36–44.
Friedman DB, Hill S, Keller JW, Merchant NB, Levy SE, Coffey RJ, Caprioli RM. Proteomics 4 (2004) 793–811.

6.5
Image analysis

Image acquisition is preferably done with scanners rather than with CCD cameras, because high resolution and a homogeneous background from the centre to the edges are required. Visible patterns are recorded with modified desktop scanners in the light transmission mode, fluorescent dyes or radioactivity is detected with special laser scanners (see page 85f).

For automated picking of selected protein spots the internal reference markers are scanned together with the protein spots. Their x- and y- coordinates will be a part of the picking list, which contains the x- and y-coordinates of the selected spots.

Image analysis is a very important step in the proteome analysis procedure, because the evaluation of the complex patterns can easily become the bottleneck.

Gel evaluation software

The evaluation of the highly complex 2-D patterns requires software, which automatically detects the spots, corrects the background, and quantifies the spot volumina. Furthermore the isoelectric points and molecular weights of the spots are interpolated after 2-D calibration. For DIGE evaluation special software is available, which can co-detect the co-migrated proteins in the different image channels.

These virtual average gels reduce the noise of the system.

Average gels: For statistically reliable quantitative data samples are often grown, prepared and run in doublets or triplets. Therefore a software tool is necessary to construct composite gels from a number of 2-D patterns of the same sample.

The patterns of different samples are compared by first defining a *reference* gel and then matching the spot positions with those of the reference. Because staining is never perfectly reproducible, a *normalization* step is needed before quantitative spot comparisons between gels can be performed.

Links to external databases are established by employing a web browser.

Qualitative and quantitative *differences* in spot patterns are then determined and displayed for the detection and measurement of up or down regulated gene products. Those proteins, which show differences between the samples

Statistics

Gel-to-gel variations need to be eliminated.

Statistic tools are mostly included in the software packages, in order to check the significance of a change in protein expression. The confidence level rises, when replicate gels have been run and evaluated. The use of pooled standard, which is only possible with DIGE gels, increases the confidence level considerably, even without replicate gels.

Spot picking list

The cellophane trick has been found by Burghardt Scheibe. The "cellophane" procedure works also for fluorescence labelled or stained gels, because cellophane does not give a fluorescent background.

Advanced *automated spot pickers* use the coordinates of the spots from the image analysis. Thus also radiolabelled and other "invisible" spots can easily be collected with high precision. For this procedure the gels have to be fixed on a glass plate or film support. Non-backed gels can be placed on a dry cellophane sheet, which is clamped and stretched into a set of plastic frames. The gels stick tightly to the cellophane and cannot shrink or swell anymore during scanning, spot-picking, and in between. A picking list is created giving the x/y coordinates of the reference markers and the protein spots selected for further analysis.

Database software

In principle, database software is used for combining the results of many 2-D experiments – a 2-D experiment is usually composed of a series of 2-D gels.

For the analysis of multiple gel experiments and intensive statistic studies database software is employed. With the definition of complex queries of the quantitative behaviour of gene products, control points in biochemical pathways and protein families can be found.

6.6
Protein spot identification

The position of a spot in the gel, marked by the isoelectric point and the molecular weight, is *not* enough information for the identification of a protein. In fact, the renaissance of 2-D electrophoresis was initiated by the new developments in *mass spectrometry*. Now the spots can be identified with much less effort and much higher throughput than in the past. In Fig. 52 a schematic overview over the spot identification methods is displayed.

It should be mentioned, that the databases have at least the same importance: Protein, peptide, genome, and/or EST databases are needed for protein identification.

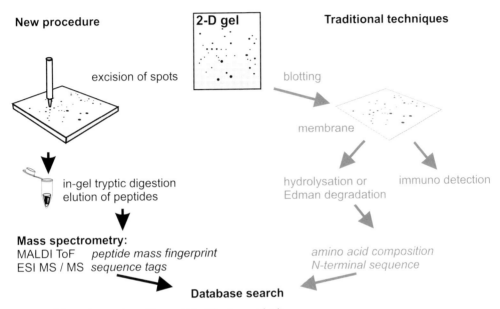

Fig. 52: Schematic overview of the spot identification methods.

Traditional spot identification methods were: Co-running of known proteins, immuno detection, amino acid composition analysis, and N-terminal sequence analysis with Edman degradation. These methods were limited to already known and/or highly abundant proteins. For Edman degradation often several 2-D gels were run, the protein spots were pooled from different gels. N-terminal sequencing is slow (10 hours for 20 amino acids) and expensive, more than 50 % of the proteins are N-terminal blocked, post-translational modifications are not detected.

For immuno detection, amino acid composition analysis and N-terminal sequence analysis blotting on nitrocellulose, activated glass fibre or PVDF membranes is needed as an intermediate step.
Amino acid composition analysis requires ca. 0.1 pmol, microsequencing at least 1 pmol protein.

With electro elution of intact proteins from polyacrylamide gels very low yields are obtained. Note, that blotting is an electro elution process.

During the development of the new mass spectrometry methods, blotting had been used in order to remove contaminating buffers and detergents. However, the overall transfer efficiency of blotting was never sufficient. Furthermore, mass spectrometry picks up chemical compounds from the membranes.

Thus, the proteins are digested inside of excised gel pieces with proteolytic enzymes, the peptide mixtures are submitted for further analysis.

Hellman U, Wernstedt C, Góñez J, Heldin C-H. Anal Biochem. 224 (1995) 451–455.

The latter technique is also preferred for chemical microsequencing: because of the poor transfer reliability, and to obtain sequence information also from N-terminally blocked proteins (Hellman *et al.* 1995).

Automated digestion for very small protein amounts is currently under development.

When spots are cut manually, there is high danger to *contaminate* the gels with keratins and other substances from the air. It is not only a cumbersome task, errors of picking wrong spots can jeopardize the results. In Fig. 53 the concept of an automated spot-picker, digester, and spotter is displayed: the system is placed into a closed cabinet; spot positions are imported from the image analysis results.

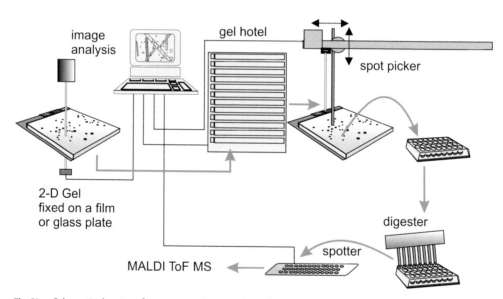

Fig. 53: Schematic drawing of an automated spot picking, digestion, and spotting system.

6.6.1
Mass spectrometry methods

The reason why mass spectrometry was not employed already many years earlier is the late development of ionization methods for peptides and proteins. For mass spectrometry analysis the sample molecules must be available as ionized gas.

Up to a few years ago only volatile small molecules could be analysed with mass spectrometry.

Two methods have survived the "evolution" of ionization procedures for peptides and proteins: *Electrospray* (ESI), invented by Yamashita and Fenn (1984) and *matrix assisted laser desorption ionization* (MALDI), developed by Karas and Hillenkamp (1988).

Yamashita MJ, Fenn B. J Phys Chem. 88 (1984) 4451–4459. Karas M, Hillenkamp F. Anal Chem. 60 (1988) 2299–2301.

Also, sensitivity, accuracy, and resolution of the instruments had been developed so far, that the data can be used in the database search with an adequate confidence level.

Protein spots of silver stained gels can now be identified.

With new mass spectrometry equipment sensitivities in the range of femto to attomol are reached.

But it is also sensitive to contaminations, like keratin etc.

MALDI ToF

MALDI ToF is mostly employed for protein identification via fingerprinting of tryptic peptides. The peptides are mixed with low molecular weight compounds, which have an absorption maximum at the wavelength of the laser – the *matrix*.

The matrix substances used for peptide analysis have been selected by Beavis and Chait (1989, 1990): For entire proteins sinapinic acid is used, for peptides α-cyano-4-hydroxy cinnamic acid or 2,5-dihydroxybenzoic acid (DHB). The mixtures are applied and dried as spots onto a metal slide.

Beavis RC, Chait BT. Rapid Commun Mass Spectrom. 3 (1989) 432–435. Beavis RC. Org Mass Spectrom. 27 (1992) 653–659.

This slide is inserted into a vacuum box. A pulsed laser beam (337 nm) is fired into the spot; the small matrix molecules absorb the energy. They move away from the target with supersonic speed. The matrix molecules take the peptides and proteins with them. While this happens, the charged matrix molecules transfer their charges on the peptide and protein molecules. MALDI produces mainly singly charged ions.

The charge transfer does not always work for all peptides, which is the main reason for missing peptides.

All ions with their different masses are accelerated in an electric field with the help of a high voltage grid to the same kinetic energy and induced into the high vacuum tube. Light ions arrive at the detector sooner than heavy ions (see Fig. 54). The recorded flight times are used to calculate the masses of the ions in m/z (mass per charge).

With a low mass rejection function the matrix ions are cancelled out. Secondary electron multiplier is mostly used as detectors, which provide sensitivity in the femtomol / attomol range.

Several technical developments have contributed to a better performance, higher mass accuracy, and enhanced resolution of the MALDI ToF technology:

Also new A/D converters (1–2 GHz) acquire more data points for higher resolution.

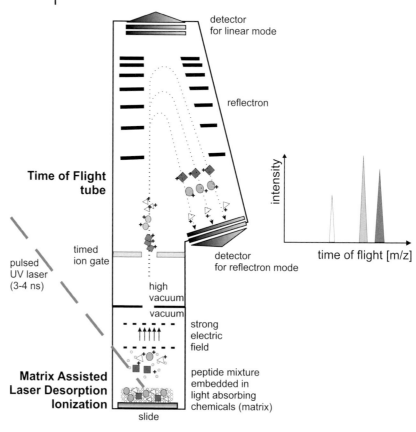

Fig. 54: MALDI ToF with quadratic field reflectron. For further explanation see text.

Also known as "time focusing". **Pulsed extraction:** The voltage of the acceleration grid is not applied permanently, but switched on with a time delay (a few nanoseconds) from the laser impulse. At this moment ions with a higher start velocity are further away from the slide than slower ions; they will thus receive a lower kinetic energy. In this way all ions will arrive at the flight tube at the same time.

Multiple spectra from a single laser shot are summed for signal averaging to achieve accurate mass determination.   *Reflectron tubes* reverse the drift direction of the ions in an electric counter field. This "ion mirror" corrects the start velocity distribution of ions of the same mass: Ions of the same mass, but higher start energy drift deeper into the reflector, thus fly a longer distance and catch up with the slower moving ions at a certain point after the reflector. The detector is located at this focusing point. This leads to a sharper signal, resulting in higher resolution.

Also the longer flight length obtained in a reflectron increases the resolution without the need for a larger instrument.

High molecular weight peptides and proteins are only detected in the linear mode; that is why most ToF instruments can be used in both modes.

The reflector reduces the sensitivity.

In a quadratic field reflectron the potential plates are arranged in such way, that the same focusing point and the same reflector voltage can be used for measurements of a wide mass / charge range in one shot. Thus the voltage has not to be stepped up; no signals are lost at the seams of a stitched mass spectrum.

Anderson UN, Colburn AW, Makarov AA, Raptakis EN, Reynolds DJ, Derrick PJ, Davis SC, Hoffman AD and Thomson S. Rev Sci Instrum 69 (1998) 1650–1660.

Post source decay (PSD):

In the field-free drift region of the mass analyzer a part of the peptides become *fragmented*, mainly by collision with background gas. These metastable fragments have the same velocity as their parent ions and would arrive at a linear detector at the same time. However, because the lighter fragments have smaller energy, in a *reflectron* they will turn back at an earlier time and will be accelerated back earlier than the larger fragments.

Note: the vacuum in a MALDI ToF is ca. 10^{-6} Torr. This means, that there are still some gas molecules left, which can collide with peptide ions.

As fragmentation occurs at the peptide bonds, the pattern shows ion series, whose masses differ by the sizes of the respective amino acids. In this way sequence information is obtained (Kaufmann *et al.* 1994). Also PTMs are detected.

Kaufmann R, Kirsch D, Spengler B. Int J Mass Spectrom Ion Processes 131 (1994) 355–385.

Because the peptides are only singly charged, this fragmentation is relatively rare; it is difficult to control.

CAF derivatisation (Chemically assisted fragmentation)

By *chemical derivatisation* of the tryptic peptides the fragmentation can be enhanced. One example is the method developed by Keough *et al.* (1999): A sulfonic acid group is added to the N-terminus, which causes a proton on the C-terminus to balance the negative charge coming from the low pK value. With MALDI the peptide will then gain a second proton from the matrix, which can bounce along the peptide backbone and initiate enhanced fragmentation. In this way strong fragment signals are obtained in the mass spectrum. As a second feature of this method the modified N-termini are not detected, because they contain a negative and a positive charge; they are neutral and will not be returned by the ion mirror. Only the C-terminal fragments are detected. These spectra are very easy to read out for amino acid sequencing and detection of phosphorylation sites.

Keough T, Youngquist RS, Lacey MP. Proc Natl Acad Sci USA. 96 (1999) 7131–7136. This modification increases the fragmentation rate and provides clearer PSD spectra.

For analysing PSD data, a peptide is selected with an *ion gate* at the entrance to the flight tube. In a quadratic field reflectron the resolution of the fragment peaks is constant over the complete mass range making it very useful for PSD analysis.

Only one spectrum has to be acquired to obtain a complete PSD ion series.

In Fig. 55 a PSD spectrum of a derivatised peptide separated in a quadratic field reflectron MALDI ToF is shown.

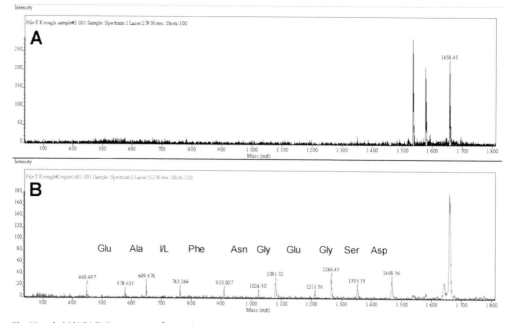

Fig. 55: A. MALDI ToF spectrum of peptides. B. PSD spectrum of a selected peptide (m/z = 1658.45). Derivatisation according to Keough *et al.* (1999), measurement with a quadratic field reflectron. With kind permission from Amersham Pharmacia Biotech, Uppsala, Sweden.

Electrospray ionization

Frequently also the term API (atmospheric pressure ionization) is used.

Here the supernatant of the tryptic digest is pressed through a metal capillary or a capillary with metal coating – for instance a cell injection needle – at high potential. The spray of fine highly charged droplets is created at atmospheric pressure in the presence of an electric field. These charged droplets are attracted to the inlet of the mass spectrometer, which is hold at lower potential.

Dry gas and heat are applied to the droplets; this causes the solvent to evaporate. The droplets decrease in size. Same charges give mutual repulses, those exceed the force of surface tension, the ions leave the droplet and are induced through a capillary or an orifice into the vacuum of the mass analyser (see Fig. 56).

Multiple protonation occurs: even 40 to 60 fold. This makes the combination with mass analysers possible, which have limited ion size capabilities like quadrupole mass filters and ion traps.

Wilm M, Mann M. Int J Mass Spectrom Ion Proc. 136 (1994) 167–180.

With *nano electrospray* 1µL sample liquid can be sprayed for more than 30 min, which allows extended analysis. Even unseparated peptide mixtures can be analysed (Wilm and Mann 1994).

Electrospray mass spectrometers can be coupled on-line to capillary electrophoresis and liquid chromatography systems.

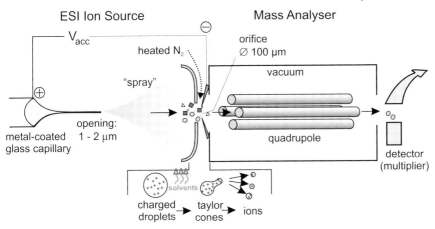

Fig. 56: Schematic drawing of electrospray ionization. Different kinds of mass analysers can be employed, including tandem mass analysers for structural analysis.

Comparison of ESI – MALDI ion sources

Electrospray spectrograms are more complex to evaluate than MALDI spectrograms, because multiply charged ions are produced. With electrospray there are usually more peptides ionized than with MALDI, leading to higher coverage. Electrospray ionization is very mild, proteins with intact tertiary structures and even noncovalent protein complexes can be analysed, the signals can be made quantitative. ESI can be combined with more possible mass analysers, including tandem mass spectrometry (MS/MS or MS^n). Mass spectra generated by ESI methods can provide quantitative information; this is not possible with MALDI.

MALDI can be employed for high throughput and automated applications for ca. 200 proteins per day, ESI is used for more thorough investigations for 10 to 20 proteins per day.

Tandem mass spectrometry

For *structural analysis* the peptides are selected with a first mass analysis, actively fragmented by collision-induced dissociation (CID) with argon or helium under low-energy conditions; these fragments are measured with a second mass analysis. This can, for instance, be done by using an ion trap: The ions are captured a three-dimensional quadrupolar field and "stored" for a certain time (0.1 to 10 ms). It is filled with a low amount of Helium gas to reduce the initial entrance speed of the ions and to allow CID. The ions are ejected from the ion trap in the sequence of their masses either by continuously raising the AC voltage or by the application of multipolar fields.

With an ion trap system further fragmentation of the fragments (MS^n) for PTM analysis can be carried out.

Roepstorff P, Fohlmann J.
Biomed Mass Spectrom. 11
(1984) 601.
Wilm M, Shevchenko A,
Houthaeve T, Breit S, Schwei-
gerer L, Fotsis T, Mann M.
Nature 379 (1996) 466–469.

Sequence information: Different ions are produced by fragmentation: y-ions with intact C-terminus and b-ions with intact N-terminus (Roepstorff and Fohlmann, 1984). The evaluation software is checking the measured mass differences of the fragments against the masses of the amino acids; it suggests possible y- and b-ion series, and finally the amino acid sequence. Even though glutamine and lysine (M = 128), as well as leucine and isoleucine (M = 113) have identical masses, the method can be used for de novo sequencing (Wilm et al. 1996).

Mørtz E, Saraneva T, Haebel S,
Julkunen I, Roepstorff P.
Electrophoresis 17 (1996)
925–931.

Comisarow MB, Marshall AG.
Chem Phys Letters. 25 (1974)
282–283.
The limitations of such instruments are mainly the purchasing price and the expertise of the operator.

Post-translational modification analysis: Phosphorylation and acetylation can be identified with the help of the software, but in many cases differential enzymatic treatment and repeated analysis has to be performed, like for the case of glycosylations (Mørtz et al. 1996).

At the high end there are *FT-ICR* (Fourier transform – Ion cyclotron resonance) mass spectrometers based on the work of Comisarow and Marshall (1974), which reach a resolution of 10^6. They can be connected to MALDI and ESI ion sources, and allow identification of proteins even in complex peptide mixtures.

More comprehensive information on mass spectrometry of peptides is found in the following books:

Siuzdak G. The expanding role of mass spectrometry for biotechnology. MCC Press, San Diego (2003).

Chapman JR Ed. Mass spectrometry of proteins and peptides. Methods in molecular biology. Vol 146. Humana Press, Totowa (2000).

6.6.2
Peptide mass fingerprinting

Chemical (CnBr) or enzymatic cleavage methods are employed. A "well defined" procedure is required. Mostly trypsin is used, which cleaves a polypeptide after each arginine and lysine.

With mass spectrometry such accurate *peptide sizes* are measured, that the results can be compared to those which are mathematically derived from the genome databases: The triplets from the DNA sequence are theoretically translated into an amino acid sequence, the peptide sizes are predicted using the knowledge of the preferred cleavage motifs of the enzyme used in the practical experiment. Fig. 57 shows a diagram of protein identification with peptide mass fingerprinting.

This is a multistep procedure including several times drying, in order to get the reactants and the enzyme into the gel by rehydration.

Prior to digestion, the buffer components and SDS are washed out from the gel piece, the cysteins are modified with vinylpyridine or iodoacetamide in order to cancel out different oxidation forms.

Note: In a MALDI ToF spectrogram always some of the peptides are missing.
Berndt P, Hobohm U,
Langen H. Electrophoresis 20

Peptide mass fingerprinting is usually performed using a MALDI ToF (see page 111), because this ionization produces only *singly charged peptides* – this makes evaluation of the mass spectrum easier. Three to four peptides measured with high mass accuracy (ca. 25 ppm) are sufficient to identify a protein (Berndt et al. 1999).

(1999) 3521–3526.

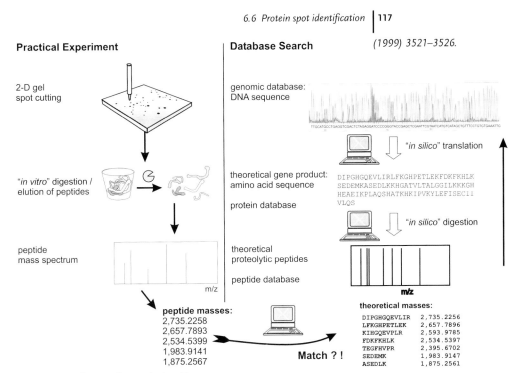

Fig. 57: Peptide mass fingerprinting. The peptide masses of the digested protein are compared to the theoretical masses of peptides, which are mathematically derived from the genomic database.

With the accuracy and the resolution needed, different isotopic peaks of one peptide are resolved, because in nature exists a certain distribution of ^{12}C and ^{13}C isotopes (100 : 1). For PMF search the monoisotopic peak containing only the most common isotope ^{12}C is used.

Peptide mass finger printing is the easiest and fastest method to identify proteins. However, in 20 to 40 % of the cases the *search software* does not find a match, or the match is uncertain or not plausible. This can be caused by several reasons, for instance: wrong predictions of the open reading frames from the genomic database, sequence errors in the genomic database, mutations, PTMs, etc. Those spots have to be further analysed with post source decay for sequence information, which will help to identify ca 10 % more proteins.

When this does not give a result, the sample has to be submitted to tandem mass spectrometry to obtain more detailed structural information: a larger part of the amino acid sequence or the post-translational modification.

Also protein mixtures – for instance more than one protein in one spot – have to be analysed with MS/MS.

6.6.3
Protein characterization

Structural data have mostly been acquired with ESI *tandem mass spectrometers* (see page 114 f.). The – usually short – amino acid sequence can be read out from the mass spectrum of the fragmented peptides. With this information the search software can browse the EST (expressed sequence tag) databases. In this way protein identification is performed with higher confidence than with the peptide mass search alone.

This technique is more complicated to handle, mostly because electrospray ionization produces multiply charged ions, which make the mass spectra very complex.

This is just one example for PTM identification; others are published in the relevant literature.

Also information on *post-translational modifications* can be extracted from the collision induced dissociation (CID) mass spectra, for instance phosphorylation. The peptide mixture is measured before and after treatment with phosphatase; the distance of the resulting shift of the peptide peak is measured.

New developed MALDI-ToF instruments together with new derivatisation chemistry allow structural analysis of peptides with post source decay (see page 113). This makes protein identification and characterization much easier: Only one instrument is required, it needs just a few more laser shots to acquire the PSD data than preparing a sample for ESI.

However, MALDI ToF data do not provide a complete coverage of peptides. Thus, for *de novo* sequencing and intensive PTM analysis ESI tandem mass spectrometers are needed.

6.7
Bioinformatics

Proteome analysis is highly depending on bioinformatics tools. They are necessary from protein spot identification and characterization to the statistical analysis of the large data sets produced by the high throughput proteome studies.

Protein and peptide *mass spectra* can mostly not be manually evaluated. Particularly the multiple charged ESI peak patterns need deconvolution algorithms.

The quality of database search results is obviously highly dependent on the software employed.

The mass spectrometric measurements need to be correlated with protein and genome sequences. In-silico translations and digestions have to be performed. Missed cleavage sites must be taken in account. Advanced database search programs include plausibility and significance checks for protein identification as a quality control. This is particularly important for the automation of protein identification.

The software must convert fragment ion series into amino acid sequences, and search the EST databases. Modified amino acids and many other post-translational modifications can be already detected and characterized without further in-vitro experiments.

Raw data must be stored in an organized way; administration of analysis results in databases is required. Appropriate visualization and interpretation tools are needed as well.

Finally, biological information must be extracted from these complex data, which is probably the most complicated job for bioinformatics.

6.8
Functional proteomics

The drawback of the proteomics analysis procedure described so far is, that information on three-dimensional folding of the proteins, protein-protein complexes etc., which are important for the function of a protein, are destroyed by denaturation of the samples during the analysis. Native multiple analysis for the screening of proteins is not possible.

Therefore the following strategy is pursued:

At first target proteins are identified, characterized and correlated with "protein families". Once some structural information are known, smaller subsets of proteins are analysed with milder separation and measuring techniques: for instance, some proteins are fished out of a cell lysate with affinity chromatography and then proteins with intact tertiary structure, or protein-protein complexes are analysed after electrospray ionization (see reviews by Lamond and Mann, 1997, and Pandey and Mann, 2000).

Lamond AI, Mann M. Trends Cell Biol. 7 (1997) 139–142.
Pandey A, Mann M. Nature 405 (2000) 837–846.

7

Instrumentation

The equipment for capillary electrophoresis and automated DNA sequencing has already been described together with the method.

In most laboratories using electrophoresis the apparatus consists of three principle pieces of equipment:

This means gel electrophoresis in the wide sense.

a) power supply
b) cooling or heating thermostat
c) separation chamber with combined gel casting system.

Concerning a): For electrophoresis DC power supplies are needed, which yield high voltages and allow to set the maximum output of current, voltage and power.

Concerning b): Many home-made systems are used without cooling or heating. Yet it has been proved that better and more reproducible separations are obtained with temperature controlled equipment.

Concerning c): The core of electrophoretic equipment is the separation chamber. A number of types exist because of the many different methods and modifications.

7.1
Current and voltage conditions

To establish electrophoretic separation conditions a few physical rules should be recalled:

This is also important when working under defined conditions.

The driving force behind electrophoresis is the product of the charge $\Theta^\pm$ (net charge) of a substance and the electric field E, measured in V/cm. For the speed of migration of a substance v in cm/s this means:

The net charge $\Theta^\pm$ can be taken to be the sum of the elementary charges, measured in As.

$$v = \frac{\Theta^\pm \times E}{R}$$

Electrophoresis in Practice, Fourth Edition. Reiner Westermeier
Copyright © 2005 WILEY-VCH Verlag GmbH & Co. KGaA, Weinheim
ISBN: 3-527-31181-5

Thus a certain field strength is necessary for an electrophoretic migration.

The frictional constant R is dependent on the molecular radius, r, (Stokes radius) in cm and the viscosity, η, of the separation medium measured in $N\ s/cm^2$.

To reach the field strength the voltage U must be applied, it is measured in volt (V) and the separation distance d in cm.

$U = E \times d$

Voltage = field strength × separation distance.

If an electric field is applied to a conducting medium (buffer), an electric current, I, will flow. It is measured in amperes (A) yet is usually given in mA for electrophoresis. The magnitude of the current depends on the ionic strength of the buffer. In electrophoresis relatively high currents are used while for isoelectric focusing they are smaller because the pH gradient has a relatively low conductivity.

The product of the voltage and the current is the power, P, given in watt (W):

$P = U \times I$

Power = voltage × current

Joule heat

The product of (electrical) power and time is energy. During electrophoresis most of the electrical energy is transformed into heat.

If this is not taken into consideration, the gel can burn through.

For this reason the temperature should be controlled during electrophoresis. Since the cooling efficiency, that is the heat dissipation, cannot be increased indefinitely, a certain intensity should not be exceeded.

Guidelines for cooled 0.5 mm thick horizontal gels:
Electrophoresis: ca. 2 W/mL gel volume;
IEF: ca.1 W/mL gel volume
IPG: see instructions

Fig. 58 demonstrates the relationship between voltage, current, power and the dimension of the electrophoretic medium. The longer the separation distance, the higher the current necessary to reach a specific field strength. At a given ionic strength, the field strength is proportional to the cross-section: the thicker the gel, the greater the current. The power is proportional to the volume of the gel.

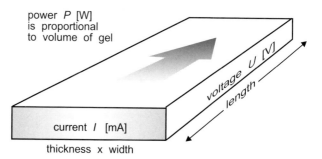

Fig. 58: Schematic diagram of the relationships between the separation medium and current, voltage and power conditions during electrophoresis.

This also means that the power and the current must be reduced if only part of a gel is used, but, for the same separation distance:

Half gel: half the current – half the power – same voltage.

One should always be aware, that the values set in the power supply are maximum values; the real values change during the run, because they are controlled by the conductivity of the buffer and the gel. It is often forgotten, that more concentrated gels have higher resistances than gels with low T values.

This is often used for flatbed techniques.

The conductivity of the system changes during the run, particularily in disc electrophoresis and isoelectric focusing experiments.

7.2
Power supply

Different models and degrees of specification exist:

1. *Simple* power supplies can be regulated by the voltage.
2. *Typical* electrophoresis power supplies can be run with constant current or constant voltage.
3. Power supplies which are also designed for isoelectric focusing supply high voltages. Their power is also stabilized so that a maximum setting can be programmed.

usually 200 V maximum

usually up to 500 or 1000 V, 200 or 400 mA

usually up to 3000 or 5000 V, 150 or 250 mA, 100 or 200 W

In isoelectric focusing the conductivity of the gel drops when the pH gradient has formed, and the buffer ions have migrated from the gel. Regulating the maximum power prevents overloading the gel with high voltages. Additional control over the focusing conditions is provided by a volt/hour integrator.

4. *Programmable* power supplies have an additional microprocessor with which different separation conditions with various steps can be recalled.

A volt/hour and an ampere/hour integrator are usually also built in.

For example:

First phase: low voltage for gentle sample entry into the gel.

Second phase: high voltage for rapid separation and to avoid diffusion.

Third phase: low voltage to avoid diffusion and migration of the zones.

For long separation distance gels, high voltages must be applied and low current. Optimal resolution and high reproducibility is obtained, when these gels are run with a "Voltage ramping".

In method 12 on page 310 a voltage ramping program is described.

7.3
Separation chambers

7.3.1
Vertical apparatus

Electrophoresis is carried out in gel slabs, which are cast in glass cassettes. The samples are applied in gel pockets with a syringe or a micropipette. The current is conducted through platinum electrodes which are located in the buffer tanks.

An example of a vertical system with temperature control which is used for gel slabs is shown in Fig. 59. To dissipate Joule heat the lower buffer is cooled by a heat exchange system.

A
B

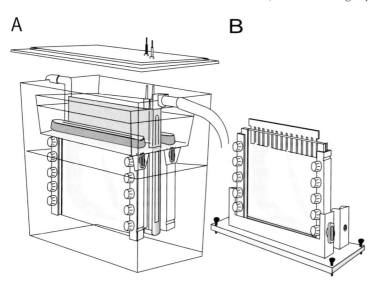

Fig 59: Vertical electrophoresis chamber. (A) Separation chamber with heat exchanger, up to 4 gels can be run in parallel. (B) Gel casting stand with comb for sample wells.

Ansorge W, De Maeyer L.
J Chromatogr. 202 (1980)
45–53.

Manual sequencing chambers

Since long gels are required for manual DNA sequencing, special chambers with thermostatable heating plates, are used (Fig. 60). The ultrathin large gels are prepared with the horizontal sliding technique of Ansorge and De Maeyer (1980). The same setup is employed for automated sequencing, however, because of the on-line detection of the fragment, the gels can be shorter.

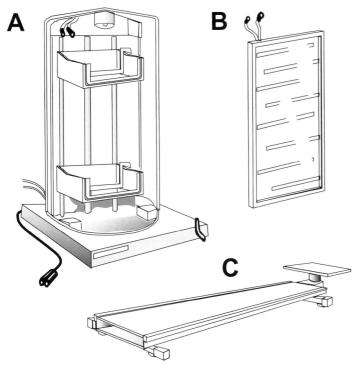

Fig. 60: Manual DNA, sequencing chamber.
– (A) Vertical chamber in safety cabinet.
(B) Thermostatable plate. – (C) Horizontal gel
casting apparatus for the sliding technique.

7.3.2
Horizontal apparatus

DNA analysis in agarose gels

For analytical and preparative separation of DNA fragments and RNA restriction fragments "submarine" chambers are usually used. The agarose separation gel is submerged under a thin layer of buffer between the lateral buffer tanks. A new development is a system using readymade agarose gels in the size of microtiter plates with 96 slots for running multiple samples.

Horizontal chambers exist in different sizes.

For electrophoresis in a *pulsed field* a controlling device is connected to the power supply, which switch the electrodes – at predefined frequency – in the north/south and the east/west directions. Diodes are built-in to the electrodes so that when they are switched off, they cannot influence the field. Since these separations can last for a long time – up to several days – the buffer must be cooled and circulated (Fig. 61).

For non-homogeneous fields the point electrodes are placed in electrode grooves set at right angles.

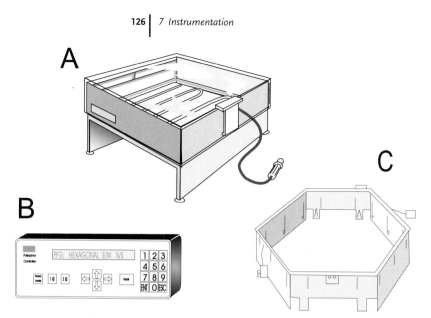

Fig. 61: System for pulsed field DNA gel electrophoresis (PFG).
– (A) PFG submarine chamber with cooling coil and buffer circulation pump
(not visible); – (B) Programmable pulse controlling device; – (C) Hexagonal
electrode for linear sample lanes.

and protein analysis in agarose
Most of the instructions in part
II are designed for this kind of
electrophoresis equipment,
because almost all methods
can be performed on it.

Protein and DNA analysis in polyacrylamide gels

Horizontal chambers with thermostatable plates and lateral buffer
tanks are very versatile (Fig. 62): They are equipped for analytical and
preparative isoelectric focusing, for several variations of immuno and
affinity electrophoresis, all zone electrophoresis techniques in restrictive and non-restrictive gels, high resolution 2D electrophoresis, as
well as semi-dry blotting. High voltages can be applied, because there
are no problems with insulation of buffer tanks; and many techniques can be applied without using a buffer tank at all.

If the gel supporting area does not need to be too large, thermo-
electric cooling or heating with Peltier elements can be used instead
of employing a thermostatic circulator (Fig. 63).

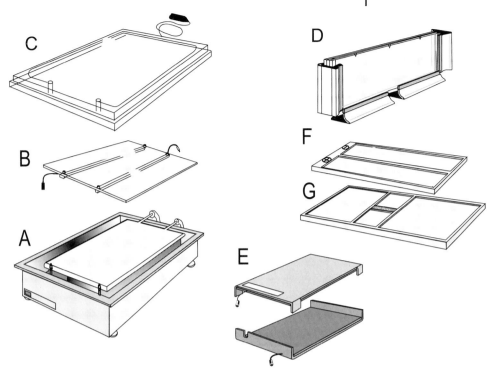

Fig. 62: Horizontal flatbed electrophoresis system. – (A) Separation chamber with cooling plate and lateral buffer tanks. – (B) Electrodes for isoelectric focusing and electrophoresis with buffer strips. – (C) Safety lid. – (D) Gel casting cassette for ultrathin homogeneous and gradient gels. – (E) Graphite plate electrodes for semi-dry blotting. – (F) PaperPool for soaking of electrode wicks. – (G) GelPool for rehydration of dry gels.

Fig. 63: Horizontal apparatus with Peltier cooling and built-in electrode system.

7.4
Staining apparatus for gels and blots

For staining blotting membranes instead of the gel tray a multicompartment tray is placed on the base and connected to the pumps (not shown in Fig 64).

Very helpful, particularly for silver staining with its many steps, is a staining apparatus. It saves time and improves reproducibility of the result considerably. With a modified tray and and program it can also be used for the development of blotting membranes. Fig. 64 shows an automated gel and blot stainer which consists of a programmable controler, a rocking tray, a 10-port valve and a fast peristaltic pump.

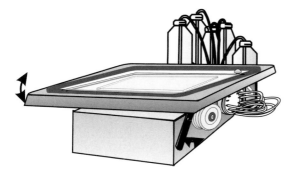

Fig. 64: Automated gel stainer for gels of different sizes with and without film support.

7.5
Automated electrophoresis

The gels only need to be placed in the separation compartment, the sample applied and the gel transferred to the development chamber after the separation.

A complete automated electrophoresis system, the PhastSystem®, is composed of a horizontal electrophoresis chamber with a Peltier element which cools and heats the separation bed, and a built-in programmable power supply and development unit (see Fig. 65). The current and temperatures for separation and staining as well as the different development procedures can be programmed and recalled for the various electrophoresis and staining methods. There is a blotting unit with graphite electrodes for electrophoretic transfers.

Special gels with support films for focusing, titration curves and electrophoresis, native and SDS-electrophoresis buffers, which are cast in agarose, as well as staining tablets and a silver staining kit exist for this system.

These combs correspond to sample applicators.

The samples are applied automatically – at a defined time – with multiple sample application combs. The separation and development steps occur very quickly because the gels are only 0.3 and 0.4 mm thick and relatively small, 4 cm × 5 cm.

A SDS-electrophoresis run in a pore gradient gel lasts 1.5 h including silver staining. Therefore, very sharp bands are obtain, and in spite of the small geometry, a high resolution is achieved.

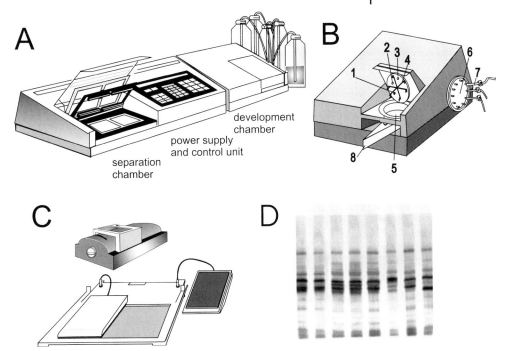

Fig. 65: System for automated electrophoresis (PhastSystem®).
– (A) The complete apparatus; – (B) the development unit: (1) opening to the
membrane pump, (2) temperature sensor, (3) level sensor, (4) rotating gel
holder, (5) development chamber, (6) 10-port valve, (7) numbered PVC tubing,
(8) closing mechanism; – (C) Blotting system to place in the separation compartment and filmremover. – (D) PhastGel medium: SDS electrophoresis after
automated silver staining (With kind permission of Professor A. Görg).

The development unit can be warmed to 50 °C. It contains tubing
for the entry and exit of the staining solutions and a membrane
pump for creating vacuum or pressure in the chamber as well as for
emptying it or pumping in solutions. There also is a gel holder for
rotating the gel in the solution. All the functions can be programmed.
The timer, temperature and level sensors regulate the execution of
the program.

*The development unit is moni-
tored by the separation and
control unit.*

The advantages of such an automated electrophoresis and development unit are numerous:

- rapid and very reproducible separations,
- electrophoresis can be carried out immediately and without
  preparation,
- problems due to "human error" are reduced to a minimum,
- reduction of the work load (ready-to-use gels and buffers, automated separation and development),

- multiple sample application,
- no handling of liquid buffers,
- clean work, can be carried out at any location,
- rapid switching between separation and development.

7.6
Instruments for 2-D electrophoresis

7.6.1
Isoelectric focusing apparatus

Immobilized pH gradient strips can be run in an horizontal electrophoresis apparatus. However, due to safety regulations the applied voltage must not exceed a limit of 3,500 V, because power supply, separation chamber, and cooling circulators are connected via external tubings and cables.

For more information on the first dimension IPG system see page 282 ff.

In order to overcome this limitation and have a time and manual work consuming procedure, new instrument has been developed, which is shown in Fig. 66 C. Here power supply, cooling system, electric contacts, sample application and rehydration trays are integrated into one system. The programmable power supply delivers up to 8,000 V. The samples are applied into the individual narrow ceramics trays, the strips are placed with the dried gel surface down onto the sample solution (Fig. 66 A). The proteins enter the gel during it is rehydrated in the sample solution. Rehydration with applied low voltage often improves the entrance of high molecular weight proteins. For very high sample concentration, however, and for cases, where cup-loading is required, an alternative ceramics tray is needed (Fig. 66 B).

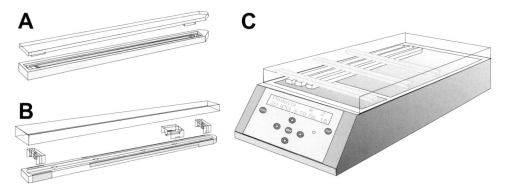

Fig. 66: Isoelectric focusing system for 2-D electrophoresis. – (A) Ceramics tray for rehydration-loading and running gel strips with immobilized pH gradients. – (B) Ceramics tray for running the gel strips upside and for cup-loading. – (C) Isoelectric focusing apparatus.

7.6.2
Multiple slab gel apparatus

For high resolution and high throughput 2-D electrophoresis in proteomics it is necessary to run many large formate SDS gels under identical conditions. In Fig. 67 a system for running 12 slab gels is shown. In the design shown below cooling of the gels is performed by continuously pumping the anodal buffer over a heat exchanger plate and force it through channels in the lateral side walls to flow between the gel cassettes. In this way the Joule heat is removed so efficiently, that a separation of 12 gels in 25 × 20 cm large cassettes is completely done within 4 to 5 hours. This limits diffusion considerably, thus the spots are higher concentrated than after over night runs, leading to higher resolution and sensitivity of detection.

Higher concentrated spots means also more protein per gel volume. This improves the enzyme kinetics for tryptic in-gel digestion for peptide mass fingerprinting.

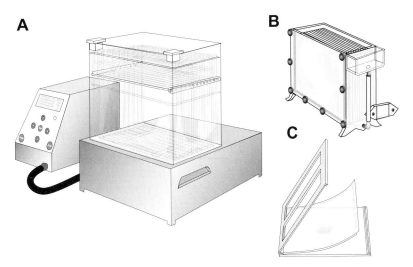

Fig. 67: SDS electrophoresis system for 2-D electrophoresis. – (A) Apparatus for multiple slab gels with integrated cooling and power supply. – (B) Multiple gel caster. – (C) Special cassette for readymade gels on film support.

7.7
Safety measures

In most electrophoresis techniques high voltages (>200 V) are used to reach the field strengths required for separation. So as not to threaten safety in the laboratory, electrophoresis should only be performed in closed separation compartments. A danger of electric shock by contact exists when the separation system is open. In addition, the system should shut off automatically in the event of a short-circuit.

The cables and plugs must be designed for constant current with high voltages.

*Versatile uses means: **either** high voltages **or** high currents. In any case, the power is limited by the heat production.*

Many separation chambers, e.g. for vertical electrophoresis or submarine techniques, are only licensed forvoltages up to 500 V. Power supplies with specifications for versatile uses can be equipped with a sensor for coded plugs, which only allows the maximum voltages for special chambers.

Electrophoresis equipment should be kept in a dry place.

The separation unit should be designed in that way, that the the plugs and sockets placed so that the current is automatically shut off if the chamber is opened by mistake. Electrophoresis systems should be designed so that the power supply is above or at least at the same height as the separation unit, to avoid buffer running into the power supply should it spill out.

7.8
Environmental aspects

Choosing an appropriate instrument system can also be influenced by environmental aspects.

Only feasible with horizontal systems.

Using buffer strips or buffer wicks is advantageous over large buffer volumes for two main reasons:

The gels, strips or wicks can be cut to size.

This applies for chemicals as well as for radioactivity.

- No excess chemicals: Only that amount of chemical is used which is needed for the run of the number of samples to be separated.
- The volume of materials and liquid disposed is much less than for buffer tanks in the conventional techniques.

Also the detection method can influence both the choice of the appropriate electrophoresis method and the instrumental system.

Here are two examples:

There are a few more advantages with the alternative polyacrylamide gel method; see Method 12 and the following on page 299 of this book.

The technique of native electrophoresis in washed and rehydrated polyacrylamide gels offers a number of possibilities, see method 5 in this book.

- *DNA electrophoresis:* As ethidium bromide is not welcome in every laboratory, the agarose-submarine-ethidiumbromide technique is – in many cases – replaced by using rehydrated thin polyacrylamide gels on carrier films with subsequent silver staining.
- *Isoelectric focusing:* In isoelectric focusing experiments, efficient fixing of the proteins and simultaneous washing out of the carrier ampholytes can only be done with trichloroacetic acid. Many laboratories with routine applications and a high gel throughput look for an alternative for this halogenated acid. Some separation problems may be solved by employing a basic or acidic native electrophoresis technique instead, because alternative fixing and staining methods can easily be used here.

Part II:
Equipment and methods

Electrophoresis in Practice, Fourth Edition. Reiner Westermeier
Copyright © 2005 WILEY-VCH Verlag GmbH & Co. KGaA, Weinheim
ISBN: 3-527-31181-5

Equipment for part II

Almost all methods described here are performed in a horizontal system with the same equipment. In methods10 and 11also procedures for vertical equipment is explained.

The small items and pieces of equipment as well as the principal stock solutions can be used for almost all the methods.

In principle the sequence of the first nine methods does not correspond to their importance or frequency of use but rather their simplicity and cost. The DNA methods are placed to the end, because all of them are new developments.

This should also help for the planning of an electrophoresis course.

Methods:

Small molecules:

1. PAGE of dyes
4. Native PAGE in amphoteric buffers

Proteins:

2. Agarose and immunoelectrophoresis
3. Titration curve analysis
4. Native PAGE in amphoteric buffers
5. Agarose IEF
6. PAGIEF in rehydrated gels
7. Horizontal SDS PAGE
8. Vertical PAGE
9. Semi-dry blotting of proteins
10. IEF in immobilized pH gradients
11. High resolution 2-D electrophoresis

Electrophoresis in Practice, Fourth Edition. Reiner Westermeier
Copyright © 2005 WILEY-VCH Verlag GmbH & Co. KGaA, Weinheim
ISBN: 3-527-31181-5

DNA:

8. Vertical PAGE
12. PAGE of double stranded DNA
13. Native PAGE of single stranded DNA
14. Denaturing gradient gel electrophoresis
15. Denaturing PAGE of DNA

Instrumentation

for methods 1 to 9, 11

only for method 9!

Multiphor II	horizontal electrophoresis unit
Mighty Small SE 250	vertical electrophoresis unit
EPS 3501 XL	programmable power supply 3500 V
Multitemp III	thermostatic circulator
Processor Plus	silver staining and blot developing apparatus
Novablot	graphite plate electrodes for blotting
Film Remover	apparatus for removing support films
IPG strip kit	for IPG-IEF of 2D electrophoresis on Multiphor
Strip positioner	for SDS PAGE on Multiphor
IPGphor, strip trays	IEF system for 2-D electrophoresis
Ettan DALT II	SDS PAGE system for 2-D electrophoresis
Gel dryer	gel drying frames and loading platform
Universal gel kit:	contains glass plates, glass plates with gaskets, clamps, gradient mixer, tubing, tubing clamps, scalpel, tape;
GelPool	Tray for rehydration of dry gels
PaperPool	Double tray for soaking of electrode wicks
Immmuno- electrophoresis kit:	contains levelling table, spirit level, glass plates, electrode wicks, holder, special scalpell, punching template, gel punch with vacuum tubing, sample application film;

field strength probe and voltmeter
hand roller,
water gauge
silicone rubber sample applicator mask
humidity chamber for the agarose techniques
stainless steel staining trays
destaining tanks
glass tray for silver staining

scissors, spatulas
assorted glass ware: beakers, measuring cylinders, erlenmeyers, test tubes etc.
magnetic stirrer bars in different sizes
graduated pipettes of 5 and 10 mL + pipetting device
(e.g. Peleus ball)
micropipettes adjustable from 2 to 1,000 μL

Consumables:
Disposible gloves, tissue paper, filter paper, Scotch and Dymo tape
   pipette tips, Eppendorf cups, test tubes with screw caps 15 and 50 mL

Evaluation:

ImageScanner	desk top scanner modified for electrophoresis gels
Typhoon	multifluorescence scanner
Computer (Windows)	
ImageMaster	1 D and 2D evaluation software

SPECIAL LABORATORY EQUIPMENT

Method	1	2	3	4	5	6	7	8	9	10	11	12	13	14	15
Glass rod		○			○										
Heating block							□					□		□	□
Heating cabinet or incubator		○			○		○			○	○				
Heating stirrer			■	■		■	■	■							
Laboratory elevator ("Laborboy")						○	○	○		○	○		○		
Microwave oven		○			○						○				
Paper cutter "Roll and Cut".				○						○					
Rocking platform							■	■	■	■	■	■	■	■	■
Small magnetic stirrer							○	○		○	○	○	○	○	
Spatula in different sizes	○	○	○	○	○	○	○	○	○	○	○	○	○	○	○
Table centrifuge		□	□	□	□	□	□	□	□	□	□	□		□	□
Forceps straight and curved						○	○	○		○	○	○	○	○	
UV lamp									■						
Ventilator			○	○	○	○	○	○		○	○	○	○	○	○
Watch glasses		○			○										
Water jet vacuum pump		○	○			○	○	○		○	○				○

□ For sample preparation, ○ for the method, ■ for detection

CONSUMABLES

Equipment for part II | 139

Method	1	2	3	4	5	6	7	8	9	10	11	12	13	14	15
Blotting membrane, nitrocellulose									○						
Cellophane								■			■				
Dymo tape		○	○	○			○					○	○	○	○
Electrode wicks	○	○	○	○			○					○	○	○	○
Filter paper	○	○ ■	○	○	■ ○		○	○	○	○	○	○	○	○	○
Focusing strips					○	○				○	○				
GelBond® Film for agarose (12.5 × 26 cm)		○			○			○							
GelBond® PAG Film (12.5 × 26 cm)				○		○	○			○	○	○			
GelBond® PAG Film (20.3 × 26 cm)										○	○				
Parafilm®	○			○				○			○				
Pipette tips	○ □	○ □	○ □	○ □	○ □	○ □	○ □		□	○ □	○ □	○ □	○ □	○ □	○ □
Plastic bags	○		○	○		○				○	○	○	○	○	○
Polyester film untreated	○		○	○		○	○				○				
PVC film (overhead)									■						
Sample application pieces					○	○	○				○				

□ For sample application, ○ for the method, ■ for detection

CHEMICALS
Chemicals

Method	1	2	3	4	5	6	7	8	9	10	11	12	13	14	15
Acetic acid (96 %)	■	■	■	■	■○	■	■	■		■○	■○	■○	■○	■○	○
Acid violet										■	■				
Acrylamide	○		○	○		○	○	○		○	○	○	○	○	○
Agarose L		○													
Agarose IEF					○										
Mixed bed ion-exchanger	○			○		○						○	○	○	○
ε-Aminocaproic acid				○					○						
Ammonia solution (25 %), NH₃						■									
Ammonium nitrate			■	■	■	■				■					
Ammonium persulphate	○		○				○					○	○	○	○
Ammonium sulphate							■				■				
Ampholine® carrier ampholyte, depending on the pH interval											□				
Antibodies		○			■				■						
Benzene sulphonic acid											■	■	■	■	■
Benzoinmethyl ether									■						
Bisacrylamide	○		○			○						○	○	○	○
Boric acid												○	○	○	○
Calcium lactate		○													

□ For sample preparation, ○ for the method, ■ for detection

CHEMICALS, CONTINUED

Method	1	2	3	4	5	6	7	8	9	10	11	12	13	14	15
CelloSeal®	○			○			○	○		○	○				
CHAPS						□					□○				
Coomassie Brilliant Blue G-250		■	■			■	■	■	■	■					
Coomassie Brilliant Blue R-350		■	■			■	■	■		■	■				
Cupric sulphate			■			■				■					
Dithiothreitol						□	□	□			□				
EDTA-Na₂		□○				□	□	□	□	□	○	○	○□	○	□
Ethanol						■	■	■	■	■	■	■	■	■	■
Ethidium bromide								■							
Ethylenglycol						○	○	○			○				
Dyes for marking the front:															
Orange G, Bromophenol Blue	□			□			□	○		○	○	○	○	○	○
Xylencyanol				□○								□	□	□	□
Pyronine, Basic Blue (cationic)															
Formaldehyde solution 37 %		■			■	■	■	■		■	■	■	■	■	■
Formamide											■		□		□
Glutardialdehyde solution 25 %						■	■	■		■					
Glycerol	○		○	○		○	○	○	○	○	○	○	○	○	○
Glycine		■	■	■											■

CHEMICALS, CONTINUED

Method	1	2	3	4	5	6	7	8	9	10	11	12	13	14	15
HEPES				○											
Hydrochloric acid, fuming							○	○		○	○				
Immobiline II Starterkit, assorted (6 × 10 mL)								○		○					
Imidazole							■	■		■	■				
Indian Ink									■						
Iodoacetamide						□	□	□			□				
Kerosene for cooling	○	○	○	○	○	○	○	○		○	○	○	○	○	○
Marker proteins for MW Calibration kit (14 400 to 94 000 Da)							○	○		○					
Collagen solution up to 300 kDa							○	○							
Peptide marker							○	○							
Marker proteins for pI, depending on the pH interval					○					○					
2-Mercaptoethanol						□				□ □	□ □				
Methanol	■	■	■	■	■	■	■	○	■	■					■
NAP columns for DNA purification, and to desalt protein				□	□	□	□	□	□	□	□	□	□	□	□
Non-ionic detergents: Nonidet NP-40, Triton X-100, ProSolv II:		□ □				□ □				□ □	□ □				

CHEMICALS, CONTINUED

Method	1	2	3	4	5	6	7	8	9	10	11	12	13	14	15
Pefabloc, protease inhibitor						□	□	□			□				
Pharmalyte® carrier ampholyte, depending on the pH interval			○		○					□	□				
PhastGel® Blue R, staining tablets Coomassie R-350		■	■	■	■	■	■		■	■					
Phosphoric acid								■		■	■	○	○	○	○
PMSF, protease inhibitor						□	□	□		□	□	○	○	○	○
Repel Silane	○		○	○	○	○	○	○		○	○		○		
SDS							○/□	○/□	○		○/□		○		
Silver nitrate		■			■	■	■	■				■	■	■	■
Soda lime pellets (CO2 trap)					○	○				○	○				
Sodium acetate						■	■	■		■	■	■	■	■	■
Sodium carbonate		■			■	■	■	■		■	■				
Sodium chloride		■							■						
Sodium dihydrogen phosphate	○								○						
Sodium hydrogen phosphate	○														
Sodiumhydroxid						○		■		○	○		□		
Sodium thiosulphate							■	■			■	■	■	■	■
Sorbitol					○	○									

CHEMICALS, CONTINUED

Method	1	2	3	4	5	6	7	8	9	10	11	12	13	14	15
Sulfuric acid conc.			■			■			■						
Sulphosalicylic acid						■									
TEMED	○		○	○		○	○	○	○	○	○	○	○	○	○
Thiourea											□○				
TMPTMA									■						
Trichloracetic acid		■	■	■	■	■				■	■				
Tricine		○				○	○	○	○	○	○	○	○	○	○
Tris		○				○	○	○	○	○	○	○	○	○	
Tungstosilicic acid		■	■		■	■				■					
Tween 20									■						
Urea						○□				○□	○□	□		○	□○
Zinc sulphate							■	■			■				

All chemicals must be of analytical (p.A.) quality. Double-distilled water should be used for all solutions.

□ For sample preparation, ○ for the method, ■ for detection

Method 1:
PAGE of dyes

In most laboratories electrophoresis is used for the separation of rela- *High molecular weight:* tively high molecular weight substances, such as proteins and nucleic *> 10 kDa* acids. Substances with low molecular weights such as polyphenols *Low molecular weight: < 1 kDa* and dyes are separated by column or thin-layer chromatography.

In the following part, a simple electrophoretic method for the sepa- *Foodstuffs, car paints, cosmetic* ration of substances of low molecular weight will be described using *dyes etc.* dyes as an example.

1
Sample preparation

Ten milligrams of each dye are dissolved in 5 mL of distilled water, 1.5 µL is applied for each run.

2
Stock solutions

Acrylamide, Bis (T = 30%, C = 3%)
29.1 g of acrylamide + 0.9 g of Bis, made up to 100 mL with distilled *Dispose of the remains ecologi-* water *cally: polymerize with an excess of APS.*

> ■ Caution!
> **Acrylamide and Bis are toxic in the monomeric form. Avoid skin contact and do not pipette by mouth.**

The solution is stable (for IEF) for one week when stored in the *But for this method, the solu-* dark at 4 °C (refrigerator): *tion can be kept for several weeks.*

Ammonium persulfate solution (APS) 40% (w/v):
dissolve 400 mg of ammonium persulfate in 1 mL of distilled water. *Stable for one week when stored in the dark at 4°C.*

Electrophoresis in Practice, Fourth Edition. Reiner Westermeier
Copyright © 2005 WILEY-VCH Verlag GmbH & Co. KGaA, Weinheim
ISBN: 3-527-31181-5

Continuous buffer.

0.75 mol/L phosphate buffer pH 7.0:
38.6 g of Na_2HPO_4
8.25 g of NaH_2PO_4
fill up to 500 mL with distilled water

3
Preparing the casting cassette

This method works best with very thin gels (0.25 mm),

Fast separation → limited diffusion

- because high fields can be applied

Chemical fixing is not possible.

- because the substances which are separated can be fixed by rapid drying.

Gasket

A scalpel is too sharp. The knife presses the edges together so the layers stay together.

Two layers of Parafilm® (50 cm wide) are superimposed and cut with a knife so that a U-shaped gasket is formed and adheres to the glass plate.

Slotformer

Sample application is done in small sample wells which are polymerized in the surface of the gel. To form the sample wells in the gel a template must be fixed to the glass plate. A cleaned and degreased glass plate is

- placed on the template pattern (in the appendix)
- fixed to the work surface.

The sample wells are placed in the middle since there are anionic and cationic dyes.

Place two layers of "Scotch tape" (one layer 50 μm) on the starting point and smooth them down so that no bubbles appear. Cut the slot former out with a scalpel to 1 × 7 mm (Fig. 1). After pressing out the holes of the slot former, remove the remains of the cellophane with methanol.

If longer separation distances are desired, the slot former is placed closer to the edge. Anionic substances are applied near the cathode and cationic ones near the anode.

This operation only needs to be carried out once.

The slot former is then made hydrophobic. A few mL of Repel Silane are spread evenly over the whole slot former with a tissue under the fume hood. When the Repel Silane is dry, the chloride ions which, result from the coating are rinsed off with water.

When one side of the gasket is coated with CelloSeal®, it adheres to the glass plate.

The Parafilm® gasket is placed along the edge of the slot former plate (Fig. 1).

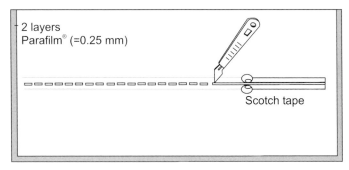

Fig. 1: Preparing the slot former.

Assembling the gel cassette

The gel is covalently polymerized on a plastic film for mechanical support and easier handling. Place the glass plate on a clean absorbent paper towel and moisten it with a small volume of water. Apply the GelBond PAG film on the glass plate with a rubber roller, placing the untreated hydrophobic side down (Fig. 2). A thin layer of water then forms between the glass plate and the film and holds them together by adhesion. The excess water, which runs out the sides is soaked up by the tissue.

To facilitate pouring the gel solution, the film should overlap the length of the glass plate by 1 mm.

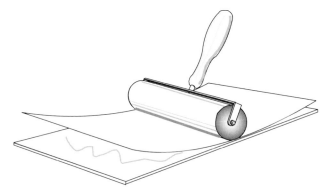

Fig. 2: Applying the support film with a roller.

Place the slot former over the glass plate and clamp the cassette together (Fig. 3).

Slot templates facing down.

Since a very thin layer of gel is needed for this method, the cassette is wedged open with paper clips and the clamps are pushed low down on the sides (Fig. 4).

This prevents the formation of air bubbles in the layer.

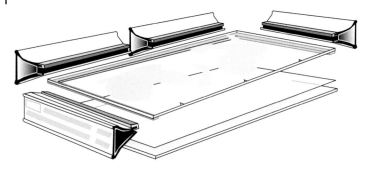

Fig. 3: Assembling the gel cassette.

4
Casting ultrathin-layer gels

A higher T value leads to less sharp bands.

Glycerol has two functions: it prevents diffusion when proteins are applied, it keeps the gel elastic during drying.

Recipe for two gels
0.33 mol/L phosphate buffer, pH 7.0 (T = 8%, C = 3%)
Mix in test tubes with screw caps (15 mL):
4.0 mL of acrylamide, Bis solution
2.5 mL of glycerol (87%)
4.0 mL of phosphate buffer
fill up to 15 mL with distilled water
7 µL of TEMED (100%)
15 µL of APS

100 µL of 60 % (v/v) isopro-panol-water are then layered on the edge of the solution. Isopropanol prevents oxygen, which inhibits polymerization, from diffusing into the gel. The gel will then present a well-defined, aesthetic upper edge.

Immediately after mixing pour in 7.5 mL per gel using a pipette or 20 mL syringe (Fig. 4).

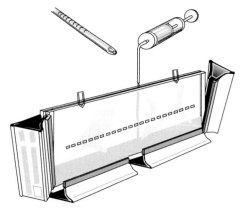

Fig. 4: Casting an ultrathin-layer polyacrylamide gel: after the calculated amount of polymerization solution has been added, remove the paper clips and push the clamps back into position.

Polymerization

• Let the gel stand for one hour at room temperature.

5
Electrophoretic separation

- Switch on the cooling system: +10 °C.

Removing the gel

- Remove the clamps and lay the sandwich on the cooling plate at 10 °C with the glass plate containing the slot former on the bottom.

  Cooling gives the gel a better consistency and it usually begins to separate from the slot former in the cassette already.

- Hold the cassette vertically and lift the GelBond film from the glass plate with a thin spatula.
- Pull the film and the gel from the slot former.
- Coat the cooling plate with 2 mL of the contact fluid kerosene.

  Water and other liquids are inadequate

- Place the gel on the cooling plate with the film on the bottom. Avoid air bubbles.
- Lay two of the electrode wicks into the compartements of the PaperPool (if smaller gel portions are used, cut them to size). Apply 20 mL of phosphate buffer to each wick (Fig. 5). Place one strip onto the anodal edge of the gel and the other strip onto the cathodal edge, each overlapping the gel edges by 5 mm (Fig. 6). Smooth out air bubbles by sliding bent tip foreceps along the edges of the wicks laying in contact with the gel.

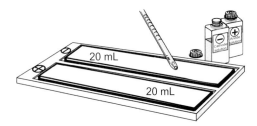

20 mL

20 mL

Fig. 5: Soaking the wicks in electrode buffer in the PaperPool.

- Quickly and carefully pipette the sample into the sample wells: 1.5 µL. Apply pure colors and mixtures of dyes but keep acid and basic dyes separate.

  Acid and basic dyes precipitate when combined!

- Clean platinum electrode wires before (and after) each electrophoresis with a wet tissue paper.
- Move electrodes so that they will rest on the outer edge of the electrode wicks. Connect the cables of the electrodes to the apparatus and lower the electrode holder plate (Fig. 6).
- Close the safety lid.

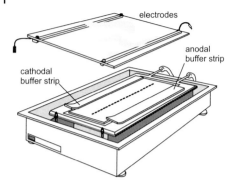

Fig. 6: Gel and electrode wicks for the separation of dyes. The samples are applied in the middle.

Running conditions:
Power supply: 400 V,   60 mA ,   20 W,   about 1 h.

After separation

- Switch off the power supply,
- Open the safety lid,
- Remove the electrode wicks,
- Remove the gel and dry immediately.

It is best to dry the gel on a warm surface, for example on a light box (switched on!).

The separation of different dyes is shown in Fig. 7. Several food colorings can be seen to consist of mixtures.

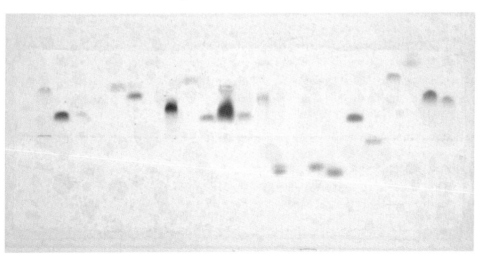

Fig. 7: Ultrathin-layer polyacrylamide gel electrophoresis of dyes. Anode at the top.

Method 2:
Agarose and immuno electrophoresis

The principles and uses of agarose gel, affinity and immunoelectro- *see pages 17 and following*
phoresis are described in part I.

Because of the large pores, agarose electrophoresis is especially
suited to the separation of lipoproteins and immunoglobulins, and to
specific detection by immunofixation.

Immunoelectrophoresis according to Grabar and Williams (1953) *Immunodiffusion is another*
and Laurell (1966) is not only used in clinical diagnostic and pharma- *possible method.*
ceutical production, it is also an official method for detection of falsi- *Ouchterlony Ö. Allergy. 6*
fications and the use of forbidden additives in the food industry. *(1958) 6.*

These methods are traditionally carried out in a Tris-barbituric acid *Susann J. The Valley of the*
buffer (veronal buffer). A few years ago the use of barbituric acid was *Dolls. Corgi Publ. London*
limited by the drug law (Susann, 1966). For this reason the following *(1966).*
methods will use a Tris-Tricine buffer.

1
Sample preparation

- Marker proteins pI 5.5 to 10.7 + 100 µL of distilled water. *Apply 6.5 µL*
- Store deep-frozen portions of meat extract from pork, rabbit, *Apply 6.5 µL*
  veal, beef. Dilute before use: 100 µL of meat extract + 300 µL
  of double-distilled water.
- *Other samples*:
  Set the protein concentration around 1 to 3 µg/mL. Dilute *Apply 6.5 µL*
  with double-distilled water. The salt concentration should not
  exceed 50 mmol/L

Desalting with a NAP-10 column may be necessary: apply 1 mL of
sample – use 1.5 mL of eluent.

Electrophoresis in Practice, Fourth Edition. Reiner Westermeier
Copyright © 2005 WILEY-VCH Verlag GmbH & Co. KGaA, Weinheim
ISBN: 3-527-31181-5

2
Stock solutions

Continuous buffer

For agarose electrophoresis also an amphoteric buffer system can be used, for example use 0.6 mol/L HEPES as in method 4. Agarose is mixed with the rehydration solution, boiled and poured on the gel.

Tris-Tricine lactate buffer pH 8.6:

   117.6 g of Tris
   51.6 g of Tricine
   12.7 g of calcium lactate
   make up to 3 L with distilled water.

Physiological salt solution (0.15 mol/L):

   9 g of NaCl, make up to 1 L with distilled water.

Bromophenol Blue solution:

   10 mg, make up to 10 mL with distilled water.

Agarose gel solution:

   1 g of Agarose L
   100 mL of Tris-Tricine lactate buffer

- Sprinkle the dry agarose on the surface of the buffer solution (to prevent the formation of lumps) and heat in a microwave oven on the lowest setting till the agarose is melted.
- Pour the solution in test tubes (15 mL each)

The test tubes can be kept in the refrigerator at 4 °C for several months.

3
Preparing the gels

a) Agarose gel electrophoresis

The "spacer" is the glass plate with the 0.5 mm thick, U-shaped silicone rubber gasket.

Agarose gel electrophoresis also works best when the samples are pipetted into small wells. To form these sample wells in the gel, a mould must be fixed on to the spacer glass plate.

Preparing the slot former

"Dymo" tape with a smooth adhesive surface should be used. When a structured adhesive surface is used, small air bubbles can be enclosed and holes appear around the wells.

The cleaned and degreased glass plate with the 0.5 mm U-shaped silicone rubber gasket is placed on the slot former template (slot former template in the appendix). On the place, which will be used as starting point, a layer of "Dymo" tape is applied (embossing tape, 250 μm thick) on the glass plate avoiding air bubbles. Cut out the slot former with a scalpel (Fig. 1). After pressing the slots once more against the glass plate, remove the remains of plastic with methanol.

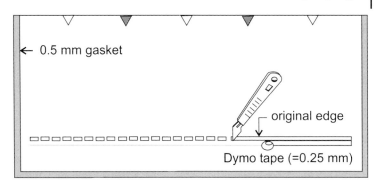

Fig. 1: Preparing the slot former.

The mold is then made hydrophobic. A few mL of Repel Silane are spread over the whole slot former with a tissue under the fume hood. When the Repel Silane is dry, the chloride ions resulting from the coating are washed off with water. *This operation only needs to be carried out once.*

Assembling the gel cassette

The gel is molded on a support film for better mechanical stability and easier handling. Place a glass plate on an absorbent tissue and moisten it with a few mL of water. Apply the GelBond film with a rubber roller placing the untreated hydrophobic side on the bottom (Fig. 2). A thin layer of water then forms between the glass plate and the film and holds them together by adhesion. The excess water which runs out is soaked up by the tissue. To facilitate the pouring of the gel solution, the film should overlap the length of the glass plate by about 1 mm. *GelBond film is used for agarose: it is a polyester film with a coating of dry agarose.*

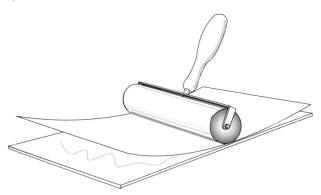

Fig. 2: Applying the support film with a roller.

The slot former is placed over the glass plate and the cassette is clamped together (Fig. 3). Before filling with the hot agarose solution, the cassette and a 10 mL glass pipette should be warmed in a heating cabinet at 75 °C.

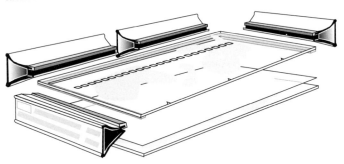

Fig. 3: Assembling the gel cassette.

- Take a 15 mL test tube out of the refrigerator and liquefy the contents in the microwave.

Avoid air bubbles; should some appear nevertheless, remove them with a long strip of polyester film.

- Remove the cassette from the heating cabinet; draw the hot agarose solution in the pipette with a rubber bulb and quickly release the solution in the cassette (Fig. 4).

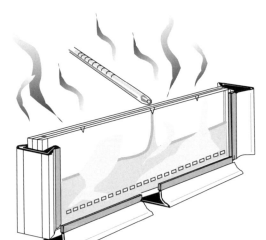

Fig. 4: Pouring the hot agarose solution into the prewarmed cassette.

This allows the gel to set slowly.

- Let the cassette stand for 1 or 2 h at room temperature.
- Remove the clamps and take out the gel.

- Place the gel on a wet filter paper and leave it overnight in a humidity chamber (Fig. 5) in the refrigerator. It can be kept up to one week under these conditions.

Only then does the definite agarose gel structure form (see page 15).

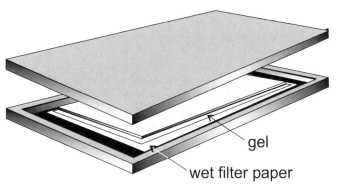

gel

wet filter paper

Fig. 5: Storing the agarose gel overnight in a humidity chamber.

b) Immunoelectrophoresis gels

Small gels are usually prepared, especially for the "rocket" technique, so as to be sparing with expensive antibodies.

In this technique, the antibodies are poured directly in the gel.

The *Grabar-Williams technique* as well will be described for small gels here.

Figure 11 gives an example in a large gel.

For these techniques it is advantageous to use the immunoelectrophoresis kit because it contains templates and foils in the right sizes for the holes and troughs.

- Cut out 8.4 cm wide pieces of GelBond film.

They can also be bought in this size.

To ensure that the gel thickness is uniform, the solution is poured on the film on a leveling table. 12 mL of gel solution are used for each plate.

- Use a spirit level to ensure that the levelling table is horizontal.
- Place a sheet of GelBond with the hydrophilic side up in the middle of the table.
- Liquefy the gel solution (in the test tube) by warming it.

When antibodies have to be added to the gel ("rocket" technique): cool the gel to 55 °C, add about 120 µL of antibody solution*, mix the solution (avoid air bubbles).

* This is an indicative value, it depends on the antigen-antibody titer.

- Pour the gel solution on the GelBond film and let the gel set.

The gels can be stored in a humidity chamber in the refrigerator.

Punching out the sample wells and troughs

Choose the size of the wells according to the sample concentration and antibody titer.

The gel puncher is connected to a water jet pump with thin vacuum tubing so that the pieces of gel can be sucked away immediately (Fig. 6 and 7).

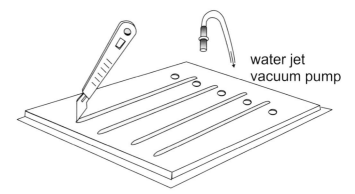

Fig. 6: Punching out the sample wells and troughs.

Grabar-Williams gel (Fig. 6)

- Place the gel (without antibodies) on the separation bed.
- Use the Grabar-Williams template.

The other holes in the template are for agar gels.

- Punch out 5 sample wells at places which will later be near the cathodic side.
- Excise the troughs with a scalpel; the gel strips remain in the gel during electrophoresis.

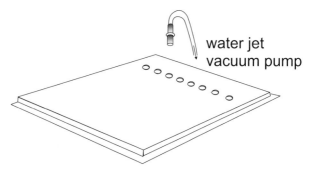

Fig. 7: Punching the sample wells for rocket immunoelectrophoresis.

"Rocket" gel (Fig. 7):
- Place the gel (with antibodies) on the separation bed.
- Use the Laurell template.
- Punch out 8 wells: use every second hole.

4
Electrophoresis

- Switch on the cooling system (10 °C).
- Place the cooling plate on the side, beside the Multiphor.
- Place the electrodes (orange plastic plates) in the inner compartment of the tanks and plug them in.
- Pour in the Tris-Tricine lactate buffer, 1 L per tank; put the cooling plate back in.
- Coat the cooling plate with 1 mL of contact fluid, kerosene.   *Do not use water, as it can*
- Place the gel with the film on the bottom on the cooling plate   *cause electrical shorting.* (Fig. 6).
- Dry the surface with filter paper (Fig. 8) because agarose gels   *This is also true for immuno-* have a liquid film on the surface.   *electrophoresis gels.*

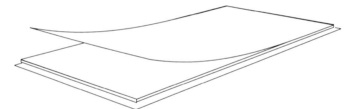

Fig. 8:   Drying the surface of the gel with filter paper.

- Place the gel on the cooling plate, the sample wells should be   *The wicks should overlap the* on the cathodal side (Fig. 9).   *gel by about 1 cm.*
- Soak 8 layers of electrode wicks in the buffer and place them on the gel to ensure good contact between the gel and the buffer.
- Rapidly pipette 5 µL of each sample in the trough.   *The samples should not diffuse.*
- Close the safety lid.
- Switch on the power supply. Running conditions: maximum 400 V, 30 mA, 30 W.
- Fixing, staining etc, should be done directly after the run.

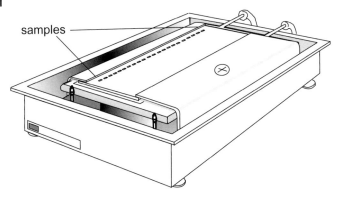

samples

Fig. 9: Agarose gel electrophoresis.

Grabar-Williams technique

In general sample application follows the scheme: sample – positive control – sample – positive control – sample.

- Pipette the samples and controls in the troughs (5 to 20 μL).
- Add one drop of Bromophenol Blue solution and start the run.

Separation conditions:

Field strength 10 V/cm (Check with a voltage probe).
Temperature 10 °C.
Time about 45 min (till Bromophenol Blue has reached the edge).

- Place the gel in the gel holder and with the "shovel" on the back of the scalpel remove the gel strips from the troughs.
- Place the gel in the humidity chamber.
- Pipette 100 μL of antibody in each trough.

The precipitin arcs then form.
The precipitates remain in the
gel.

- Let the solution diffuse for about 15 h at room temperature.
- Staining is only done after the non-precipitated antigens and antibodies have been washed out (see below).

Laurell technique

- Pipette the samples in the wells of the antibody containing gel as quickly as possible to prevent diffusion.

In general a series of dilutions (4 samples) of an antibody solution is run at the same time so as to obtain a concentration calibration curve.

- Start the electrophoresis immediately after sample application.

Separation conditions:

Field strength   10 V/cm (Check with voltage probe)
Temperature   10 °C
Time   3 h

- Staining should be done after the non-precipitated antigens and antibodies have been washed out (see below).

The buffer is titrated to pH 8.6 where the specific antibodies in the gel have a minimal net charge and no electrophoretic mobility (Caution! Antibodies can have different optimum pH values depending on their origin; for rabbit antibodies pH 7.8 is usually used).

The sample proteins (antigens) are charged at pH 8.6 and migrate in the gel.

The determination of antigens with a very low or no net charge at this pH is problematic (e.g. IgG). However, the isoelectric point of these proteins can be lowered by acetylation or carbamylation and their electrophoretic mobility thus increased.

At first the antigen molecules are in excess, so the antigen-antibody complexes are soluble and migrate towards the anode. The equivalence concentration is reached along both sides of the migrating antigen track and a rocket-shaped precipitin line is formed continuously starting at the bottom.

It is important that the antibodies be of good quality, otherwise the lines are not well defined or several "rockets" are formed in another.

The area of the "rocket" is directly proportional to the antigen concentration. Exact concentration determination is done by measuring the area of the "rocket". But in many cases it is sufficient to measure the height of the rocket.

5
Protein detection

Coomassie staining (Agarose electrophoresis)

Solutions with H_2O_{dist}

- *Fixing:* 30 min in 20% (w/v) TCA;
- *Washing:* 2 × 15 min in 200 mL of fresh solution each time: 10% glacial acetic acid, 25% methanol;
- *Drying:* cover the gel with 3 layers of filter paper and place a glass plate and a weight (1 to 2 kg) over them (Fig. 10). Remove after 10 min and finish drying in a heating cabinet;

First humidify the filter paper that lies directly on the agarose.

- *Staining:* 10 min in 0.5% (w/v) Coomassie R-350 in 10% glacial acetic acid, 25% methanol: dissolve 3 tablets of PhastGel Blue (0.4 g dye each) in 250 mL;
- *Destaining:* in 10% glacial acetic acid, 25% methanol till the background is clear:
- *Drying:* in the heating cabinet.

Immuno fixing of agarose electrophoresis

All the other proteins and the excess antibody are washed out of the gel with a NaCl solution.

With this fixing method only the protein bands which have formed an insoluble immunoprecipitate with the antibody are fixed.

Antibody solution: 1:2 (or 1:3 depending on the antibody titer) dilute with double-distilled water and apply on the surface of the gel with a glass rod or a pipette: about 400 to 600 µL;

- *Incubation:* 90 min in the humidity chamber in an incubator or a heating cabinet at 37 °C;

First humidify the filter paper that lies directly on the agarose.

- *Pressing:* 20 min with 3 layers of filter paper, a glass plate and a weight (Fig. 10);
- *Washing:* in physiological sodium chloride solution (0.9% NaCl w/v) overnight;
- *Drying:* see Coomassie staining;
- *Staining and destaining:* as for Coomassie staining.

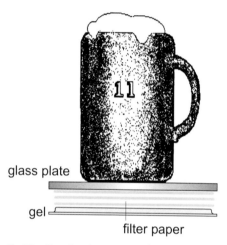

glass plate

gel

filter paper

Fig. 10: Pressing the agarose gel.

It is important to know the antibody titer, since for example, hollow bands can occur when the antibody solution is too concentrated: in the middle one antibody binds to one antigen and no precipitate results.

Coomassie staining (immuno electrophoresis)

In *"rocket"* electrophoresis as in Grabar-Williams electrophoresis, the gel contains antibodies, which must be washed out before staining.

- *Pressing:* 20 min with 3 layers of filter paper, a glass plate and weight (Fig. 10);
- *Washing:* in physiological sodium chloride solution (0.9% NaCl w/v) overnight;
- *Drying:* see Coomassie staining;
- *Staining and destaining:* as for Coomassie staining.

First humidify the filter paper that lies directly on the gel.

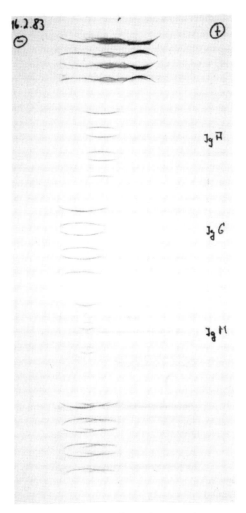

Fig. 11: Large immunoelectrophoresis gel with the Grabar-Williams technique.

Silver staining

Willoughby EW, Lambert A.
Anal Biochem. 130 (1983)
353–358.

If the sensitivity of Coomassie staining is not sufficient, silver staining can be performed on the dry gel (Willoughby and Lambert, 1983):

- *Solution A:* 25 g of Na_2CO_3, 500 mL of double-distilled water;
- *Solution B:* 1.0 g of NH_4NO_3, 1.0 g of $AgNO_3$, 5.0 g of tungsto-silisic acid, 7.0 mL of formaldehyde solution (37%), make up to 500 mL with double distilled water.
- *Staining:* Mix 35 mL of solution A and 65 mL of solution B just before use, put the gel in the resulting whitish solution and incubate while agitating until the desired intensity is reached. Briefly rinse with double-distilled water;
- *Stop* with 0.05 mol/L glycine. Remove the remains of metallic silver from the gel and the support film with a cotton swab;
- *Air dry.*

Method 3:
Titration curve analysis

The principles of titration curve analysis are described in part 1.

The technical execution with washed and rehydrated polyacrylamide gels will be explained here. This variant excludes the influences of ionic catalysts which are found in the gel after polymerization.

When it is not necessary to obtain very precise results, the carrier ampholyte can be polymerized directly into the gel.

see page 64 f

APS and TEMED can influence the formation of the pH gradient.

This saves washing, drying and rehydration.

1
Sample preparation

- Marker proteins pI 4.7 to 10.6 or
- Marker proteins pI 5.5 to 10.7 + 100 µL of double distilled water.   *Apply 50 µL*
- Meat extract from pork, rabbit, veal, beef, store in frozen portions. Dilute before use: 100 µL of meat extract + 300 µL of H_2O_{dist}.   *Apply 50 µL*

Other samples:

Set the protein concentration around 1 to 3 mg/mL.   *Apply 50 µL*

Dilute with double-distilled water. The salt concentration should not exceed 50 mmol/L: apply 1 mL of sample solution – use 1.5 mL of eluent.

Electrophoresis in Practice, Fourth Edition. Reiner Westermeier
Copyright © 2005 WILEY-VCH Verlag GmbH & Co. KGaA, Weinheim
ISBN: 3-527-31181-5

2
Stock solutions

Acrylamide, Bis (T = 30%, C = 3%)

29.1 g of acrylamide + 0.9 g of Bis, make up to 100 mL with double-distilled water or

Dispose of the remains ecologically, polymerize with an excess of APS.

reconstitute PrePAG Mix (29.1:0.9) with 100 mL of double-distilled water.

> ■ Caution!
> **Acrylamide and Bis are toxic in the monomeric form. Avoid contact with the skin, do not pipette by hand.**

Can still be used for SDS gels for several weeks.

Can be stored for one week in a dark place at 4 °C (refrigerator).

Ammonium persulfate solution (APS) 40% (w/v):

Can be stored for one week in the refrigerator (4 °C).

Dissolve 400 mg of ammonium persulfate in 1 mL of double-distilled water.

0.25 mol/L Tris-HCl buffer:

3.03 g of Tris + 80 mL of double-distilled water, titrate to pH 8.4 with 4 mol/L HCl and make up to 100 mL with double-distilled water.

The buffer is removed during washing.

This buffer corresponds to the anode buffer of horizontal electrophoresis as in method 7. It is used to set the pH value of the polymerization solution, since this works best in a slightly basic medium.

Glycerol 87% (w/v):

Glycerol fulfills several roles: in the polymerization solution it improves the fluidity and makes the solution denser at the same time so that overlayering is easier. In the last washing solution, glycerol ensures that the gel does not roll up upon drying.

3
Preparing the blank gels

Preparing the casting cassette

The gel is later cut in half so two separations can be run simultaneously. If only one gel is needed, the second half can be stored in the humidity chamber in the refrigerator (2 weeks at most).

In titration curve analysis the sample is applied in long narrow grooves which run along the whole pH gradient which is electrophoretically established beforehand. To form the sample grooves a mould must be fixed on a glass plate. Since square gels are used for this method, two gels are poured together with the glass plate of the gel kit and rehydrated later.

Two 10 cm long pieces of "Dymo" tape (embossing tape, 250 μm thick) are cut and applied on the cleaned and degreased glass plate with a 0.5 mm thick U-shaped gasket. As shown in Fig. 1, a piece is cut away with the scalpel in such a way that 1 to 2 cm Dymo tape remain. After pressing the tape once more against the glass plate, the remains of the tape are removed with methanol.

Dymo tape with a smooth adhesive surface should be used. When the adhesive surface is structured, small air bubbles which inhibit polymerization can be enclosed.

This mold is then made hydrophobic. A few mL of Repel Silane are spread over the whole slot former with a tissue under the fume hood. When the Repel Silane is dry, the chloride ions which result from the coating are rinsed off with water.

This operation only needs to be carried out once.

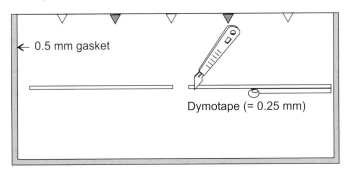

0.5 mm gasket

Dymotape (= 0.25 mm)

Fig. 1: Cutting the narrow adhesive tape to form the sample troughs in the gel.

Assembling the gel cassette

- Remove the GelBond PAG film from the package.

  The hydrophobic side can be identified with a few drops of water.

- Pour a few mL of water on the glass plate and place the support film on it with the hydrophobic side down. Press the support film onto the glass plate with a roller (Fig. 2).

  This facilitates filling the mold later on.

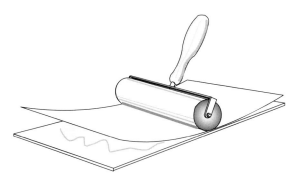

Fig. 2: Rolling on the support film.

The "spacer" is the glass plate with the 0.5 mm thick U-shaped silicone rubber gasket.

- The spacer is then placed on the glass plate with the gasket facing downwards and the cassette is clamped together (Fig. 3).

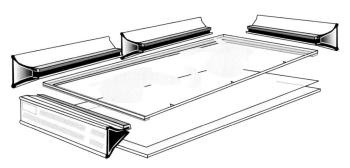

Fig. 3: Assembling the gel cassette.

Two types of gels can be prepared:

– Directly used gels
– Rehydratable gels

Tab. 1: Composition for monomer solution for gels with $T = 4.2\%$

	Directly used gel	Rehydratable gel
Acrylamide/Bis solution 30 % T / 3 % C	2.7 mL	2.7 mL
Glycerol (85 %)	–	1 mL
0.25 mol/L Tris-HCl buffer	–	0.5 mL
Sorbitol	2 g	–
Ampholine/Pharmalytes	1.3 mL	–
TEMED	10 μL	10 μL
with H_2O_{dist} make up to	20 mL	20 mL
APS (40 %)	40 μL	20 μL

Filling the gel cassette

When everything else is ready, APS is added to the polymerization solution.

Never pipette the toxic monomer solutions by mouth!

The cassette is filled with a 10 mL pipette or a 20 mL syringe (Fig. 4). Draw the solution into the pipette (a 10 mL scale plus the headspace make 18 mL) and place it in the middle notch. Slowly release the solution it will be directed into the cassette by the strip of film sticking out.

Air bubbles which might occur can be dislodged out with a long strip of polyester film.

Introduce 100 μL of 60 % v/v isopropanol-water into each notch. Isopropanol prevents oxygen, which inhibits polymerization, from diffusing into the gel. The gel will then have a well defined aesthetic edge.

Polymerization

Let the gel stand for one hour at room temperature.

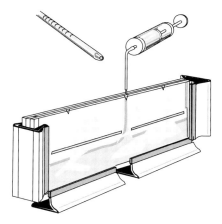

Fig. 4: Pouring the polymerization solution.

Removing the gel from the cassette

After the gel has polymerized for one hour remove the clamps and lift the glass plate off the film with a spatula. Peel the gel of the spacer by slowly pulling on a corner of the film.

Washing the gel

Wash the gel in H_2O_{dist} by shaking it three times for 20 min. Add 2 % glycerol to the last washing solution.

This washes the remains of monomers, APS and TEMED out of the gel.

Drying the gel overnight at room temperature. Then cover the gels with a protecting film and store them frozen (–20 °C) in sealed plastic bags.

Heat drying damages the swelling capacity, which is why the gels should be stored frozen.

4
Titration curve analysis

Reswelling the rehydratable gel

Place the GelPool on a horizontal table. Select the appropriate reswelling chamber of the GelPool. Clean it with distilled water and tissue paper. Pipet the rehydration solution (one gel):

400 µL ethylenglycol (= 7.5% v/v)
390 µL of Ampholine or Pharmalyte (= 3% w/v)
make up to 5.2 mL with H_2O_{dist}.

Ethylenglycol increases the viscosity of the solution, so the curves become smoother.
For carrier ampholytes concentrations see page 55.

Set the edge of the gel-film – with the dry gel surface facing downward – into the rehydration solution (Fig. 5 A) and slowly lower the film down. At the same time move the gel-film to and fro, in order to achieve an even distribution of the liqid and to avoid trapping airbubbles.

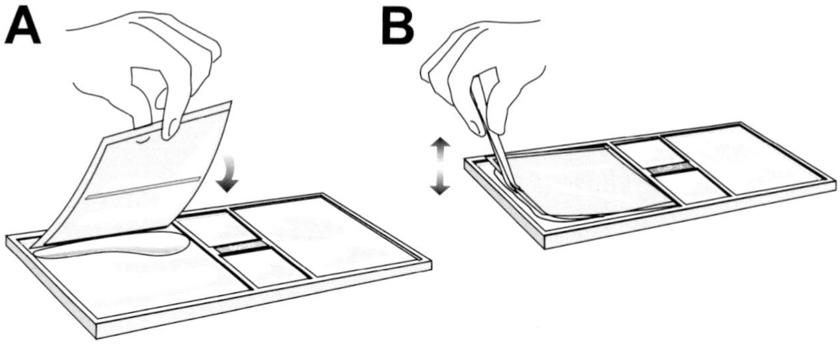

Fig. 5: Rehydration of a titration curve gel in the GelPool.

Lift the film at the edges with foreceps, and slowly lower them down, in order to maintain an even distribution of the liquid (Fig. 5 B) and to remove airbubbles.

Check, whether the gel can be moved around on its reswelling liquid.

■ Note:
Repeat this measure several times during the first 15 min, to prevent the gel from sticking to the GelPool.

60 min later the gel has reswollen completely and can be removed from the GelPool.

• Dry the sample grooves with filter paper.

Run without sample.

a) Formation of the pH gradient

• Switch on the cooling system: +10 °C.

Do not use water! This can cause electrical shorting.

• Pipette 1 mL of contact fluid, kerosene, on to the middle of the cooling plate.

This method must be carried out at a defined temperature since the pH gradient and the net charge are temperature dependent.

• Place the gel in the middle of the gel with the support film on the bottom. The sample troughs should lie perpendicular to the electrodes (Fig. 6).

If two gels are run in parallel the current and power must be doubled.

• The power supply settings are the following:

1500 V,   7 mA,   7 W

Electrode strips are not needed for washed gels.

The electrodes are placed directly on the edge of the gel (Fig. 6).

• Plug in the cable. Take care that the long anode cable is hooked to the front.
• Close the safety lid.
• Switch on the power supply.

see chapter 3.

A continuous pH gradient from 3.5 to 9.5 will form in about 60 min.

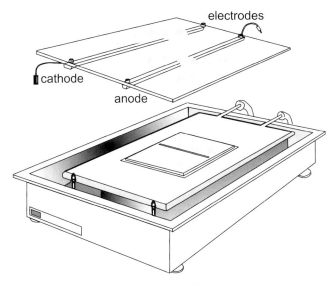

Fig. 6: Placing the titration gel on the cooling plate for the formation of the pH gradient.

b) Native electrophoresis in the pH spectrum

- Once the power is switched off, the safety lid opened and the electrodes are removed, the gel is turned by 90°: The alkaline side of the gel is now directed towards the side of the cooling tubing!

This standard is used when placing the gels and presenting results to simplify the interpretation and comparison of results.

This means that the proteins in the left side of the gel are positively charged and migrate towards the cathode while those in the right side are negatively charged and migrate towards the anode (Fig. 9 and 35 in chapter 3.7).

- Using a pipette introduce 50 µL of sample solution(1 mg/mL) into the trough.

Work quickly to prevent diffusion from destroying the gradient.

- Proceed with the electrophoresis immediately:
  1200 V,   10 mA,   5 W.

The proteins with different mobilities now migrate towards the anode or the cathode with velocities which depend on the actual pH of the medium.

Titration curves are formed.

- Stop after 20 min and stain the gel.

5
Coomassie and silver staining

Colloidal Coomassie staining

Diezel W, Kopperschläger G, Hofmann E. Anal Biochem. 48 (1972) 617–620.
Blakesley RW, Boezi JA. Anal Biochem. 82 (1977) 580–582.

The result is quite quickly visible with this method. Few steps are necessary, the staining solutions are stable and there is no background staining. Oligopeptides (10 to 15 amino acids) which are not properly fixed by other methods can be revealed here. In addition, the solution is almost odorless (Diezel *et al.* 1972; Blakesley and Boezi, 1977).

Preparation of the staining solution:

Dissolve 2 g of Coomassie G-250 in 1 L of distilled water and add 1 L of sulfuric acid (1 mol/L or 55.5 mL of conc H_2SO_4 per L) while stirring. After stirring for 3 h, filter (paper filter) and add 220 mL of sodium hydroxide (10 mol/L or 88 g in 220 mL). Finally add 310 mL of 100% TCA (w/v) and mix well, the solution will turn green.

Fixing and staining: 3 h at 50 °C or overnight at room temperature in the colloidal solution;

Washing out the acid: soak in water for 1 or 2 h, the green color of the curves will become blue and more intense.

Fast Coomassie staining

Use distilled water for all solutions

Stock solutions:

TCA: 100% TCA (w/v) 1 L
A: 0.2% (w/v) $CuSO_4$ + 20% glacial acetic acid
B: 60% methanol

1 tablet = 0.4 g of Coomassie Brilliant Blue R-350

dissolve 1 Phast Blue R tablet in 400 mL of double distilled water, add 600 mL of methanol and stir for 5 to 10 min.

Staining:

- *Fixing:* 10 min in 200 mL of 20% TCA;
- *Washing:* 2 min in 200 mL of washing solution (mix equal parts of A and B);
- *Staining:* 15 min in 200 mL of 0.02% (w/v) R-350 solution at 50 °C while stirring (Fig. 7);
- *Destaining:* 15 to 20 min in washing solution at 50 °C while stirring;
- *Preserving:* 10 min in 200 mL of 5% glycerol;
- *Drying:* air-dry.

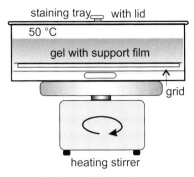

Fig. 7: Appliance for hot staining.

5 minute silver staining of dried gels

This method can be applied to stain dried gels already stained with Coomassie (the background must be completely clear) to increase sensitivity, or else unstained gels can be stained directly after pretreatment (Krause and Elbertzhagen, 1987). A significant advantage of this method is that no proteins or peptides are lost during the procedure. They often diffuse out of the gel during other silver staining methods because they cannot be irreversibly fixed in the focusing gels.

Krause I, Elbertzhagen H. In: Radola BJ, Ed.. Elektrophorese-Forum '87. This edition. (1987) 382–384.

According to the silver staining method for agarose gels (Kerenyi and Gallyas 1972, Willoughby and Lambert, 1983).

Pretreatment of the unstained gels:

- fix for 30 min in 20% TCA,
- wash for 2 × 5 min in 45% methanol. 10% glacial acetic acid,
- wash for 4 × 2 min in distilled water,
- impregnate for 2 min in 0.75% glycerol,
- air-dry.

Solution A: 25 g of Na_2CO_3, 500 mL of double distilled water;

Solution B: 1.0 g of NH_4NO_3, 1.0 g of $AgNO_3$, 5.0 g of tungstosilisic acid, 7.0 mL of formaldehyde solution (37%), make up to 500 mL with double-distilled water.

Silver staining:

Mix 35 mL of solution A with 65 mL of solution B just before use. Immediately soak the gel in the resulting whitish suspension and incubate while agitating until the desired intensity is reached. Briefly rinse with distilled water.

Stop with 0.05 mol/L glycine. Remove remains of metallic silver from the gel and support film with a cotton swab.

Air-dry.

6

Interpreting the curves

The schematic curves of three proteins A, B and C are shown in Fig. 8. These results can be interpreted as follows for:

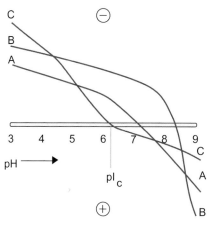

Fig. 8: Schematic representation of titration curves.

Ion-exchange chromatography:

The intensity of the net charge and thus the binding to the separation medium present most differences there.

- Flat curves point to bad separation in ion-exchangers, since a change in the pH in the buffer only has a limited influence of the net charge.
- The separation capacity is best when elution is carried out at a pH at which the curves lie far apart on the y-axis: in this example at pH 3.5 for a cation-exchanger. If an anion-exchanger is used and the elution is done at pH 7.7, protein B will not bind and will elute in the exclusion volume.

A mixture of amphoteric buffers is used for elution in this case and the sample is fractionated according to the pIs.

- If the intersections with the x-axis (pI) lie far apart, chromato-focusing will be effective.

Zone electrophoresis

It is better to choose another pH value.

- If the buffer with a pH value at which the curves intersect is chosen, these proteins can only be separated on the basis of their frictional resistance in a restrictive medium.

The mobilities are very different there.

- For electrophoresis in a support free medium or for separations in non-restrictive media, choose a pH value at which the curves lie clearly apart on the y-axis: pH 5.3 or 7.7 in this example.

Isoelectric focusing

- The protein C intersects the x-axis at a flat angle, this means that the mobility gradient is low at the pI. High field strength is necessary to focuse C.   *In addition the focusing time must be increased.*

- The curve of protein C lies along the x-axis above the pI. For IEF the sample must be applied at a pH < pI, otherwise C will migrate very slowly or not at all.   *The protein only has a low charge here.*

- If one or more proteins do not migrate out of the troughs or produce a smeared zone in a certain area, this means that the protein is unstable at this pH or that a few proteins aggregate (not shown).   *The sample should not be applied in this interval.*

Method 4:
Native PAGE in amphoteric buffers

Polyacrylamide gel electrophoresis of proteins under native conditions is described at several occasions in part I.

See pages 34 and following, as well as 42 and 65.

A series of buffers systems have been developed for the native separation of proteins (Jovin *et al.* 1970; Ornstein, 1964; Davis, 1964; Maurer, 1968). Disc electrophoresis in a basic buffer according to *Ornstein and Davis* is the technique most used. A homogeneous glycine-acetic acid buffer at pH 3.1 is used for the separation of basic proteins, but the gel preparation is relatively complicated.

Catalysts such as ammonium persulfate and TEMED are necessary for the polymerization of acrylamide. These substances dissociate in the gel and form ions which can have a marked influence on the buffer system calculated beforehand.

For these reasons many buffers could only be used in agarose gels earlier, but electroendosmosis was a problem then.

When gels are polymerized on support films and the catalysts are washed out, the gels can be dried and rehydrated in the desired buffer and the problem thus avoided.

These gels can be prepared in the laboratory or purchased in the dried form.

An amphoteric buffering substance which establishes a constant pH value in the gel near its pI can be used as for titration curve analysis (steady-state value). If the buffering substance is not influenced by other ions or electrolytes, it is not charged and cannot migrate.

See also the principles of "rocket" immunoelectrophoresis page 19.

A washed and dried homogeneous gel is simply rehydrated in the buffer solution and the electrodes are connected directly to the gel for the separation of low molecular weight non-amphoteric substances.

No buffer reservoirs are necessary.

A discontinuity in the buffer system and in the gel porosity appreciably increases the sharpness of the bands and the resolution during the separation of proteins. The amphoteric compound is only needed in the gel.

The buffer strips contain leading or terminating ions and an acid or a base to increase the conductivity and ionization of the substances.

Two systems have been chosen for the following experiments:

a) a separation method for cationic dyes.
b) a cationic electrophoresis for proteins.

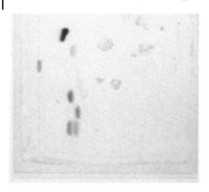

Fig. 1: Electrophoresis of cationic dyes in 0.5 mol/L HEPES, method a.

1
Sample preparation

Method a (dyes)

Apply 1.5 µL

- Dissolve 10 mg of each dye in 5 µL of distilled water.

Method b (proteins)

Apply 6.5 µL
Apply 6.5 µL

- Marker proteins pI 5.5 to 10.7 + 100 µL of distilled water.
- Meat extracts from pork, rabbit, veal and beef frozen in portions. Dilute before use: 100 µL of meat extract + 300 µL of distilled water.

Other samples:

Apply 6.5 µL

- Set the protein concentration around 1 to 3 mg/mL. Dilute with distilled water. The salt concentration should not exceed 50 mmol/L.

It may be necessary to desalt with a NAP-10 column: apply 1 mL of sample – use 1.5 mL of eluent.

2
Stock solutions

Acrylamide, Bis solution (T = 30%, C = 2%):
29.4 g of acrylamide + 0.6 g of bisacrylamide, make up to 100 mL with distilled water (H_2O_{Bidist}).

C = 2% in the resolving gel solution prevents the separation gel from peeling off the support film and cracking during drying.

Acrylamide, Bis solution (T = 30%, C = 3%):
29.1 g of acrylamide + 0.9 g of bisacrylamide, make up to 100 mL with distilled water (H_2O_{Bidist}).

This solution is used for slightly concentrated plateaus with C = 3%, because the slot would become unstable if the degree of polymerization were lower.

> ■ Caution!
> **Acrylamide and bisacrylamide are toxic in the monomeric form. Avoid skin contact and dispose of the remains ecologically (polymerize the remains with an excess of APS).**

Ammonium persulfate solution (APS) 40% (w/v):
Dissolve 400 mg of ammonium persulfate in 1 mL H_2O_{Bidist}.

It can be stored for one week in the refrigerator (4°C).

0.25 mol/L Tris-HCl buffer:
3.03 g of Tris + 80 mL of distilled water, titrate to pH 8.4 with 4 mol/L HCl, make up to 100 mL with H_2O_{Bidist}.
   This buffer corresponds to the anode buffer of horizontal electrophoresis as in method 7. It is used to set the pH of the polymerization solution, since polymerization works best in a slightly basic medium.

The buffer is removed during the washing step.

100 mmol/L arginine solution:
1.742 g of arginine, make up to 100 mL with distilled water.

Store at +4°C

300 mmol/L acetic acid :
1.8 mL of acetic acid (96%), make up to 100 mL with distilled water.

300 mmol/L ε-amino caproic acid:
18 g of ε-amino caproic acid, make up to 500 mL with distilled water.

Store at +4°C

Pyronine solution (cationic dye marker) 1% (w/v):
1 g of pyronine, make up to 100 mL with distilled water.

Glycerol 85%:
Glycerol fulfills many functions: it improves the fluidity of the polymerization solution and makes it denser at the same time so that the final overlayering is easier. In the last washing step, it ensures that the gel does not roll up when it is dried.

3
Preparing the empty gels

Slot former

The "spacer" is the glass plate with the 0.5 mm thick U-shaped silicone rubber gasket fixed to it.

Sample application is done in small wells which are molded in the surface of the gel during polymerization. To form these slots a mould must be fixed on a glass plate (spacer).

- Method a:
  The casting mold for ultrathin gels (0.25 mm) from method 1 is used.

"Dymo" tape with a smooth adhesive surface should be used. Small air bubbles can be enclosed when the adhesive surface is structured, these inhibit polymerization and holes appear around the slots.

- Method b:
  The cleaned and degreased glass plate with 0.5 mm U-shaped spacer is placed on the template (slot former template in the appendix). A layer of "Dymo" tape (embossing tape, 250 μm thick) is applied, avoiding air bubbles, at the starting point. Several layers of "Scotch tape" (one layer 50 μm) can be used instead (Fig. 2). The slot former is cut out with a scalpel. After pressing the individual slot former pieces against the glass plate, the remains of sticky tape are removed with methanol.

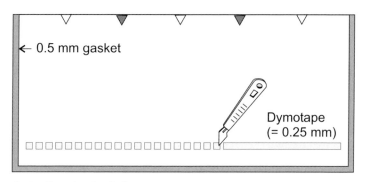

Fig. 2: Preparing the slotformer.

This treatment only needs to be carried out once.

The casting mold is then made hydrophobic by spreading a few mL of Repel Silane over the whole slot former with a tissue under the fume hood. When the Repel Silane is dry, the chloride ions which result from the coating are rinsed off with water.

Assembling the casting cassette

For mechanical stability and to facilitate handling, the gel is covalently polymerized on a support film. The glass plate is placed on an absorbent tissue and wetted with a few mL of water. The GelBond PAG film is applied with a roller with the untreated hydrophobic side down (Fig. 3). A thin layer of water then forms between the film and the glass plate and holds them together by adhesion. The excess

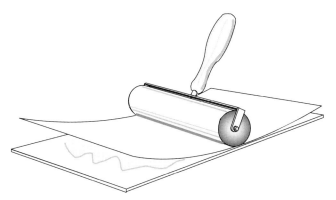

Fig. 3: Rolling on the support film.

water which runs out is soaked up by the tissue. To facilitate pouring in the gel solution, the film should overlap the length of the glass plate by about 1 mm.

The finished slot former is placed on the glass plate and the cassette is clamped together (Fig. 4).

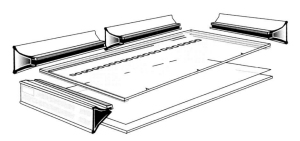

Fig. 4: Assembling the casting cassette.

Polymerization solutions

- *Method a: gel recipe for 2 gels (T = 8 %, C = 3 %):*

  Introduce and mix in test tubes with screw caps (15 mL):
  4.0 mL of acrylamide, Bis solution 30 %T, 3 %C
  0.5 mL of Tris-HCl make up to 15 mL with distilled water
  7 µL of TEMED (100 %)
  15 µL of APS

- *Method b: discontinuous gel*
  Cool the casting cassette in the refrigerator at 4 °C for about 10 min: this delays the onset of polymerization. This step is necessary because the stacking gel with large pores and the resolving gel with small pores are cast in one piece. The polymerization solutions which have different densities take 5 to 10 min to settle.

Increasing T leads to less sharp bands.

See method 1 for the casting technique. These gels are also washed and dried.

In the summer in a warm laboratory, the gel solutions should also be brought to 4 °C.

APS is only added shortly before filling the cassette.

Tab. 1: Composition of the polymerization solutions

	Stacking gel (4 %T, 3 %C)	Resolv. Gel (10 %T, 2 %C)
Acrylamide/Bis 30 %T, 3 %C	1.3 mL	–
Acrylamide/Bis 30 %T, 2 %C	–	5.0 mL
0.25 mol/L Tris-HCl buffer	0.2 mL	0.3 mL
Glycerol (85 %)	2 mL	0.3 mL
TEMED	5 μL	7 μL
with H_2O_{dist} make up to	10 mL	15 mL
APS (40 %)	10 mL	15 μL

Filling the cooled gel cassette

Never pipette the toxic monomers by mouth!

The cassette is filled with a 10 mL pipette or a 20 mL syringe (Fig. 5). Draw the solution into the pipette with a pipetting device. The stacking gel plateau is poured first, and then the resolving gel solution which contains less glycerol and is less dense. Pour the solutions in slowly. The gel solution is directed into the cassette by to the piece of film sticking out.

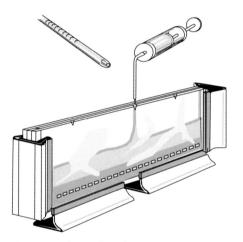

Fig. 5 Introducing the polymerization solutions in the gel cassette.

If air bubbles are trapped in the solution, they can be dislodged with a long strip of polyester film.

100 μL of 60 % (v/v) isopropanol are then layered in each filling notch. Isopropanol prevents oxygen, which inhibits polymerization, from diffusing into the gel. The gel will then present a well-defined, aesthetic upper edge.

Polymerization

Let the gel stand at room temperature.

Removing the gel from the casting cassette

After the gel has polymerized over night, the clamps are removed and the glass plate ently lifted off the film with a spatula. The gel can slowly be pulled away from the spacer by grasping a corner of the film.

There is a slow "silent polymeri-zation" after the gel has solidi-fied, which should be completed before the gel is washed.

Washing the gel

The gels (0.25 mm gels for dyes and 0.5 mm gels for proteins) are washed three times for 20 min by shaking in double-distilled water. The last washing solution should contain 2% glycerol. If a "Multi-Wash" apparatus is used the gel is deionized by pumping double-dis-tilled water through the mixed bed ion-exchanger cartridge for 30 min (Fig. 6).

The remains of monomers, APS and TEMED are removed from the gel.

- Before drying the gel, soak it for about 15 min in a 10% gly-cerol solution.
- Air-dry the gel overnight.

Then store them in the refrig-erator covered with polyester film.

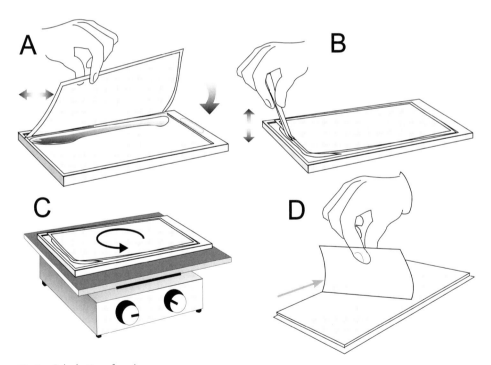

Fig. 6: Rehydration of a gel.
(A) Placing the dry gel into the GelPool; (B) Lifting the gel for an even distribu-tion of the liquid. (C) Rehydration on a rocking platform (not always necessary).
(D) Removing the excess buffer from the gel surface with filter paper.

4
Electrophoresis

Rehydration in amphoteric buffers

The gels can be used in one piece, or – depending on the number of samples – cut into smaller portions with scissors (when they are still dry).

- Lay GelPool onto a horizontal table; select the appropriate reswelling chamber, pipet rehydration solution into the chamber, for

  a complete gel: 25 mL
  a half gel: 13 mL

The rest of the gel should be sealed airtight in a plastic bag and stored in a freezer.

Very even rehydration is obtained when performing it on a shaker at a slow rotation rate (Fig. 6C). If no shaker is used, lift gel edges repeatedly.

- Set the edge of the gel-film – with the gel surface facing down – into the rehydration buffer (Fig. 6) and slowly lower it, avoiding air bubbles.
- Using forecps, lift the film up to its middle, and lower it again without catching air bubbles, in order to achieve an even distribution of the liquid (Fig. 6 B). Repeat this during the first 10 min.

When the gel surface is dry enough, this is indicated by a noise like a whistle.

60 min later the gel has reswollen completely and is removed from the GelPool. Dry sample wells with clean filter paper, wipe buffer off the gel surface with the edge of a filter paper (Fig. 6D).

Method a (dyes):

Rehydration solution (0.5 mol/L HEPES)

The very thin gel needs only 10 mL of reswelling solution.

- 1.15 g of HEPES, make up to 10 mL with $H_2O_{Bidist.}$
- Rehydrate for about 1 h.
- Dry the sample wells with filter paper.

Water or other liquids are not suitable, as they can cause electrical shorting.

- Wet the cooling plate with 1 mL of contact fluid kerosene.
- Place the gel, film side down, on the cooling plate.

See methods 3 and 6
Avoid air bubbles.

In this method the electrodes can be placed directly on the edge of the gel; no buffer wicks are necessary.

1.5 μL of each

- Quickly introduce the samples in the wells:
- Power supply settings: 400 V, 60 mA, 20 W, about 1 h.

After separation

It is important to dry the gel immediately for mechanical fixing of the zones.

- Switch off the power supply and open the safety lid.

Immediately place the dye gel on a warm surface, a light box for example. A separation is shown on Fig. 1.

Method b (PAGE of cationic proteins)

For the separation of lipophilic proteins (e.g. alcohol soluble fractions from cereals), it suggested to add 0.1 to 0.5 % (w/v) ProSolv II and 1 to 3 mol/L urea to the rehydration buffer.

Add urea and detergents only for these applications with lipo-philic membrane proteins.

Tab. 2: Rehydration solution (0.6 mol/L HEPES)

The dye pyronine permits the front to be seen.

0.6 mol/L HEPES	3.5 g
1 mmol/L acetic acid	83 µL (from the stock solution)
10 mmol/L arginine	2.5 mL (from the stock solution)
0.001 % pyronine	8 µL (from the stock solution)

make up to 25 mL with distilled water.

Rehydration time: about 1 h.

- Dry out the sample wells and the gel surface with the edge of a clean filter paper.

  Cathode buffer:
  30 mmol/L acetic acid:
  2 mL (from the stock
  solution)

  Anode buffer:
  113 mmol/L ε-amino caproic
  acid: 7.5 mL
  (from the stock solution)
  5 mmol/L acetic acid: 333 µL
  (from the stock solution)

  to → 20mL with H_2O_{Bidist}

  to → 20mL with H_2O_{Bidist}

- Wet the cooling plate with 1 mL of contact fluid kerosene.

  Water or other liquids are not suitable, as they can cause electrical shorting.

- Place the gel (surface up) onto the center of the cooling plate: the side containing the sample wells must be oriented towards the anode (Fig. 8).

  Multiphor II: cathodal side of the wells matching line no. 12

- Lay two of the electrode wicks into the compartements of the PaperPool (if smaller gel portions are used, cut them to size). Apply 20 mL of the anode and cathode buffer respectively to the wicks (Fig. 7). Place the anode strip onto the anodal edge

  Always apply anode wick first, in order to prevent buffer contamination of the trailing ions.

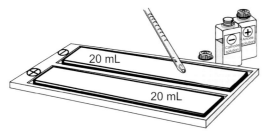

20 mL

20 mL

Fig. 7: Soaking the wicks with the electrode buffers.

of the gel, matching the grid on the cooling plate between the lines 13 and 14. Place the cathode strip onto the cathodal edge, matching the grid between 3 and 4 (Fig. 8).

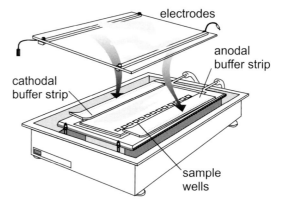

Fig. 8: Cationic native electrophoresis: sample application at the anode.

6.5 μL in each.

- Smooth out air bubbles by sliding bent tip forceps along the edges of the wicks laying in contact with the gel (first anode, then cathode).
- Quickly fill the sample wells:

Clean platinum electrode wires before (and after) each electrophoresis with a wet tissue paper.

- Move electrodes so that they will rest on the outer edge of the electrode wicks. Connect the cables of the electrodes to the apparatus and lower the electrode holder plate (Fig. 8). Close the safety lid.

Running conditions at 10 °C (maximum settings):

Phase 1 is for a mild sample entry and effective stacking.

Tab. 3: Power supply program

	U (V)	I (mA)	P (W)	t (min)
complete gel:	500 V	10 mA	10 W	10 min
	1200 V	28 mA	28 W	50 min
half gel:	500 V	5 mA	5 W	10 min
	1200 V	14 mA	14 W	50 min

> **Note:**
> **The field strength can be increased at low conductivity: this results in a fast separation compared to conventional native electrophoresis.**

The separation will then take more time.

If composite proteins such as chloroplast proteins are separated, the voltage should not exceed 300 V.

After separation

- Switch off the power supply and open the safety lid.
- Remove the electrode wicks and either place them on the edge of the cooling plate or let them slip into the tank.
- Protein gels are stained or blotted.

See method 9 for blotting.

5
Coomassie and silver staining

Colloidal Coomassie staining

The result is quite quickly visible with this method. Few steps are necessary, the staining solutions are stable and there is no background staining. Oligopeptides (10 to 15 amino acids) which are not properly fixed by other methods can be revealed here. In addition, the solution is almost odorless (Diezel *et al.* 1972; Blakesley and Boezi, 1977).

Diezel W, Kopperschläger G, Hofmann E. Anal Biochem. 48 (1972) 617–620.
Blakesley RW, Boezi JA. Anal Biochem. 82 (1977) 580–582.

Preparation of the staining solution:

Dissolve 2 g of Coomassie G-250 in 1 L of distilled water and add 1 L of sulfuric acid (1 mol/L or 55.5 mL of conc H_2SO_4 per L) while stirring. After stirring for 3 h, filter (paper filter) and add 220 mL of sodium hydroxide (10 mol/L or 88 g in 220 mL). Finally add 310 mL of 100 % TCA (w/v) and mix well, the solution will turn green.

Fixing and staining: 3 h at 50 °C or overnight at room temperature in the colloidal solution;

Washing out the acid: soak in water for 1 or 2 h, the green color of the curves will become blue and more intense.

Fast Coomassie staining
Stock solutions:

    Use distilled water for all solutions
    TCA: 100 % TCA (w/v) 1 L
    A: 0.2 % (w/v) $CuSO_4$ + 20 % glacial acetic acid
    B: 60 % methanol

dissolve 1 Phast Blue R tablet in 400 mL of double distilled water, add 600 mL of methanol and stir for 5 to 10 min.

1 tablet = 0.4 g of Coomassie Brilliant Blue R-350

Staining:

- *Fixing:* 10 min in 200 mL of 20 % TCA;
- *Washing:* 2 min in 200 mL of washing solution (mix equal parts of A and B);
- *Staining:* 15 min in 200 mL of 0.02 % (w/v) R-350 solution at 50 °C while stirring (Fig. 7);
- *Destaining:* 15 to 20 min in washing solution at 50 °C while stirring;
- *Preserving:* 10 min in 200 mL of 5 % glycerol;
- *Drying:* air-dry.

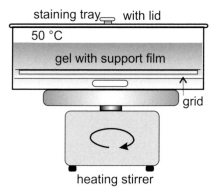

Fig. 9: Appliance for hot staining of gels.

5 minute silver staining of dried gels

Krause I, Elbertzhagen H. In: Radola BJ, Ed. Elektrophorese-Forum '87. This edition. (1987) 382–384.

According to the silver staining method for agarose gels (Kerenyi and Gallyas 1972, Willoughby and Lambert, 1983).

This method can be applied to stain dried gels already stained with Coomassie (the background must be completely clear) to increase sensitivity, or else unstained gels can be stained directly after pretreatment (Krause and Elbertzhagen, 1987). A significant advantage of this method is that no proteins or peptides are lost during the procedure. They often diffuse out of the gel during other silver staining methods because they cannot be irreversibly fixed in the focusing gels.

Pretreatment of the unstained gels:

- fix for 30 min in 20 % TCA,
- wash for 2 × 5 min in 45 % methanol. 10 % glacial acetic acid,
- wash for 4 × 2 min in distilled water,
- impregnate for 2 min in 0.75 % glycerol,
- air-dry.

Solution A: 25 g of Na_2CO_3, 500 mL of double distilled water;

Solution B: 1.0 g of NH_4NO_3, 1.0 g of $AgNO_3$, 5.0 g of tungstosilisic acid, 7.0 mL of formaldehyde solution (37 %), make up to 500 mL with double-distilled water.

Silver staining:

Mix 35 mL of solution A with 65 mL of solution B just before use. Immediately soak the gel in the resulting whitish suspension and incubate while agitating until the desired intensity is reached. Briefly rinse with distilled water.

Stop with 0.05 mol/L glycine. Remove remains of metallic silver from the gel and support film with a cotton swab.

Air-dry.

Fig. 10 shows an electropherogram of different protein samples which were analyzed as described here.

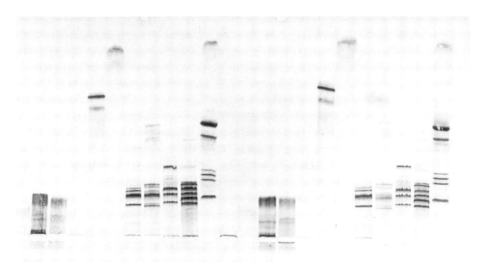

Fig. 10:   Cationic native electrophoresis of marker proteins and meat extracts in 0.6 mol/L HEPES. Staining with PhastBlue R-350 (Cathode on top).

For *alkaline* separations, no amphoteric buffer has been found yet, which meets the buffer capacity requirements. Very good results are obtained with washed and dried gels rehydrated in the conventional Tris-HCl buffer for basic native electrophoresis. In the cathode Tris-glycine is used in this case.

Method 5:
Agarose IEF

The principles of isoelectric focusing in carrier ampholyte pH gradients are described in part I. There are several reasons for the use of agarose as separation medium.

See pages 51 and following.

- No toxic monomer solutions are needed;
- There are no catalysts which can interfere with the separation;
- No polymerization solutions have to be prepared and stored;
- The matrix has large pores;

- Therefore the separation times are shorter;
- The staining times are shorter;immunofixation can be carried out in the gel.

No acrylamide, no Bis.
No TEMED, no APS.
Less work.
High molecular weight proteins are not a problem.

Less frictional resistance.
The dried gel is stained. See page 18.

There are however, a few problems with agarose:

- The gels are not absolutely free of electroendosmosis, since not all the carboxyl and sulfate groups are removed during purification.

- For these reasons agarose gels are not suited for use at extreme pH ranges (acid or basic).

- Gels containing urea are difficult to prepare, because urea disrupts the agarose network.

This leads to a cathodic drift of the gradient and water transport.

Agarose electrophoresis works best in the middle of the pH range.

Use of rehydratable agarose gels (Hoffman et al. 1989).

1
Sample preparation

- Marker proteins pI 4.7 to 10.6 or
- Marker proteins pI 5.5 to 10.7 + 100 µL of double-distilled water. *Apply 10 µL*
- Meat extracts from pork, rabbit, veal and beef frozen in portions. Dilute before use: 100 µL of meat extract + 300 µL of double-distilled water. *Apply 10 µL*

Electrophoresis in Practice, Fourth Edition. Reiner Westermeier
Copyright © 2005 WILEY-VCH Verlag GmbH & Co. KGaA, Weinheim
ISBN: 3-527-31181-5

Apply 10 µL

- Other samples:
Set the protein concentrations around 1 to 3 mg/mL. Dilute with double-distilled water. The salt concentration should not exceed 50 mmol/L.

It may be necessary to desalt with a NAP-10 column:
apply 1 mL – use 1.5 mL of eluent.

2
Preparing the agarose gel

Not to be confused with GelBond PAG film.

Unlike polymerization of acrylamide.

Leifheit H-J, Gathof AG, Cleve H. Ärztl Lab. 33 (1987) 10–12.

Agarose gels are cast on GelBond film: a polyester film coated with a dry agarose layer.

Agarose gels can be cast several ways: oxygen from the air does not inhibit gelation.

Here the gel is cast in a vertical prewarmed cassette, since this produces an uniform gel layer. The following method is based on the experience of Dr. Hans-Jürgen Leifheit, Munich (1987) whom we would like to thank for his helpful advice.

Making the spacer plate hydrophobic

The "spacer" is the glass plate with the 0.5 mm thick U-shaped silicone rubber gasket.

This treatment only needs to be carried out once.

A few mL of Repel Silane are spread over the inner face of the spacer plate with a tissue under the fume hood. When the Repel Silane is dry, the chloride ions which result from the coating are rinsed off with water.

Assembling the casting cassette:

Identify the hydrophilic side with a few drops of water.

- Remove the GelBond film from the package.

This facilitates filling the cassette later on.

Pour a few mL of water on the glass plate and place the support film on it with the hydrophobic side down. Press the film against the glass with a roller (Fig. 1), the film should overlap the length of the glass plate by about 1 mm.

The spacer plate, with the gasket on the bottom, is then placed on the glass plate and the cassette is clamped together (Fig. 2).

This ensures that the solution does not solidify immediately.

Before pouring the hot agarose solution, prewarm the cassette and a 10 mL pipette to 75 °C in a heating cabinet.

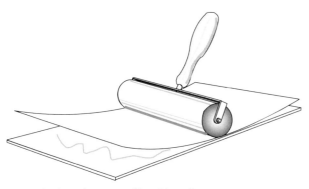

Fig. 1: Applying the support film with a roller.

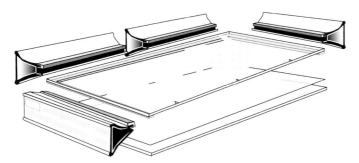

Fig. 2: Assembling the gel cassette.

Preparation of the agarose solution (0.8% agarose)

It is important to store agarose in a dry place because it is very hygroscopic.

If the agarose is humid, too little will be weighed.

- In a 100 mL beaker:
  2 g of sorbitol
  19 mL of distilled water
  0.16 g of agarose IEF

  Sorbitol improves the mechanical properties of the gel and since it is hygroscopic, it works against the electro-osmotic water flow.

- Mix and boil the solution – covered with a watch glass – until the agarose has completely dissolved either in a microwave oven at the lowest setting or while stirring slower on a magnetic heating stirrer.

  Rapid stirring damages the mechanical properties of agarose.

- Degas to remove CO_2.
- Place the beaker in a heating cabinet for a few minutes to cool the solution down to 75 °C.

  It should not be too hot for the carrier ampholytes.

- Add 1.3 mL of Ampholine pH 3.5 to 9.5 or Pharmalyte pH 3 to 10 and stir with a glass rod.

  Most of the commercially available solutions contain 40 % (w/v) carrier ampholytes. The product "Pharmalytes™" are produced with a different chemistry, the concentration can therefore not be specified. However, they are used with the same volumes like a 40 % solution.

Avoid air bubbles. Should some be trapped nevertheless, dislodge them with a long strip of polyester film.

- Remove the cassette from the heating cabinet; draw the hot agarose solution in the prewarmed pipette and quickly release it in the cassette (Fig. 3).

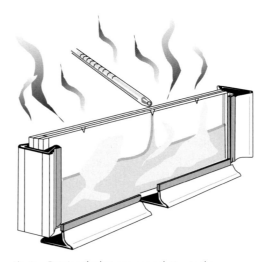

Fig. 3: Pouring the hot agarose solution in the prewarmed cassette.

The gel slowly sets.

- Let the cassette stand for 1 or 2 h at room temperature.
- Remove the clamps and take out the gel.

Only then does the final agarose gel structure form (see page 15).

- Place the gel on a piece of moist tissue and store it overnight in a humidity chamber (Fig. 4) in the refrigerator, it can be kept for up to a week.

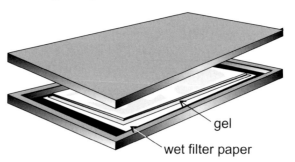

gel

wet filter paper

Fig. 4: Agarose gel in the humidity chamber.

Preparation of electrode solutions

In Tab 1 a list of electrode solutions is given for respective pH intervals.

Tab. 1: Electrode solutions for IEF in agarose gels

pH gradient	anode	cathode
3.5 – 9.5	0.25 mol/L acetic acid	0.25 mol/L NaOH
2.5 – 4.5	0.25 mol/L acetic acid	0.40 mol/L HEPES
4.0 – 6.5	0.25 mol/L acetic acid	0.25 mol /L NaOH
5.0 – 8.0	0.04 mol/L glutamic acid	0.25 mol/L NaOH

3
Isoelectric focusing

- The gel is placed with the film on the bottom on the cooling plate at 10 °C, using about 1 mL kerosene (Fig. 6).
- Dry the surface with filter paper (Fig. 5) because agarose gels have a liquid film on the surface.

IEF must be performed at a defined constant temperature because the pH gradient and the pIs are temperature dependent.

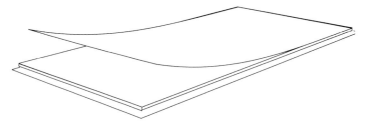

Fig. 5: Drying the surface of the gel with filter paper.

Electrode solutions:

To reduce cathodic drift apply electrode strips made from filter paper between the electrodes and both edges of the gel and let them soak in the electrode solutions. For a pH gradient from 3.5 to 9.5 these are:

Anode:	Cathode:
0.25 mol/L acetic acid	0.25 mol/L NaOH

These are also used for pH 4.0 to 6.5.

- Cut the electrode strips shorter than the gel (< 25 cm);
- Soak the strips thoroughly in the corresponding solutions;
- Blot them with dry filter paper for about 1 min to remove excess liquid;

They should not protrude the sides.

- Place the electrode strips along the edge of the gel;
- Shift the electrodes along the electrode holder so that they lie over the electrode strips (see Fig. 6);
- Plug in the cable, make sure that the long anodic connecting cable is hooked to the front.

Make sure that the acid strip lies under the anode and the basic strip under the cathode.

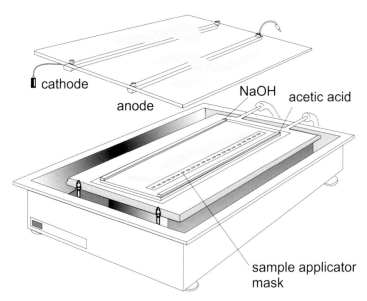

cathode

anode

NaOH acetic acid

sample applicator mask

Fig. 6: Agarose IEF with electrode strips and sample application strip.

Separation conditions

The values for current and power are valid for a whole gel. For IEF of half a gel use half the values for mA and W.

The pH gradient is formed during this step.

- *Prefocusing:* 30 min at 1400 V, 30 mA, 8 W,

It may be necessary to do a step test.

- The optimum point for sample application in agarose IEF depends on the characteristic of the sample (see also method 6, page 204). But most samples can be applied where mentioned here.

  Place the sample applicator strip (mask) on the gel surface 2 cm away from the anode.

Do not use any filter paper or cellulose plates for agarose IEF!

- *Sample application:* apply 10 µL in each hole of the sample applicator .

This also helps high molecular weight proteins, e.g. IgM to enter the gel

- *Desalting:* 30 min at max. 150 V; the other settings remain.

This is valid for gradients from pH 3 to 10.

- *Separation:* 60 min at 1500 V, 30 mA, 8 W.

During the run it may be necessary to interrupt the separation and blot the electrode strips with filter paper.

For narrower gradients, e.g. pH 5 to 8, IEF should last about 2 h, since the proteins with a low net charge must cover long distances.

- Switch off the power supply, open the safety lid, remove the electrodes and blot the excess water from the strips;
- Continue the separation.

The proteins are then stained, immunofixed or blotted. Should problems occur, consult the trouble-shooting guide in the appendix.

4
Protein detection

Coomassie Blue staining

- *Fixing:* 30 min in 20% (w/v) TCA;
- *Washing:* 2 ×15 min in 200 mL of fresh solution each time: 10% glacial acetic acid, 25% methanol;
- *Drying:* place 3 layers of filter paper on the gel and a 1 to 2 kg weight on top (Fig. 7). Remove everything after 10 min and finish drying in the heating chamber;
- *Staining:* 10 min in 0.5% (w/v) Coomassie R-350 in 10% glacial acetic acid, 25% methanol: dissolve 3 PhastGel Blue R tablets (0.4 g dye each) in 250 mL;
- *Destaining:* in 10% glacial acetic acid, 25% methanol till the background is clear;
- *Drying:* in the heating cabinet.

Prepare all solutions with distilled water.

First moisten the piece of filter paper lying directly on the agarose.

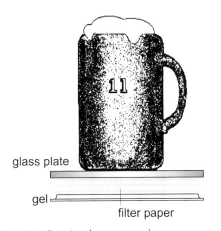

glass plate

gel

filter paper

Fig. 7: Pressing the agarose gel.

Immunofixation

All the other proteins and excess antibodies are washed out of the gel with a NaCl solution.

Only the protein bands, which have formed an insoluble immunoprecipitate with the antibody are stained with this method.

Antibody solution: dilute 1:2 (or 1:3 depending on the antibody titer) with double-distilled water and spread it over the surface of the gel with a glass rod or a pipette: use about 400 to 600 µL;

First moisten the piece of filter paper lying directly on the gel.

- *Incubation:* 90 min in the humidity chamber in a heating chamber or incubator at 37 °C;
- *Pressing:* 20 min under 3 layers of filter paper, a glass plate and a weight (Fig. 7);
- *Washing:* in a physiological salt solution (0.9% NaCl w/v) overnight;
- *Drying:* see Coomassie staining;
- *Staining and destaining* as for Coomassie staining.

It is important to know the antibody titer, since hollow bands can appear in the middle when the antibody solutions are too concentrated: one antigen binds to one antibody and no precipitate is formed.

Silver staining

If the sensitivity of Coomassie staining is not sufficient, silver staining is carried out on the dry gel (Willoughby and Lambert, 1983):

Solution A: 25 g of Na_2CO_3, 500 mL of distilled water;

Solution B: 1.0 g of NH_4NO_3, 1.0 g of $AgNO_3$, 5.0 g of tungstosilicic acid, 7.0 mL of formaldehyde solution (37%), make up to 500 mL with distilled water.

Staining: mix 35 mL of solution A with 65 mL of solution B just before use. Soak the gel immediately in the resulting whitish solution and incubate while agitating until the desired intensity is reached. Rinse briefly with distilled water;

Stop   with 0.05 mol/L glycerol. Remove remains of metallic silver from the gel and the back of the support film with a cotton swab.

Drying: air-dry.

Method 6:
PAGIEF in rehydrated gels

The principles of isoelectric focusing in carrier ampholyte pH gradi- *See pages 51 and following.*
ents are described in part I.

The use of washed, dried and rehydrated polyacrylamide gels is
advantageous for a number of reasons:

- A few carrier ampholytes inhibit the polymerization of gels, *Especially strongly basic*
  especially the reaction on the film; this means that the thin gel *ampholytes.*
  may swim around like a jellyfish in an aggressive staining solution.
- Gels containing ampholytes are slightly sticky, and thus not *Because of inhibited polymeri-*
  very easy to remove from the casting cassette. *zation as well.*
- Blank gels can be washed: APS, TEMED and unreacted acryl- *During IEF the otherwise*
  amide and Bis monomers can thus be washed out of the gel. *indispensable electrode wicks,*
  This allows focusing to occur with fewer interferences, as is *soaked in acid or basic buffer,*
  readily apparent in the straight bands in the acid pH range as *can be left out: this allows*
  well. *separation up to the platinum*
  electrode!
- Zymogramm techniques work better in absence of acrylamide *Enzymes maintain their*
  monomers and catalysts. *activities.*
- Gels can easily be prepared in large quantities and kept dry. *With fresh acrylamide*
  monomer solution.
- Chemical additives, which allow the separation of many pro- *For example non-ionic deter-*
  teins but which would inhibit polymerization, can be added to *gents such as Triton or NP-40.*
  the gel without any problems.
- Sample wells can be formed in the gel without disturbing the *See method 1 for preparation*
  pH gradient. *of a "slot former" for example.*
- Blank, ready-made gels can be bought. *Handling acrylamide mono-*
  mers is not necessary and a lot
  of time and effort are saved.

1
Sample preparation

Apply 10 μL

Apply 10 μL

Apply 10 to 20 μL

- Marker proteins pI 4.7 to 10.6 or
- Marker proteins pI 5.5 to 10.7 + 100 μL of distilled water.
- Meat extracts from pork, rabbit, veal and beef frozen in portions. Dilute before use: 100 μL of meat extract + 300 μL of double-distilled water.
- *Other samples:*
  Set the protein concentration around 1 to 3 mg/mL. Dilute with distilled water. The salt concentration should not exceed 50 mmol/L.

It might be necessary to desalt with a NAP-10 column: apply 1 mL sample solution – use 1.5 mL of eluent.

2
Stock solutions

Acrylamide, Bis (T = 30 %, C = 3 %):
29.1 g of acrylamide + 0.9 g of Bis, make up to 100 mL with distilled water or

Dispose of the remains ecologically: polymerize with an excess of APS.

PrePAG Mix (29.1 : 0.9), reconstitute with 100 mL of distilled water.

■ Caution!
Acrylamide and Bis are toxic in the monomeric form. Avoid skin contact and do not pipette by mouth.

It can still be kept for several weeks for SDS gels.

When stored in a dark place at 4 °C (refrigerator) the solution can be kept for one week.

Ammonium persulfate solution (APS) 40 % (w/v):

Stable for one week when stored in the refrigerator (4 °C).

Dissolve 400 mg of ammonium persulfate in 1 mL of distilled water.

0.25 mol/L Tris-HCl buffer:
3.03 g of Tris + 80 mL of distilled water, titrate to pH 8.4 with 4 mol/L HCl, make up to 100 mL with distilled water.

The buffer is removed during the washing step.

This buffer corresponds to the anodic buffer for horizontal electrophoresis as in method 7. It is used to set the pH of the polymerization solution, since it works best at a slightly basic pH.

Glycerol 87 % (w/v):
Glycerol fulfills several functions: it improves the fluidity of the polymerization solution and makes it denser at the same time so that the final overlayering is easier. In the last washing solution, it ensures that the gel does not roll up during drying.

3
Preparing the blank gels

Assembling the casting cassette

*   Remove the GelBond PAG film from the package.

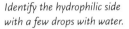

Identify the hydrophilic side with a few drops with water.

Fig. 1: Applying the support film with a roller.

Pour a few mL of water on the glass plate and place the support film on it with the hydrophobic side on the bottom. Press the film on the glass plate with a roller so that the film overlaps the length of the glass plate by about 1 mm.

This facilitates filling the cassette later on.

Since the gel should bind to the GelBond PAG film after polymerization but not to the spacer plate, the spacer is made hydrophobic with Repel Silane. It is then placed on the film with the gasket on the bottom and the cassette is clamped together (Fig. 2).

The "spacer plate" is the glass plate with the 0.5 mm thick U-shaped silicone rubber gasket.

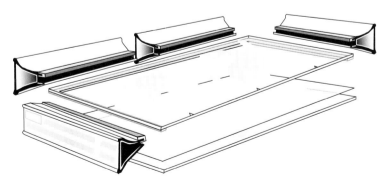

Fig. 2: Assembling the casting cassette.

- Composition of the polymerization solution for a gel with $T = 4.2\%$:
  2.7 mL of acrylamide/Bis solution
  0.5 mL of 0.25 mol/L Tris-HCl
  1 mL of glycerol
  10 µL of TEMED
  make up to 20 mL with distilled water.

Filling the casting cassette

The clock is running now: this solution polymerizes after at most 20 min.

When everything else is ready, add 20 µL of APS to the polymerization solution.

The cassette is filled with a 10 mL pipette or a 20 mL syringe (Fig. 3). Fill the pipette completely (the 10 mL scale plus the headspace make 18 mL) and place the tip of the pipette in the middle notch of the spacer. Release the solution slowly. The gel is directed in to the cassette by the piece of film sticking out.

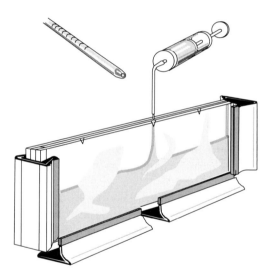

Fig. 3: Pouring the polymerization solution.

Air bubbles which might be trapped can be dislodged with a long strip of polyester film.

Polymerization.

100 µL of 60% v/v isopropanol-water are then pipetted into the 3 notches. Isopropanol prevents oxygen which inhibits polymerization from diffusing in the gel. The gel then presents a well defined aesthetic edge.

Let the gel stand for one hour at room temperature.

Removing the gel from the casting cassette

After the gel has polymerized for one hour, remove the clamps and carefully lift the glass plate from the support film with a spatula. Slowly peel the gel away from the spacer by pulling on a corner of the film.

Washing the gel

Wash the gel it three times in a tray with shaking in 300 mL distilled water for 20 min. The last washing step should contain 2 % glycerol.

This step washes the remains of monomers, APS and TEMED out of the gel.

Drying the gel

Dry the gels overnight at room temperature with a fan. Then cover them with a protecting film, and store them frozen (−20 °C) in sealed plastic bags.

Heat drying damages the swelling capacity, which is why the gels should be stored frozen.

4
Isoelectric focusing

Rehydration solution (Ampholine, Pharmalyte):

Example for an optimized broad gradient pH 3 to 10.

0.79 mL of ethylenglycol (= 7.5 % v/v)
200 µL of Ampholine + 580 µL Pharmalyte
make up to 10.5 mL with distilled water.
7.5 % (v/v) of ethylenglycol in the gel increases viscosity,
so the bands are straighter.

Most of the commercially available solutions contain 40 % (w/v) carrier ampholytes. The product "Pharmalytes^{TM}" are produced with a different chemistry, the concentration can therefore not be specified. However, they are used with the same volumes like a 40 % solution.

Narrow gradients: When narrow gradients (e.g. pH 5 to 7) are employed, add 10 % (v/v) carrier ampholytes of a wide range (e.g. 3 to 10). See also "6 Perspectives" on page 208 ff.

Focus narrow gradients for a longer time: separation phase prolongued up to 3 hours.

Reswelling the gel

The gels can be used in one piece or cut into two halves or smaller portions – depending on the number of samples to be separated.

 Place the GelPool on a horizontal table. Select the appropriate reswelling chamber of the GelPool, depending on the gel size. Clean it with distilled water and tissue paper. Pipet the appropriate volume of rehydration solution, for

We suggest to degas the solution in order to remove CO_2 for sharper bands in the alkaline region.

At the same time move the gel-film to and fro, in order to achieve an even distribution of the liqid and to avoid trapping airbubbles.

complete gel:	10.4 mL
half gel:	5.2 mL

Set the edge of the gel-film – with the dry gel surface facing downward – into the rehydration solution (Fig. 4A) and slowly lower the film down.

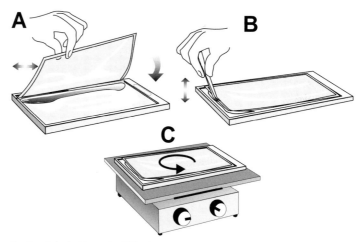

Fig. 4: Rehydrating a dry IEF gel.
(A) Laying the gel surface into the tray of the GelPool, which contains the exact amount of rehydration solution, which is needed.
(B) Lifting the edges of the gel with foreceps to prevent sticking of the gel to the tray surface.
(C) Rehydrating on a rockling platform (not absolutely necessary).

Lift the film at the edges with foreceps, and slowly lower them down, in order to maintain an even distribution of the liquid (Fig. 4B) and to remove air bubbles.

■ Note:
Repeat this measure several times during the first 15 min, to prevent the gel from sticking to the GelPool. Check, whether the gel can be moved around on its reswelling liquid.

If the solution contains urea or nonionic detergents, reswelling takes ca. 3 hours.

60 min later the gel has taken up the complete volume of the solution and can be removed from the GelPool.

Separation of proteins

IEF must be performed at a constant defined temperature since the pH gradient and the pIs are temperature dependent.

• Cool the cooling plate to 10 °C and place the gel on it, with the film on the bottom, using a few mL of kerosene (Fig. 5).

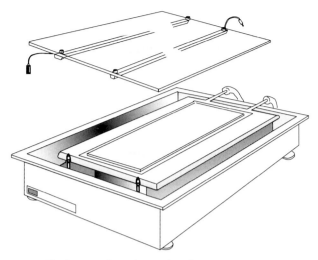

Fig. 5: The focusing electrodes are directly placed on the edge of the gel.

The following settings for separation have been optimized:

Tab. 1: Power supply program for PAGIEF

Settings (max.):	Voltage	Current	Power	Time	
complete gel:	700 V	12 mA	8 W	20 min	*pre-IEF*
	500 V	8 mA	8 W	20 min	*sample entry*
	2000 V	14 mA	4 W	90 min	*separation*
	2500 V	14 mA	18 W	10 min	*band sharpening*

■ Note:
For half gels 1/2 mA and 1/2 W have to be set!

- Place the electrodes directly on the edges of the gel (Fig. 5).
- Electrode wicks do not have to be used for washed gels.
- Plug in the cables; make sure that the long anodic connecting cable is hooked to the front.
- First focus without sample. This already distributes the ampholytes in the pH gradient without the proteins migrating. In addition there are proteins which are only stable in an acidic or basic medium, before prefocusing the gel has a pH value of about 7.0 all over.

Prefocusing is not necessary for all samples, since APS and TEMED are already washed out and most proteins are not sensitive to pH; it is best to try out.

Sample application

During application it should be taken into consideration that most proteins have an optimum application point. If necessary it should be determined by applying the sample at several places (step test, Fig. 6). The samples are applied with small pieces of filter paper or silicone rubber applicator mask.

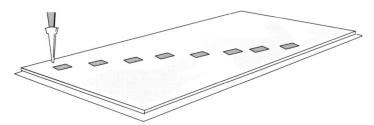

Fig. 6: Step trial test to determine the best application point.

The time for separation in table 1 is sufficient for most of the proteins. Large proteins may require slightly longer to reach their pI. Highly viscous gels (with 20 % glycerol or 8 mol/L urea solution) also require a longer focusing time.

This is valid for pH gradients from 3 to 10. When gels with a narrower pH range are used, e.g. pH 5 to 7, focusing should be performed during approximately 4 hours, since the proteins with lower net charge must migrate longer distances.

- After IEF, switch off the power supply and open the safety lid.
- The proteins are now stained or blotted. Should problems arise, consult the trouble-shooting guide in the appendix.

5
Coomassie and silver staining

Colloidal Coomassie staining

Diezel W, Kopperschläger G, Hofmann E. Anal Biochem. 48 (1972) 617–620.
Blakesley RW, Boezi JA. Anal Biochem. 82 (1977) 580–582.

The result is quite quickly visible with this method. Few steps are necessary, the staining solutions are stable and there is no background staining. Oligopeptides (10 to 15 amino acids) which are not properly fixed by other methods can be revealed here. In addition, the solution is almost odorless (Diezel *et al.* 1972; Blakesley and Boezi, 1977).

Preparation of the staining solution:

Dissolve 2 g of Coomassie G-250 in 1 L of distilled water and add 1 L of sulfuric acid (1 mol/L or 55.5 mL of conc H_2SO_4 per L) while stirring. After stirring for 3 h, filter (paper filter) and add 220 mL of sodium hydroxide (10 mol/L or 88 g in 220 mL). Finally add 310 mL of 100 % TCA (w/v) and mix well, the solution will turn green.

Fixing and staining: 3 h at 50 °C or overnight at room temperature in the colloidal solution;

Washing out the acid: soak in water for 1 or 2 h, the green color of the curves will become blue and more intense.

Fast Coomassie staining

Stock solutions:

Use distilled water for all solutions
TCA: 100 % TCA (w/v) 1 L
A: 0.2 % (w/v) $CuSO_4$ + 20 % glacial acetic acid
B: 60 % methanol

dissolve 1 Phast Blue R tablet in 400 mL of double distilled water, add 600 mL of methanol and stir for 5 to 10 min.

1 tablet = 0.4 g of Coomassie Brilliant Blue R-350

Staining:

- *Fixing:* 10 min in 200 mL of 20 % TCA;
- *Washing:* 2 min in 200 mL of washing solution (mix equal parts of A and B);
- *Staining:* 15 min in 200 mL of 0.02 % (w/v) R-350 solution at 50 °C while stirring (Fig. 7);

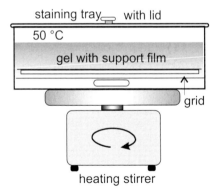

Fig. 7: Appliance for hot staining.

- *Destaining:* 15 to 20 min in washing solution at 50 °C while stirring;
- *Preserving:* 10 min in 200 mL of 5 % glycerol;
- *Drying:* air-dry.

5 minute silver staining of dried gels

Krause I, Elbertzhagen H. In: Radola BJ, Ed. Elektrophorese-Forum '87. This edition. (1987) 382–384.
According to the silver staining method for agarose gels (Kerenyi and Gallyas 1972, Willoughby and Lambert, 1983).

This method can be applied to stain dried gels already stained with Coomassie (the background must be completely clear) to increase sensitivity, or else unstained gels can be stained directly after pretreatment (Krause and Elbertzhagen, 1987). A significant advantage of this method is that no proteins or peptides are lost during the procedure. They often diffuse out of the gel during other silver staining methods because they cannot be irreversibly fixed in the focusing gels.

Pretreatment of the unstained gels:

- fix for 30 min in 20 % TCA,
- wash for 2 × 5 min in 45 % methanol. 10 % glacial acetic acid,
- wash for 4 × 2 min in distilled water,
- impregnate for 2 min in 0.75 % glycerol,
- air-dry.

Solution A: 25 g of Na_2CO_3, 500 mL of double distilled water;

Solution B: 1.0 g of NH_4NO_3, 1.0 g of $AgNO_3$, 5.0 g of tungstosilisic acid, 7.0 mL of formaldehyde solution (37 %), make up to 500 mL with double-distilled water.

Silver staining:

Mix 35 mL of solution A with 65 mL of solution B just before use. Immediately soak the gel in the resulting whitish suspension and incubate while agitating until the desired intensity is reached. Briefly rinse with distilled water.

Stop with 0.05 mol/L glycine. Remove remains of metallic silver from the gel and support film with a cotton swab.

Air-dry.

The most sensitive silver staining procedure for IEF

Because complete removal of the carrier ampholytes from the gel with simultaneous fixing of the proteins is not so easy, isoelectric focusing gels show very often a high background after highly sensitive silver staining. Most silver staining protocols for IEF are less sensitive than for native and SDS polyacrylamide gels. The highest demand on sensitivity – particularly in the basic part of the pH gradient – comes from those laboratories, who perform isoelectric focusing of oligoclonal IgG in human cerebrospinal fluid.

Wurster U. Isoelectric focusing of oligoclonal IgG on polyacrylamide gels (PAG plates pH 3.5 – 9.5) with automated silver staining. Amersham Pharmacia Biotech Application Note (1998).

Ammoniacal silver staining has been found to be the best method, however it requires many steps and is time consuming. U. Wurster (1998) has modified the procedure in order to make it compatible with automated gel staining. For this type of silver staining procedure the stainless steel tray, not the teflon coated tray is used.

Staining solutions:

A. *Fixing sol. I:* 20 % TCA

B. *Fixing sol. II:* 50 % methanol, 10 % acetic acid.

C. *Fixing solution III:* 5 % methanol, 7 % acetic acid.

D. *Fixing solution IV:* 2.5 % glutardialdehyde.

E. *Ammoniacal silver reagent:*

   a) dissolve 180 mg silver nitrate in 0.75 mL H_2O_{dist}.

   b) 3 mL 1 mol/L NaOH + 1.1 mL NH_3 (25 % stock solution), to 150 mL with H_2O_{dist}.

   c) With the help of a pipette slowly add solution a) $AgNO_3$ to the vigorously vortexed solution b) ammoniacal sodium hydroxide.

Any brownish precipitate should immediately disappear, otherwise the concentration of the ammonia solution is too low and should be checked.

F. *Developing solution:* For the preparation of 0.05 % (w/v) citric acid dissolve 125 mg monohydrate in 250 mL H_2O_{dist}. Immediately before use dilute 15 mL of the citric acid solution to 150 mL and add 200 μL of formaldehyde.

G. *Stopping solution:* 10 % ethanol, 1 % acetic acid.

H. *Preserving solution:* 1 % acetic acid, 10 % glycerol.

Tab. 2: Ammoniacal silver staining protocol (250 mL per step for manual, 150 mL per step for automated staining)

Step	Solution	IN-port	OUT-port	Time (min)
1	20 % TCA	7	7	45
2	H_2O_{dist}	0	7	0.5
3	50 % methanol, 10 %acetic acid	0	9	40
4	5 % methanol, 7 % acetic acid	8	9	20
5	Glutardialdehyde 2.5 %	2	2	30
6	H_2O_{dist}	0	7	10
7	H_2O_{dist}	0	7	30
8	H_2O_{dist}	0	7	30
9	H_2O_{dist}	0	7	360
10	H_2O_{dist}	0	7	360 (Hold)*
11	H_2O_{dist}	0	7	60
12	Silver Solution	3	3	40
13	H_2O_{dist}	0	7	0.5
14	H_2O_{dist}	0	7	8
15	Developing solution	4	9	5
16	Stop solution	5	9	5
17	Stop solution	5	9	5
18	Stop solution	5	9	5
19	Stop solution	5	9	5
20	Preserving solution	6	9	60

Staining sensitivity is optimal, when the silver staining and development solutions are used freshly.

The stainer stays in the HOLD* position after over-night water washing and must be restarted the next morning. Thus the silver solution can be prepared fresh which gives a lighter background. Alternatively the silver solution may be provided the day before together with all other solutions. In that case the HOLD option must be canceled and staining will proceed automatically until the end.

The stained gels are air-dried over night. The sticky surface is then covered by a rolled-on Mylar sheet. The gels can be punched, filed and stored for several years without fading.

6

Perspectives

Electrode solutions: For applications, which require long separation times, basic gradients, and/or with presence of high molar urea, it is necessary to use filter paper strips with electrode solutions. In table 1a list of electrode solutions is given for 0.5 mm thin gels for the respective pH intervals.

Tab. 3: Electrode solutions for IEF in polyacrylamide gels

pH Gradient	Anode	Cathode
3.5 – 9.5	0.5 mol/L H_3PO_4	0.5 mol/L NaOH
2.5 – 4.5	0.5 mol/L H_3PO_4	2 % Ampholine pH 5–7
2.5 – 4.5	0.5 mol/L H_3PO_4	0.4 mol/L HEPES
3.5 – 5.0	0.5 mol/L H_3PO_4	2 % Ampholine pH 6–8
4.0 – 5.0	0.5 mol/L H_3PO_4	1 mol/L glycine
4.0 – 6.5	0.5 mol/L acetic acid	0.5 mol/L NaOH
4.5 – 7.0	0.5 mol/L acetic acid	0.5 mol/L NaOH
5.0 – 6.5	0.5 mol/L acetic acid	0.5 mol/L NaOH
5.5 – 7.0	2 % ampholine pH 4–6	0.5 mol/L NaOH
5.0 – 8.0	0.5 mol/L acetic acid	0.5 mol/L NaOH
6.0 – 8.5	2 % ampholine pH 4–6	0.5 mol/L NaOH
7.8 – 10.0	2 % ampholine pH 6–8	1 mol/L NaOH
8.5 – 11.0	0.2 mol/L histidine	1 mol/L NaOH

The advantages of the gels have already been explained at the beginning.

Rehydrated gels: A number of possible applications exist for rehydrated focusing gels. A few of these and the necessary rehydration solutions will be presented here.

For cathodic drift see Righetti and Drysdale (1973).

Basic, water-soluble proteins: basic carrier ampholytes partially inhibit the polymerization of polyacrylamide gels. In addition basic gradients are more prone to cathodic drift than others, partly because of acrylamide monomers and remains of APS in the gel. These problems do not arise when washed and rehydrated gels are used. Practice has shown, that electrode solutions must be applied to stabilize the gradient.

Proteins and enzymes sensitive to oxidation: 2-mercaptoethanol is sometimes used to prevent oxidation in the gel, but it inhibits polymerization. In addition ammonium persulfate possesses oxidative characteristics in the gel (Brewer, 1967). Once again, these problems do not occur with rehydrated gels.

Brewer JM. Science 156 (1967) 256–257.

Heat-sensitive enzymes and enzyme-substrate complexes: 37 % DMSO in the gel is generally used during cryo-isoelectric focusing (at –10 to –20 °C) (Righetti, 1977, and Perella *et al.* 1978). In this case the gel is reconstituted with 37 % DMSO (v/v) and 2 % Ampholine pH > 5 (w/v).

Ampholytes with pH < 5 precipitate at these low temperatures.

Hydrophobic proteins: the non-ionic detergents required to solubilize hydrophobic proteins inhibit the co-polymerization of the gel and support film when they are added to the polymerization solution. Support free gels with non-ionic detergents bend during the run and can even tear at high field strengths. These problems do not occur with rehydrated gels: they can be reconstituted with nonionic or zwitterionic detergents such as LDAO or CHAPS in any desired concentration.

Complex mixtures of proteins: if proteins from cell lysates or tissue extracts must be completely separated, the aggregation or precipitation of the proteins should be prevented (Dunn and Burghess 1983). An 8 mol/L urea solution, 1 to 2 % non-ionic detergent, 2.5 % carrier ampholyte and about 1 % DTT are usually used. During polymerization the problems mentioned can occur. In addition in these highly viscous solutions the mobility of the proteins is reduced, so the cathodic drift has more influence.

Dunn MJ, Burghes AHM. Electrophoresis 4 (1983) 97–116.

Separations in these highly viscous gels take a longer time then usual. Practice has shown, that electrode solutions must be applied to stabilize the gradient.

Immunofixation: when IEF polyacrylamide gels with a T value of 5 % or less have a thickness of 0.5 mm or less, antibodies are able to diffuse into the matrix. Thus, in some cases, TCA can be avoided for fixation, and the noninteresting proteins can be washed out of the matrix. This is particularly advantageous for the detection of oligoclonal IgG bands in serum and cerebrospinal fluid. After the nonprecipitated antibodies and proteins have been washed out, silver staining can be employed to detect the antigen-antibody complexes.

This detection method is only possible since thin and soft polyacrylamide gels on film-support have been introduced. In order to save costs for antibodies, small gels are preferred for this technique.

Gianazza E, Chillemi F, Gelfi C, Righetti PG. J Biochem Biophys Methods. 1 (1979) 237–251.
Gianazza E, Chillemi F, Righetti PG. J Biochem Biophys Methods. 3 (1980) 135–141.

Detection of oligopeptides: Small peptides cannot be fixed with an acid and are not easy to stain. Gianazza *et al.* (1979 and 1980) have described a method of drying ultrathin IEF gels on filter paper and detecting the small peptides with specific amino acid stains or an iodine stain.

Method 7:
Horizontal SDS-PAGE

The principles of SDS-electrophoresis and its fields of application are *See pages 37 and following.* described in part I.

1
Sample preparation

The native state of proteins is shown in Fig. 1: the tertiary structure of *For SDS electrophoresis* a polypeptide coil – slightly simplified – and a protein (IgG) with a *proteins must be converted* quarternary structure consisting of many subunits – highly simpli- *from their native form into* fied. This spatial arrangement is conditioned by intra- and intermo- *SDS-protein micelles.* lecular hydrogen bonds, hydrophobic interactions and disulfide bridges, which are formed between cysteine residues.

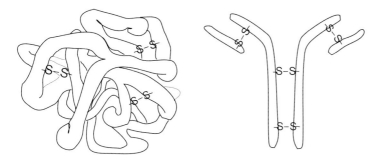

tertiary structure   quarternary structure

Fig. 1: Native structure of proteins.

Electrophoresis in Practice, Fourth Edition. Reiner Westermeier
Copyright © 2005 WILEY-VCH Verlag GmbH & Co. KGaA, Weinheim
ISBN: 3-527-31181-5

SDS-treatment

The addition of an excess of SDS to protein solutions has the following effect:

- individual charge differences of the proteins are masked,
- hydrogen bonds are cleaved,
- hydrophobic interactions are canceled,
- aggregation of the proteins is prevented,

See also pages 81 and following for qualitative and quantitative evaluation.

In the process 1.4 g of SDS are bound per gram of protein. All the micelles have a negative charge, which is proportional to the mass.in addition the polypeptides are unfolded (removal of the secondary structure) and ellipsoids are formed.

The Stokes radii of the micelles are then proportional to the molecular weight (M_r), and a separation according to the molecular weight is obtained by electrophoresis.

Some proteins, e.g. casein require 2 % SDS or more.

1 to 2% (w/v) SDS are used in the sample solution and 0.1% in the gel.

The method of sample treatment is very important for the quality of the separation and its reproducibility; the various possibilities will therefore be described in detail.

Stock buffer (pH 6.8):

6.06 g Tris + 0.4 g SDS
make up to 80 mL with double-distilled water
titrate to pH 6.8 with 4 mol/L HCl
make up to 100 mL with double-distilled water.

For some proteins it is better to use a more basic buffer, e.g. Tris-HCl pH 8.8.

This in fact is the stock buffer for the stacking gel – and also for the sample buffer – in the original publication of Laemmli. This buffer has become a standard in SDS-electrophoresis.

Non-reducing SDS treatment

A proper molecular weight estimation is not possible here: for example albumin (68 kDa) shows an apparent molecular weight of 54 kDa.

Many samples, e.g. physiological fluids such as serum or urine, are simply incubated with a 1% SDS buffer without reducing agent, because one does not want to destroy the quaternary structure of the immunoglobulins. The disulfide bonds are not cleaved by this treatment and so the protein is not fully unfolded (Fig. 2).

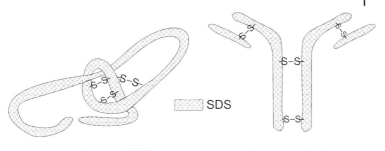

incomplete unfolding   quarternary structure

Fig. 2: Proteins treated with SDS without reducing agent.

Nonred SampB (Non reducing sample buffer)

1.0 g of SDS
3 mg of EDTA
10 mg of Bromophenol Blue

2.5 mL of stock buffer
make up to 100 mL with
double-distilled water.

EDTA is used to inhibit the oxidation of DTT in case the sample should be reduced. Dr. J. Heukeshoven, personal communication.

Diluting the samples (examples)

For *Coomassie staining* (sensitivity 0.1 to 0.3 mg per band):
Serum: 10 µL + 1.0 mL of sample buffer
Urine: 1 mL + 10 mg of SDS
(it should not be further diluted with sample buffer)
For *silver staining* (sensitivity down to 0.2 ng per band):
Serum: dilute the above sample 1:20 with Nonred SampB.
Urine: 1 mL + 10 mg of SDS (dilute 1:3 for some proteinurias).

Apply 7 µL
Apply 20 µL

Apply 3 µL
Apply 3 µL

- Incubate for 30 min at room temperature, *do not boil*! Boiling of non reduced proteins can lead to protein fragmentation.

Reducing SDS treatment

Proteins are totally unfolded and a separation according to molecular weight is possible, only when a reducing agent like dithiothreitol (DTT) is added (Fig. 3). The smell is reduced by the use of the less volatile DTT instead of 2-mercaptoethanol.

Prepare DTT and Red SampB in the appropriate quantities shortly before use.

Another advantage of using DTT instead of 2-mercapto-ethanol is, that when reduced and nonreduced sample should be separated in the same gel, DTT does not diffuse into the traces of non reduced proteins.

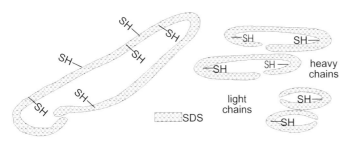

Fig. 3: SDS treated and reduced proteins.

Dithiothreitol stock solution (2.6 mol/L DTT):
dissolve 250 mg of DTT in 0.5 mL double-distilled water.

Add a bit of Orange G for better differentiation.

Red SampB (reducing sample buffer 26 mol/L):
10 mL of Nonred SampB + 100 µL of DTT solution

Diluting the samples (examples):

Apply 5 µL, 1 mg of BSA per application

For Coomassie staining (sensitivity: 0.1 to 0.3 mg per band)
LMW marker + 415 µL of Red SampB

Apply 5 µL

HMW marker + 200 µL of double-distilled water, do not boil!

Apply 5 µL

CMW marker: LMW marker + 315 µL of Red SampB + 100 mL of desalted collagen.

Apply 5 µL

Meat extracts: 100 µL + 900 µL of H_2O_{dist}, divide:
50 µL + 237.5 µL of Red SampB.

Apply 3 µL

For *silver staining* (sensitivity down to 0.2 ng per band):
dilute the above sample 1:20 with Red SampB.

- Boil the samples for 3 min (heating block).

The oxidation of DTT is inhibited by EDTA , so this treatment is not necessary for all samples: it is best to perform a comparison test.

- After cooling:
  Add 1 µL DTT solution per 100 µL of sample solution. The reducing agent can be oxidized during heating. The SH groups are better protected by the addition of reducing agent which prevents refolding and aggregation of the subunits.

In practice the (repeated) renewed addition of reducing agent is often forgotten: this results in additional bands in the high molecular weight area ("ghost bands") and precipitation at the application point.

Reducing SDS treatment with alkylation

In addition, the appearance of artifact lines during silver staining is prevented because iodo-acetamide traps the excess DTT. Alkylation with iodoacetamide works best at pH 8.0 which is why another sample buffer with a higher ionic strength (0.4 mol/L) is used. This high molarity is not a problem for small sample volumes.

Dr. J. Heukeshoven, personal communication.

The SH groups are better and more durably protected by ensuing alkylation with iodoacetamide (Fig. 4). Sharper bands result; in proteins containing many amino acids with sulfur groups a slight increase in the molecular weight can be observed.

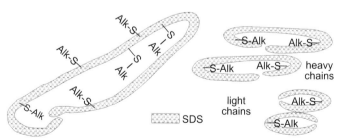

Fig. 4: Reduced and alkylated proteins treated with SDS.

Iodoacetamide solution 20% (w/v):
20 mg iodoacetamide + 100 μL H_2O_{dist}

The percentile weight error is negligible.

Sample buffer for alkylation (pH 8.0: 0.4 mol/L):
4.84 g Tris + 1.0 g SDS + 3 mg EDTA → 80 mL with H_2O_{dist},
titrate to pH 8.0 with 4 mol/L HCl,
make up to 100 mL with H_2O_{dist},
add 10 mg of Bromophenol or Orange G,
10 mL of sample buffer (Alk) + 100 μL of DTT solution.

The additional dilution of the sample by the alkylation solution can be taken into account either during the sample preparation or the application: apply 10% more sample volume.

After boiling the reduced sample:
Add 10 μL of iodoacetamide solution per 100 μL of sample solution and incubate for 30 min at room temperature.

2
Stock solutions for gel preparation

Acrylamide, Bis solution (T = 30%, C = 2%):
29.4 g of acrylamide + 0.6 g of bisacrylamide, make up to 100 mL with H_2O_{dist}

*C = 2% in the gradient gel solution prevents the **separation** gel from peeling off the support film and cracking during drying.*

Acrylamide, Bis solution (T = 30%, C= 3%):
Add 100 mL of H_2O_{dist} to the PrePAG mix (29.1:0.9).

*This solution is used for slightly concentrated **plateaus** with C = 3%, because the slot would become unstable if the degree of polymerization were lower.*

■ Caution!
  Acrylamide and bisacrylamide are toxic in the monomeric form. Avoid skin contact and dispose of the remains ecologically (polymerize the remains with an excess of APS).

Gel buffer pH 8.8 (4 × conc):
18.18 g of Tris + 0.4 g of SDS, make up to 80 mL with H_2O_{dist}.
Titrate to pH 8.8 with 4 mol/L HCl; make up to 100 mL with H_2O_{dist}.

Ammonium persulfate solution (APS):
Dissolve 400 mg of APS in 1 mL of H_2O_{dist}

Can be stored for one week in the refrigerator (4°C).

Cathode buffer (2×conc):
7.6 g of Tris + 36 g of glycine + 2.5 g of SDS, make up to 250 mL with H_2O_{dist}

Do not titrate with HCl!

Anode buffer (2×conc):
7.6 g of Tris + 2.5 g SDS + 200 mL of H_2O_{dist}. Titrate to pH = 8.4 with 4 mol/L HCl; make up to 250 mL with H_2O_{dist}

Economy measure: the cathode buffer can also be used here.

3
Preparing the casting cassette

The "spacer plate" is the glass plate with the 0.5 mm thick silicone rubber gasket glued on to it.

Gels with a completely smooth surface can be used in horizontal SDS-electrophoresis for sample application methods similar to IEF. When the gels are hand-made, it is possible to polymerize sample wells in the surface of the gel. For this a "slot former" is made out of the spacer plate:

Preparing the slot former

"Dymo" tape with a smooth adhesive surface should be used. Small air bubbles can be enclosed when the adhesive surface is structured, these inhibit polymerization and holes appear around the slots.

To make sample application wells, a mould must be fixed on to the glass plate. The cleaned and degreased spacer plate is placed on the template ("slot former" template in the appendix) on the work surface. A layer of "Dymo" tape (embossing tape, 250 µm thick) is placed on the area which will be used as starting point avoiding air bubbles. The slot former is cut out with a scalpel (Fig. 5). After pressing the individual sample wells once again against the glass plate the remains of tape are removed with methanol. Several superimposed layers of "Scotch tape" quality, one layer = 50 µm) can be used instead.

In some applications, the sample must be applied over the entire gel width for subsequent blotting for producing test strips. In these cases the "Dymo" tape is glued over the entire width, and only cut in the center for the application of markers (see method 8 Blotting).

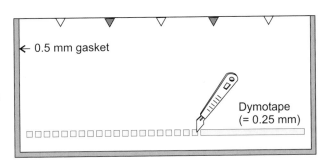

Fig. 5: Preparing the slot former.

The next step is to make the casting cassette hydrophobic. This is done with a few mL of Repel Silane which are spread over the whole slot former with a tissue. This operation should be carried out under the fume hood. When the Repel Silane is dry, the chloride ions which result from the coating are rinsed off with water.

Assembling the casting cassette

The gel is polymerized on a support film for mechanical support and easier handling. A glass plate is placed on an absorbent tissue and wetted with a few mL of water. The GelBond PAG film is applied with a roller, the untreated, hydrophobic side down (Fig. 6). This cre-

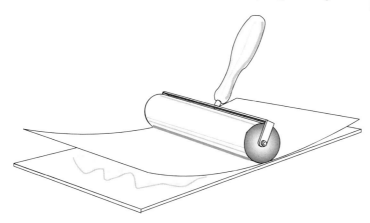

Fig. 6: Applying the support film with a roller.

ates a thin film of water between the support film and the glass plate, which holds them together by adhesion.

The excess water, which runs out is soaked up by the tissue. To facilitate pouring the gel solution, the film should overlap the long edge of the glass plate by about 1 mm.

The slot former is placed over the glass plate and the cassette is clamped together (Fig. 7).

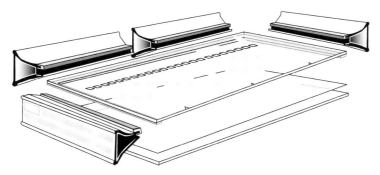

Fig. 7: Assembling the casting cassette.

The casting cassette is cooled to 4 °C in the refrigerator for about 10 min; this delays the onset of polymerization. This last step is essential, since it takes the poured gradient about 5 to 10 min to settle horizontally.

In a warm laboratory – during the summer for example – the gel solution should also be cooled to 4 °C.

4

Gradient gel

Pouring the gradient

It is not recommended to use sucrose for ultrathin gels because the solution would be too viscous.

a) Assembling the casting cassette

The gradient is prepared with a *gradient mixer*. It is made of two communicating cylinders. The front cylinder, the *mixing chamber*, contains the denser solution and a magnetic stirrer bar. The back cylinder, the *reservoir*, contains the lighter solution. The terms "dense" and "light" mean that a *density gradient* is coupled to the acrylamide solution: the dense solution contains about 25% glycerol and the light one 0%. The difference in densities prevents the solutions from mixing in the cassette and allows the gradient to settle horizontally.

Because of the difference in densities, opening the channel between the two chambers would cause the denser solution to flow back into the reservoir.

The *compensating bar* in the reservoir corrects for the difference in density and for the volume of the magnetic stirrer.

A "very dense" plateau does not mix with the gradient.

Tab. 1: Composition of the monomer solutions for 2 gradient gels 8 – 20%*T* and a sample application plateau with 5%*T*.

Pipette into 3 test tubes	Plateau 5%*T* , 3%*C*	Dense 8%*T* , 2%*C*	Light 20%*T*, 2%*C*
Glycerol (85 %)	3.2 mL	4.3 mL	–
Acrylamide, Bis 30%*T*, 3%*C*	1.25 mL	–	–
Acrylamide, Bis 30%*T*, 2%*C*	–	4.0 mL	10 mL
Gel buffer	1.9 mL	3.75 mL	3.75 mL
Bromophenol Blue (0.7 % w/v) in H_2O_{dist}	–	–	100 µL
Orange G (1 % w/v) in H_2O_{dist}	50 µL	–	–
TEMED	4 µL	7.5 mL	7.5 mL
bring to final volume with H_2O_{dist}	7.5 mL	15 mL	15 mL

In the gradient described here, the dense solution contains the low proportion of acrylamide, the light solution contains the high acrylamide concentration. In consequence the slot former is placed in the lower part of the cassette (Fig. 8 and 9).

This is in contrast to the conventional gradient pouring technique for verrtical chambers, but has a number of advantages:

During separation, the part containing glycerol slowly swell because glycerol is hygroscopic.

- The part with the small pores keeps its pore size without glycerol.

- A high proportion of glycerol in the sample application area prevents the gel from drying out, increases the stability of the sample slots, improves the solubility of high molecular proteins and compensates for the effect of the salts in the sample. *In many cases dialysis of the sample can be avoided.*
- Only one acrylamide stock solution is needed for a gel concentration of up to $T = 22.5\%$.
- Because the highly concentrated acrylamide solution is on top, the settling of the gradient is not disturbed by thermal convection. *The higher the acrylamide concentration, the more heat is produced during polymerization.*
- In the part of the gradient with the small pores, where the influence of the matrix on the zones in the most important, the perfect levelling of the gradient is important. Glycerol free solutions are less viscous. *The viscosity of the solution plays an important role for ultrathin gels.*

To pour a linear gradient (Fig. 8) both cylinders of the gradient mixer are left open. The laboratory platform ("Laborboy") is set so that the outlet lies 5 cm above the upper edge of the gel. *For reproducible gradients the outlet of the gradient mixer must always be on the same level above the edge of the gel cassette.*

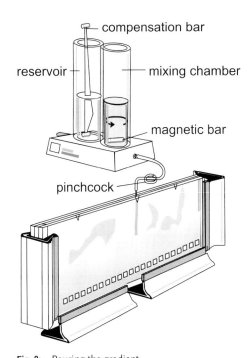

Fig. 8: Pouring the gradient.

Before filling the connecting channel between the *reservoir* and the *mixing chamber* as well as the *pinchcock* are shut.

The stirring bar is then placed in the mixing chamber and the optimum speed set on the magnet stirrer.

b) Casting the gel

The dense and the light solutions are poured directly into the gradient mixer in the following steps:

- pour the light solution into the reservoir,
- briefly open the valve to fill the connecting channel,
- pour the dense solution into the mixing chamber.

The glass plate with the film is oriented towards the user.

When this is done, the cassette can be removed from the refrigerator and connected to the gradient mixer with the slot former oriented towards the gradient mixer.

- Add APS to the plateau solution (Tab. 2).

Tab. 2: Catalysts volumes.

Gel solution	Volume	APS (40 %)
Very dense (Plateau)	3.5 mL	5 µL
Dense	7.0 mL	6 µL
Light	7.0 mL	4 µL

The notches of the slot former are oriented towards the gradient mixer.

- Pipette the 3.5 mL of the very dense solution into the cassette (Fig. 9);
- pipette APS into the reservoir;
- disperse APS while introducing the compensating stick into the reservoir;

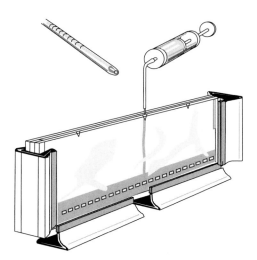

Fig. 9: Pouring the gel solution for the plateau.

- pipette the APS into the mixing chamber and stir briefly but *to disperse the catalyst* vigorously with the magnetic stirrer;
- place the outlet in the middle notch of the slot former;
- set the magnetic stirrer at moderate speed; *do not generate air bubbles*
- open the outlet valve;
- open the connecting valve.

The gradient mixer should be empty when the fluid level has *Rinse out the mixer with* reached the top of the cassette. *double-distilled water*

100 μL of 60 % (v/v) isopropanol-water are then layered in each *immediately afterwards.* filling notch. Isopropanol prevents oxygen from diffusing into the gel. The gel will then present a well-defined, aesthetic upper edge.

The gradient takes about 10 min to level out before polymerization starts. The gel should be solid after about 20 min.

The gel must be completely polymerized before electrophoresis, *At least for 3 hours, ideally* because the electrophoretic mobilities of the protein-SDS micells are *overnight, at room tempera-* greatly influenced by the sieving properties of the gel.Although, *ture.* when the gel seems to have perfectly polymerized, it can not be used immediately. The matrix only becomes regular after a slow "silent polymerization". If it is used too early, the bands are shaky, not straight and even as usual.

5
Electrophoresis

Preparing the separation chamber

- Turn on the cooling system: + 15 °C;
- Unclamp the gel cassette and place the gel with the slot former on the bottom on the cooling plate of the Multiphor;
- Place the cassette vertically to ease the glass plate from the GelBond film with a thin spatula;
- Grasp the film at a corner where the acrylamide concentration is high and pull it away from the slot former.

Placing the gel on the cooling plate
Wet the cooling plate with 1–2 mL of kerosene. *Water is not suitable, as it can cause electric shorting.*

- Place the gel onto the center of the cooling plate with the film on the bottom; the side with the slots must be oriented towards the cathode (–). Avoid air bubbles.
- Shift the gel so that the anodal edge coincides exactly with line *Wear gloves.* "5" on the scale on the cooling plate (Fig. 11).
- Lay two of the electrode wicks into the compartments of the *Be sure to use **very** clean wicks,* PaperPool. If smaller gel portions are used, cut them to size. *SDS would dissolve any traces of contaminating compounds.*

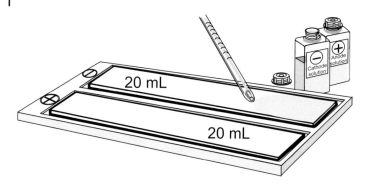

Fig. 10: Soaking the electrode wicks with the respective solutions in the PaperPool.

less volume for shorter strips

less volume for shorter strips

Always apply cathode wick first to avoid contamination of cathode buffer with leading ions.

- Mix 10 mL of the cathode buffer (2 × conc) with 10 mL distilled water and apply it onto the respective strip (Fig.10).
- Mix 10 mL of the anode buffer (2 × conc) with 10 mL distilled water and apply it onto the respective strip.
- Place the cathode wick onto the cathodal edge of the gel; the edge of the wick matching "3.5" on the cooling plate; the anodal wick over the anodal edge, matching "13.5".

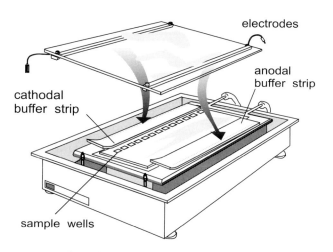

Fig. 11: Appliance for semi-dry SDS electrophoresis with electrode wicks.

Sample application

Ready-made gels are marked by cutting off a corner of the film on the side with narrow pores.

- Rapidly pipette the samples in the slots.

If a gel with a smooth surface, that is without sample wells, is used, the side of the gel with the large pores should be marked. The gel is then, as described above, placed on the cooling plate so that the samples are also applied on line "5".

The side with large pores must be placed towards the cathode!

The samples are applied 1 cm from the cathode strips and at least 1.5 cm from the sides. There are several possibilities of application (Fig. 12):

- Directly pipette droplets (1). — *Not more than 3 μL.*
- Place the sample application pieces on the gel and then apply the sample (2). — *It is possible to apply 3 to 20 μL.*
- Use a sample application strip and pipette the samples in the slots (3). — *It is possible to apply 30 to 40 μL.*
- Remove some Raschigg rings from the condenser of a rotatory evaporator, place them on the gel, apply the samples (4). — *It is possible to apply up to 100 μL.*

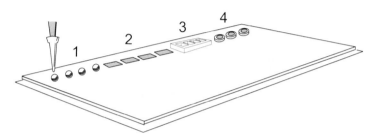

Fig. 12: Possibilities of sample application on SDS gels with an even surface.

Electrophoresis

- Clean platinum electrode wires before (and after) each electrophoresis run with a wet tissue paper. Move electrodes so that they will rest on the outer edges of the electrode wicks. — *The buffer ions must be in between the electrodes.*
- Connect the cables of the electrodes to the apparatus and lower the electrode holder plate (Fig. 11). — *Be sure, that the electrodes have complete contact on the wicks.*
- Close the safety lid.

Separation conditions:

1000 V, 50 mA, 35 W, 1 h 30 min. — *with normal power supply*

Tab. 3: Programmable power supply

Phase	U	I	P	t	
1	200 V	50 mA	30 W	10 min	*gentle sample entry*
2	600 V	50 mA	22 W	1 h 20 min	
3	100 V	5 mA	5 W	1 h	*hold against diffusion*

- Switch off the power supply;
- Open the safety lid;
- Remove the electrode strips from the gel and dispose them;
- Remove the gel from the cooling plate.

6
Protein detection

Quick Coomassie staining

The gel is placed face down on the grid.

- *Staining:* hot staining by stirring with a magnetic stirring bar in a stainless steel tank: 0.02% Coomassie R-350: Dissolve 1 PhastGel Blue tablet in 1.6 L of 10 % acetic acid; 15 min, 50 °C (see Fig. 13).

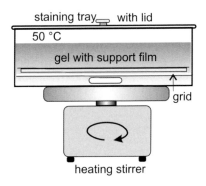

Fig. 13: Device for hot staining.

- *Destaining:* in a tray on a rocking table in 10% acetic acid for 2 h at room temperature.
- *Preserving:* in a solution of 25 mL of glycerol (87% w/v) + 225 mL of distilled water for 30 min.
- *Drying:* air-dry (room temperature).

Colloidal staining

Neuhoff V, Stamm R, Eibl H. Electrophoresis 6 (1985) 427–448.

This method has a high sensitivity (ca. 30 ng per band), but takes overnight. There is no background destaining (Neuhoff *et al.* 1985).

Preparation of the staining solution:

The staining solution can be used several times.

Slowly add 100 g of ammonium sulfate to 980 mL of a 2% H_3PO_4 solution till it has completely dissolved. Then add Coomassie G-250 solution (1 g per 20 mL of water). Do not filter the solution! Shake before use.

Staining

- *Fixing* 1 h in 12% (w/v) TCA, for gels on support films it is best to place the gel surface on the bottom (on the grid of the staining tank), so that additives with a higher density (e.g. the glycerol for gradient gels) can diffuse out of the gel. *The use of a stirrer is recommended for this step.*
- *Staining* overnight with 160 mL of staining solution [0.1% (w/v) Coomassie G-250 in 2% H_3PO_4, 10% (w/v) ammonium sulfate see above] plus 40 mL of methanol (add during staining).
- *Wash* for 1 to 3 min in 0.1 mol/L Tris, H_3PO_4 buffer, pH 6.5.
- *Rinse* (max. 1 min) in 25% (v/v) aqueous methanol.
- *Stabilize* the protein-dye complex in 20% aqueous ammonium sulfate.

Reversible imidazole-zinc negative staining

This staining procedure for SDS gels according to Hardy *et al.* (1996) produces non stained bands against a white background. A sensitivity down to 15 ng per band has been reported. With an EDTA mobilization buffer the zinc-imidazole complex can be dissolved prior to electrophoretic blotting. *Hardy E, Santana H, Sosa AE, Hernandez L, Fernandez-Patron C, Castellanos-Serra L. Anal Biochem. 240 (1996) 150–152.*

- *Fix* the gel in 200 mol/L Imidazole containing 0.1 % SDS: 2.72 g Imidazole + 0.2 g SDS, dissolve in 200 mL distilled water. 15 min with continuous shaking.
- *Rinse* with distilled water.
- *Stain* (negative) with 200 mmol/L Zinc sulphate: 5.74 g $ZnSO_4$, dissolve in 200 mL H_2O_{dist}. Shake for 30 to 60 s until white background has developed.
- *Rinse* with distilled water.
- *Store* in 200 mL of new fixing solution diluted 1 : 10 with distilled water.
- *Mobilize* proteins with 50 mmol/L EDTA, 25 mmol/L Tris; pH 8.3: 0.61 g Tris + 3.72 g EDTA-Na_2, dissolve in 200 mL distilled water, adjust to pH 8.3 with a few grains of Tris when necessary. 6 min 200 mL with vigorous shaking. *Because of this feature this technique is sometimes called the "Sleeping Beauty / Prince" staining procedure.*

Silver staining

Tab. 4: Silver staining acc. to Heukeshoven and Dernick (1986)

Step	Solution	V [mL]	t [min]
Fixing	200 mL ethanol + 50 mL acetic acid with $H_2O_{dist} \rightarrow 500$ mL	2×250	2×15
Sensitizer	75 mL ethanol*) 17 g Na-acetate 1.25 mL glutardialdehyde (25% w/v) 0.50 g $Na_2S_2O_3 \times 5\ H_2O$ with $H_2O_{dist} \rightarrow 250$ mL	250	30 or overnight
Washing	H_2O_{dist}	3×250	3×5
Silvering	0.625 g $AgNO_3$**) 100 µL formaldehyde (37%) with $H_2O_{dist} \rightarrow 250$ mL	250	20
Developer	7.5 g Na_2CO_3 120 µL formaldehyde (37%) with $H_2O_{dist} \rightarrow 300$ mL	1×100 1×200	1 3 to 7
Stopping	3.65 g EDTA-$Na_2 \times 2H_2O$ with $H_2O_{dist} \rightarrow 250$ mL	250	10
Washing	H_2O_{dist}	3×250	3×5
Preserving	25 mL glycerol (87% w/v) with $H_2O_{dist} \rightarrow 250$ mL	250	30
Drying	air-drying (room temperature)		

* First dissolve NaAc in water, then add ethanol. Add
the thiosulfate and glutaraldehyde just before use.
** Dissolve $AgNO_3$ in water, add the formaldehyde
before use.

If larger volumes are prepared in advance: fixing solution should
be made fresh, silver nitrate should be kept as 10 × concentrated
stock solution in a dark bottle, aldehydes are added before use.

■ Note:
*for automated staining only 125 mL solution per
step is required.*

Blue Toning

In general, silver stained bands can not easily be quantified, because
the staining curve is very steep. Frequently the bands show different
colors; highly concentrated fractions show hollow bands ore bands
with yellow centers. These images severely interfere with semiquanti-
tative and qualitative evaluations using densitometers or scanners.

The evaluation of these results can be improved by "blue toning" according to Berson (1983): After silver staining the gel must be washed thoroughly with distilled water and then it is immersed for 2 minutes in a freshly mixed bath of:

Berson G. Anal Biochem. 134 (1983) 230–234. This process is adopted from the photoshops to make blue slides out of photo negatives.

140 mL H_2O_{dist} + 20 mL of 5% $FeCl_3$ + 20 mL of 3 % oxalic acid + 20 mL of 3.5 % potassium hexacyanoferrate.

Blue toning is slightly improving the sensitivity and gives uniformly stained bands.

Place the gel in water and then in glycerol solution before drying.

7
Blotting

The specific detection of proteins can be carried out after their electrophoretic transfer from the gel to an immobilizing membrane (blotting membrane). Either different samples are separated on the gel and the membrane is analysed with one antibody solution, or one antigen is loaded over the entire gel width and the membrane is cut into strips for probing in different patient sera.

To blot gels which have been stained the proteins must be solubilized in SDS buffer again (by soaking the gels in the SDS buffer). In most cases, for subsequent immuno-detection, the antigen-antibody reactivity remains despite additional denaturation with staining reagents (Jackson and Thompson, 1984). If blotting is carried out immediately after electrophoresis or with handmade gels and when the entire gel must be blotted, the gel is poured on the hydrophobic side of a GelBond PAG film from which it can easily be removed after electrophoresis.

Jackson P, Thompson RJ. Electrophoresis 5 (1984) 35–42. Reversible imidazole-zinc staining, however, works better. See method 9, page 225 and following.

8
Perspectives

Most problems can be solved with the methods for SDS electrophoresis presented here. Should difficulties occur nevertheless, refer the trouble-shooting guide in part III. A few other conceivable problems are described in part I.

See pages 34 and following.

Gel characteristics: The gel described here has an acrylamide gradient from T = 8 to 20% and a sample application plateau with T = 5%. If a higher resolution in a narrower molecular weight range is required, a flatter gradient (e.g. T = 10 to 15%) or a homogeneous resolving gel (e.g. T = 10%) are used. For T = i %, the volumes of acrylamide-Bis

This also enables complete separation of complex protein mixtures with a molecular weight range of 5 to 400 kDa.

solution used for 15 mL of polymerization solution can easily be calculated as follows:

$$V \text{ mL} = i \times 0.5 \text{ mL}$$

Front is visualized by Orange G or Bromophenol Blue marker dyes.

Different gel compositions will influence the separation time. For flatter gradients and homogenous gels, it is advisable to end the separation when the front has reached the anode.

SDS electrophoresis in washed and rehydrated gels

Obviously, this recipe can only work in presence of APS, Temed and monomers of acrylamide and Bis.

The technique is almost only feasible with plastic-film-supported gels.

A series of experiments with rehydrating washed and dried gels – like in method 4 of this book – for SDS electrophoresis has shown, that the standard Tris-HCl / Tris-glycine buffer system can not be applied for these gels: the separation quality is very poor.

Good results are, however, are obtained with the Tris-acetate / Tris-tricine buffer system (as shown in Fig. 15 A):

> *Rehydration buffer:*
> 0.3 mol/L Tris / acetate pH 8.0, 0.1% SDS.

20 mL in wick

> *Cathode buffer:*
> 0.8 mol/L Tricine, 0.08 mol/L Tris, 0.1 % SDS.

20 mL in wick

> *Anode buffer:*
> 0.6 mol/L Tris / acetate pH 8.4, 0.1 % SDS.

SDS disc electrophoresis in a rehydrated and selectively equilibrated gel

An efficient stacking of the proteins in the first phase of electrophoresis is necessary here.

For the separation of samples with very high protein concentrations in one fraction (e.g. in pharmaceutical quality control) and / or very complex protein mixtures, complete separations of all fractions is only achieved, when all four discontinuities of disc electrophoresis are applied (see page 32 of this book). In conventional ready made

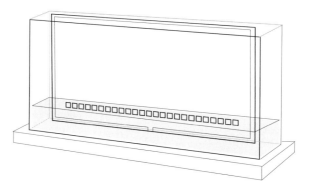

Fig. 14: Selective equilibration of the stacking gel zone in the appropriate buffer for disc electrophoresis.

gels, only the discontinuities in the gel matrix and between the leading ion in the gel and the trailing ion in the cathodal buffer can be applied.

When a gel is supported by a rigid plastic film, its stacking zone can selectively be equilibrated in a stacking gel buffer with a different pH and a lower buffer concentration, short before use. This is performed in a vertical equilibration chamber (Fig. 14). *For a 0.5 mm thin gel equilibration takes 15 min.*

If a laboratory made gel is used with Tris-HCl pH 8.8 – like described in this chapter –, and Tris-glycine is used in the cathode buffer, best results are obtained with a stacking buffer containing 1.25 mol/L Tris-HCl pH 6.7 and 0.1 % SDS. *For a Tris-acetate buffer pH 8.0 and tricine in the cathode, use a stacking buffer containing 0.1 mol/L Tris-acetate pH 5.6 and 0.1 % SDS.*

Peptide separation

Very good separations of low molecular weight peptides are obtained in SDS polyacrylamide electrophoresis with gels of relatively high acrylamide concentration, and employing a buffer with pH 8.4 (the pK-value of the basic group of the tricine, which is used in the cath- *See page 40*

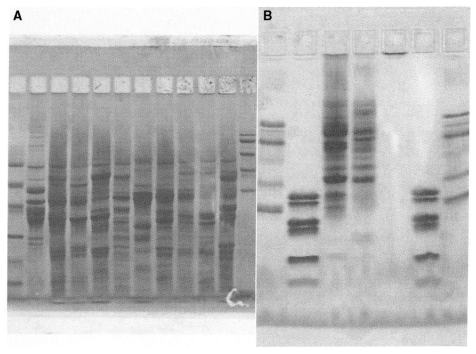

Fig. 15: (A) SDS electrophoresis of legume seed extracts and molecular weight standards in a rehydrated discontinuous gel with a 5 %T stacking zone and a 10 %T resolving zone. (B) Separation of peptide and low molecular weight markers in a rehydrated 15 %T gel containing 30 % (v/v) ethylenglycol and the buffer described in the text. Staining with Coomassie Brilliant Blue R-350.

Pharmacia TF 112: SDS-PAGE of low molecular weight proteins using PhastGel high density (1987).

ode buffer) as well as high molarity of Tris according to Schägger and von Jagow (1987). Pharmacia had introduced a PhastGel® High Density, a 20 % T gel containing 30 % ethylenglycol for peptide separations (Pharmacia, 1987).

By rehydration of a washed and dried gel with $T = 15$ % in a SDS buffer with 0.7 mol/L Tris-acetate pH 8.4 and 30 % ethylenglycol, a gel with very high resolving power in the low molecular weight area (1 to 40 kDa) is obtained within one hour. This gel should be run with the Tris-acetate anode buffer and the Tris-tricine cathode buffer according to the recipes given above. Electrophoresis takes 2 hours 45 minutes. A separation result obtained in such a gel is shown in Fig. 15 B.

Method 8:
Vertical PAGE

Polyacrylamide gel electrophoresis in a vertical setup is the standard technique in many laboratories. For the separation of proteins mostly discontinuous SDS electrophoresis according to Lämmli (1970) is performed, for DNA separations the continuous TBE (Tris borate EDTA) buffer is employed. Exact descriptions of gel casting and running conditions for gels of different sizes are found in the Hoefer Protein Electrophoresis Applications Guide (1994).

Hoefer Protein Electrophoresis Applications Guide (1994) 18–54.

Conventional procedure:

For a discontinuous system, the resolving gel is polymerized at least one day before use. After pouring the monomer solution into the cassette it is overlaid with water-saturated butanol to achieve a straight upper edge.

In the standard procedure the resolving and stacking gels are polymerized at different times.

One hour before electrophoresis the butanol solution is removed, the edge is rinsed several times with a gel buffer solution to remove unpolymerized monomers, is dried with filter paper, the stacking gel solution is poured on top of the resolving gel, and the comb is inserted. After removal of the combs the wells are first rinsed and then filled using the upper buffer. The samples must contain at least 20 % v/v glycerol or sucrose and they are underlaid with a syringe or a fine-tipped pipette.

In practice the interface between stacking and resolving gel frequently leads to problems: lateral edge effects, protein precipitate at the resolving gel edge, loss of the stacking gel during staining (probable loss of large proteins which can not migrate into the resolving gel).

Modified procedure:

Practice has shown, that the quality of results is not reduced, when the original procedures are modified as described below.

Reproducibility is even improved, when some steps are simplified.

In the following chapter simplified procedures are described how to cast individual and multiple 0.75 mm thin gels – and to run them – for a vertical minigel system: Mighty Small chamber with 6 × 8 cm gels, see Fig. 1.

The principles described can be transferred to larger and thicker gels with no problems.

Electrophoresis in Practice, Fourth Edition. Reiner Westermeier
Copyright © 2005 WILEY-VCH Verlag GmbH & Co. KGaA, Weinheim
ISBN: 3-527-31181-5

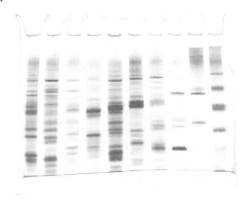

Fig. 1: Vertical SDS electrophoresis of legume seed extracts and markers, modified procedure; silver staining.

In this way, a lot of work can be saved: overlaying with butanol water, washing and drying the edge, casting the stacking gel before use.
Furthermore, no edge and corner effects occur, the stacking gel does not fall off the resolving gel.

Instead of polymerizing the resolving gels and stacking gels separately, they are polymerized together. In order to prevent mixing of these monomer solutions during pouring them into the gel cassette, the resolving gel solution must contain a certain amount of glycerol. Glycerol in the monomer solution has no negative effects on the polymerization and the separation.

This saves work and time.

Gradient gels are cast without a pump with very good reproducibility.

1
Sample preparation

For all vertical techniques, sample solutions must contain at least 20 % glycerol to prevent mixing with the cathode buffer.

*) In the original Lämmli procedure the stacking gel buffer is added; better results are achieved, when the resolving gel buffer is added here.

For SDS PAGE, the sample preparation is performed exactly as described in Method 7 (page 183 ff) with the only difference, that – because of underlaying – the sample buffer must contain 25 % (v/v) glycerol:

Nonred SampB (Non reducing sample buffer):
1.0 g of SDS + 3 mg of EDTA + 10 mg of Bromophenol Blue + 2.5 mL of gel buffer*) + 25 mL glycerol (85 %), make up to 100 mL with distilled water.
For native protein PAGE and DNA PAGE 25 % glycerol is just added to the samples.

2
Stock solutions

Acrylamide, Bis solution (T = 40%, C= 3%):
38.8 g of acrylamide + 1.2 g of Bis, make up to 100 mL with H_2O_{dist}.
   For vertical gels a 40 % T stock solution is used rather than a 30 % T solution (for the horizontal gels).

   ■ Caution!
   Acrylamide and Bis are toxic in the monomeric form. Avoid skin contact and dispose of the remains ecologically.

Because glycerol must be added to some highly concentrated starting solutions, higher concentrated acrylamide stock solutions are required. Polymerize the remains with an excess of APS.

Stacking gel buffer pH 6.8 (4 × conc):
6.06 g of Tris + 0.4 g of SDS, make up to 80 mL with H_2O_{dist}. Titrate to pH 6.8 with 4 mol/L HCl; make up to 100 mL with H_2O_{dist}

pH 6.8 for stacking gel is used only to achieve optimal polymerization conditions for samples wells (buffers will diffuse during storage).

Resolving gel buffer pH 8.8 (4 × conc):
18.18 g of Tris + 0.4 g of SDS, make up to 80 mL with H_2O_{dist}. Titrate to pH 8.8 with 4 mol/L HCl; make up to 100 mL with H_2O_{dist}.

Ammonium persulfate solution (APS):
Dissolve 400 mg of APS in 1 mL of H_2O_{dist}.

Can be stored for one week in the refrigerator (4 °C).

Push-up Solution:
11 mL Glycerol + 3.5 mL resolving gel buffer + 0.5 mL Orange G solution (1 %).

Only needed for multiple gel casting.

Cathode buffer (10 × conc):
7.6 g of Tris + 36 g of glycine + 2.5 g of SDS, make up to 250 mL with H_2O_{dist}.

Do not titrate with HCl!

Anode buffer (10 × conc):
7.6 g of Tris + 2.5 g SDS + 200 mL of H_2O_{dist}. Titrate to pH = 8.4 with 4 mol/L HCl; make up to 250 mL with H_2O_{dist}.

Economy measure: the cathode buffer can also be used here.

3
Single gel casting

A gel cassette consists of a glass plate, a notched aluminum oxide ceramics plate, two spacers and a comb (Fig. 2).
   The gels should be prepared at least one day before use.

Aluminum oxide ceramics dissipates the heat much more efficiently than glass.

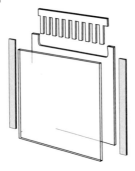

Fig. 2: Gel cassette for a vertical gel.

■ Important:
In order to prevent drying of the stacking gels, the gel cassettes should be taken out from the casting stand after one hour of polymerization. They are placed into a plastic bag, to which a few milliliters of gel buffer – diluted 1 : 4 with water – are added.

Several weeks refrigerated; at room temperature maximum 10 days shelflife.

When polyacrylamide gels contain a buffer with a pH value above 7, the shelflives of these gels are limited to a couple of weeks, because they hydrolize after some time and loose their sieving properties. When gels are not needed every day, it is recommended to cast only one or two gels, when needed.

The comb is inserted after pouring the stacking gel solution.

For the preparation of one or two gels a casting stand is used with a rubber gasket bottom to seal the the gel cassette. The cassettes are held together with two clamps (Fig. 3).

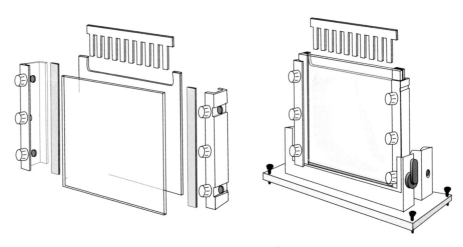

Fig. 3: Preparation of the gel casting stand for one or two gel cassettes.

Discontinuous SDS polyacrylamide gels

The difference of the pore sizes between stacking and resolving gel *See page 220 ff.*
and the buffer ions are sufficient for a good stacking effect.

The following recipe is an example for 0.75 mm thin 12 % T gels and
can easily be recalculated for other gel thicknesses and concentrations.

Tab. 1: Composition of the gel solutions for two discontinuous gels

	Resolving gel 12%T / 3%C	Stacking gel 5%T / 3%C
Glycerol (85 %)	2.0 mL	–
Acrylamide, Bis 40%T, 3%C	2.4 mL	0.5 mL
Resolving buffer	2.0 mL	–
Stacking buffer	–	1.0 mL
TEMED (100 %)	4 µL	2 µL
with H_2O_{dist} fill up	→ 8 mL	→ 4 mL
APS (40 %)	8 µL	4 µL

First pipette 3.4 mL resolving gel solution into the cassette. Then *Because of the difference in*
carefully apply 1.2 ml of stacking gel solution like an overlay. Insert *densities of the solutions they*
the comb without trapping air bubbles. *do not mix, a sharp interface*
is obtained.

Porosity gradient gels

The casting procedure is similar to the technique used for horizontal *See page 219 ff.*
gradient gels: no pump is employed.

The gradient is prepared with a *gradient maker* (s. Fig. 4). It is
made of two communicating cylinders.

The front cylinder, the mixing chamber, contains the denser solu- *In the gradient presented here,*
tion and a magnetic stirrer bar. The back cylinder, the *reservoir*, con- *the dense solution contains the*
tains the lighter solution. The dense solution contains about 25% gly- *higher proportion of acrylamide*
cerol and the light one 10%. The stacking gel solution – without *and gives the part of the gel*
glycerol – is overlaid over the gradient and copolymerized. *with the smaller pores.*

The *compensating* bar in the reservoir corrects for the difference in
density and for the volume of the magnetic stirrer.

For reproducible gradients the outlet of the gradient maker must
always be on the same level above the upper edge of the gel cassette.

The following recipe is an example for a gradient from 8% to 20 % T
in 0.75 mm thin gels and can easily be adjusted to other gel thicknesses and concentrations.

Tab. 2: Composition of the gel solutions for two gradient gels.

Pipette into 3 test tubes	Dense solution 20%T / 3%C	Light solution 8%T / 3%C	Stacking gel 5%T / 3%C
Glycerol (85 %)	1.0 mL	0.5 mL	–
Acrylamide, Bis 40%T, 3%C	2.0 mL	0.8 mL	0.5 mL
Resolving Buffer	1.0 mL	1.0 mL	–
Stacking Buffer	–	–	1.0 mL
TEMED (100 %)	2 μL	2 μL	2 μL
with H$_2$O$_{dist}$ fill up	→ 4 mL	→4 mL	→ 4 mL
APS (40 %)	4 μL	5 μL	4 μL

To pour a linear gradient (Fig. 4) both cylinders of the gradient maker are left open. The laboratory platform ("Laborboy") is set so that the outlet lies 5 cm above the upper edge of the gel. Before filling, the valve in the connecting channel between the *reservoir* and the *mixing chamber* as well as the *pinchcock* are shut. The stirring bar is then placed in the mixing chamber.

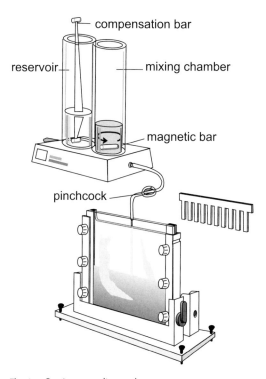

Fig. 4: Casting a gradient gel.

■ Note:

The dense solution must contain less APS, in order to start the polymerization from the upper edge.

Thermal convection can distort the gradient, when polymerization starts from the bottom.

- Pour 1.7 mL of the light solution into the reservoir,
- briefly open the valve to fill the connecting channel,
- pour 1.7 mL of the dense solution into the mixing chamber.,
- pipette APS into the reservoir, mix with compensating bar,
- pipette the APS into the mixing chamber and stir briefly but vigorously with the magnetic stirrer;, *to disperse the catalyst*
- place the tip of the tubing into the cassette,
- set the magnetic stirrer at moderate speed, *do not generate air bubbles*
- open the connecting valve,
- open the outlet valve (pinchcock);,
- when the gradient maker is empty, carefully apply 1.2 ml of stacking gel solution like an overlay,
- insert the comb without trapping air bubbles.

Rinse the gradient maker with distilled water immediately.

4
Multiple gel casting

The multiple casting stand is used with the silicon plugs inserted, thus there is almost no dead volume. First a plastic sheet is laid into the casting stand. The gel cassettes (see "2 Single gel casting") are placed into the casting stand – the ceramics plates to the back – with the combs already inserted (s. Fig. 5). It is strongly recommended to

The plastic sheets are used for filling the stand completely, and they make removing of the cassettes easier. Parafilm® works much better than the wax paper.

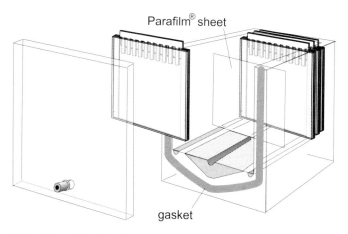

Fig. 5: Assembling the gel cassettes into a stand for multiple gel casting.

When 1 or 1.5 mm gels are cast, only 10 respectively 8 gels can be prepared in one stand.

lay sheets of Parafilm® between the gel cassettes for their easy separation after the polymerization. After 12 gel cassettes have been inserted, two plastic sheets are added. The gasket is coated with a thin film of CelloSeal®. The cover plate is clamped to the stand.

Note: *When the cassettes are packed in the correct way, the liquid flow can be observed through the front plate.*

The solutions flow into the casting stand from below with a 35 cm long tubing connected to a gradient maker, also when no gradient is prepared. The laboratory platform ("Laborboy") is set so that the outlet lies 30 cm above the inlet of the casting stand. Before filling, the valve in the connecting channel between the reservoir and the mixing chamber as well as the pinchcock are shut.

Multiple discontinuous SDS polyacrylamide gels

See page 39 ff.

The difference of the pore sizes between stacking and resolving gel and the buffer ions are sufficient for a good stacking effect. Because of the difference in densities of the solutions they do not mix during casting; a sharp interface is obtained.

The following recipe is an example for twelve 0.75 mm thin 12 % T gels and can easily be recalculated for other gel thicknesses and concentrations.

Tab. 3: Composition of the gel solutions for twelve discontinuous gels

	Resolving gel 12%T / 3%C	Stacking gel 5%T / 3%C
Glycerol (85 %)	15 mL	–
Acrylamide, Bis 40%T, 3%C	18 mL	2.5 mL
Resolving buffer	15 mL	–
Stacking buffer	–	5.0 mL
TEMED (100 %)	30 µL	10 µL
with H_2O_{dist} fill up	→ 60 mL	→ 20 mL
APS (40 %)	60 µL	20 µL

- Pour 20 mL stacking gel solution into the cylinder of the gradient maker and let them flow into the stand.
- When the first air bubble leaves the outlet, immediately close the pinchcock.

The liquid flows slowly because of the glycerol content.

- Fill 40 mL of the resolving gel solution into the cylinder and open the pinchcock again.
- When ca. 1/2 half of the liquid has flown out, pour the rest into the cylinder.
- When the liquid level – of the stacking gel solution – has reached the edge of the ceramics plate, close the pinchcock.
- Empty the gradient maker by pouring the mixing chamber out into a beaker.

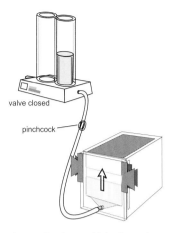

valve closed

pinchcock

Fig. 6: Casting multiple discontinuous gels.

- Disconnect the tube from the outlet, connect it to a 1000 µL micropipette, open the pinchcock, press 1 mL of "Push-up solution" into the tube, close the pinchcock.

This measure keeps the tube clear of polymerization solution.

Rinse the gradient maker with distilled water immediately.

Do not forget to remove the gel cassettes from the casting stand after 1 hour of polymerization, and place them – with a few mL of gel buffer – into sealed plastic bags.

Otherwise the stacking gel starts to dry.

Multiple SDS polyacrylamide gradient gels

To pour a linear gradient both cylinders of the gradient maker are left open. As the solutions flow into the cassettes from below, the gradient maker is used with the light solution in the mixing chamber and the dense solution in the reservoir (see Fig. 7).

The compensation bar is not placed into the reservoir, because the stirrer bar compensates for the difference in densities.

■ Note:
The dense solution must contain less APS, in order to start the polymerization from the upper edge.

Thermal convection can distort the gradient, when polymerization starts from the bottom.

Tab. 4: Composition of the gel solutions for twelve gradient gels

Pipette into 3 test tubes	Dense solution 20%T / 3%C	Light solution 8%T / 3%C	Stacking gel 5%T / 3%C
Glycerol (85 %)	6.25 mL	3.0 mL	–
Acrylamide, Bis 40%T, 3%C	12.5 mL	5.0 mL	2.5 mL
Resolving Buffer	6.25 mL	6.25 mL	–
Stacking Buffer	–	–	5.0 mL
TEMED (100 %)	12.5 µL	12.5 µL	10 µL
with H_2O_{dist} fill up	→ 25 mL	→25 mL	→ 20 mL
APS (40 %)	20 µL	25 µL	20 µL

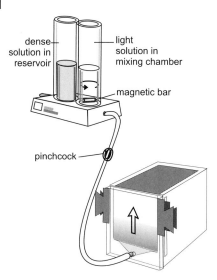

dense— solution in reservoir

light solution in mixing chamber

magnetic bar

pinchcock

Fig. 7: Casting multiple gradient gels.

The recipe is an example for twelve 0.75 mm thin gels with a gradient from 8 – 20 % *T* (see Tab. 4); the solutions can easily be recalculated for other gel thicknesses and concentrations.

- Pour 20 mL stacking gel solution into the mixing chamber of the gradient maker and let them flow into the stand,
- when the first air bubble leaves the outlet, immediately close the pinchcock,
- place the stirring bar into the mixing chamber,
- pour the 25 mL dense solution into the reservoir,
- briefly open the valve to fill the connecting channel,
- pour the 25 mL light solution into the mixing chamber.,

remove the compensating bar before casting

- pipette the APS into the reservoir and mix with the compensating bar,

to disperse the catalyst

- pipette the APS into the mixing chamber and stir briefly but vigorously with the magnetic stirrer,

do not generate air bubbles

- set the magnetic stirrer at moderate speed,
- open the connecting valve,
- open the outlet valve (pinchcock),

This measure keeps the tube clear of polymerization solution.

- When the gradient maker is empty, disconnect the tube from the outlet, connect it to a 1000 µL micropipette filled with 1 mL of "Push-up solution", open the pinchcock, press the solution into the tube, and close the pinchcock again.

Rinse the gradient maker with distilled water immediately afterwards.

Do not forget to remove the gel cassettes from the casting stand after 1 hour of polymerization, and place them – with a few mL of gel buffer – into sealed plastic bags.

5
Electrophoresis

- Clamp the gel cassettes to the core with the notched ceramics plates facing to the center, the long side of the clamps on theb glass plate.

When a chamber for two gels is used and only one gel is run, clamp a glass plate or a ceramics plate to the opposite side of the core, to prevent a short circuit between the electrodes in the anodal buffer.

- Remove the combs.
- Mix 15 mL of cathode buffer (10×conc) with 135 mL deionized water and pour 75 mL into each cathodal compartment.
- Mix 15 mL of anode buffer (10×conc) with 135 mL deionized water and pour 150 mL into the anodal compartment.
- Load the samples using a pipette with standard tips: the tip is set on the edge of the ceramics plate and the sample is pushed slowly into the well (see Fig. 8).
- Place the safety lid on and connect to the power supply.

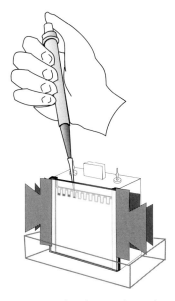

Fig. 8:   Loading the samples on the vertical gel.

Running conditions:

When the chamber is run in a cold room, cooling is much less effective.

It is recommended to cool the gels with a thermostatic circulator: quicker separations are obtained, because higher field strength can be applied, and the results are better reproducible.

Direct cooling at 10 °C:

Two gels 0.75 mm:	280 V	65 mA	18 W	1 h
One gel 0.75 mm:	280 V	33 mA	9 W	1 h

No direct cooling:

Two gels 0.75 mm:	280 V	40 mA	12 W	1 h 30 min
One gel 0.75 mm:	280 V	20 mA	6 W	1 h 30 min

After the run:

- Switch off the power supply.

Open the clampscarefully and slowly remove the cassettes from the core.

- Pour the anodal buffer out of the anodal compartment before opening the clamps, in order to avoid spilling of the buffer.

A knife or a spatula damage the glass and ceramics plates.

- Open the cassettes only with a spacer or a plastic wedge.

- Clean and dry the glass and ceramics plates carefully.

6
SDS electrophoresis of small peptides

As already mentioned in part I and in method 7, standard SDS PAGE shows a poor resolution of small peptides.

The discontinuous gel and buffer system acc. to Schägger and Von Jagow (1987) provides a very good resolution of small peptides from 1 to 20 kDa. By using, a high gel buffer concentration (1 mol/L Tris), high crosslinking (6% C), pH 8.45 in both stacking and resolving gel, and replacing glycine in the cathodal buffer by tricine, an improved destacking of the small peptides is achieved.

With the ethylenglycol in the gel, the buffer concentration can be reduced to 0.75 mol/L, which results in a quicker separation.

In the following part the recipe for buffers and the gels for a mini vertical system is described, which is derived from the method by Schägger and Von Jagow and the PhastGel® High Density; in the latter 30 % ethylenglycol is added to the polymerization solution.

Stock Solutions:

Note, that for the relatively low volume of anodal buffer a high Tris concentration is needed.

Anodal buffer 10 × conc (2 mol/L Tris, HCl pH 8.9, 1 % SDS):
48.4 g Tris + 2.0 g SDS; make up to 160 mL with H_2O_{dist}; titrate to pH 8.9 with 4 mol/L HCl; to 200 mL with H_2O_{dist}

Cathodal buffer 10 × conc (0.2 mol/L Tris, 1.6 mol/L Tricine, 1% SDS):
4.84 g Tris + 56 g tricine + 2 g SDS; to 200 mL with H_2O_{dist}

Gel buffer 4 × conc (3.0 mol/L Tris / HCL pH 8.45):
36.3 g Tris + 0.4 g SDS; make up to 80 mL with H_2O_{dist}; titrate to pH 8.45 with 4 mol/L HCl; to 100 mL with H_2O_{dist}

Monomer solution (40 % T / 6 % C) for resolving gel:
37.6 g acrylamide + 2.4 g Bis; to 100 mL with H_2O_{dist}

Store up to 3 months at room temperature in the dark.

Tab. 5: Gel solutions for two gels for the analysis of small peptides

	Resolving gel 16%T / 6%C	Stacking gel 5%T / 3%C
Ethylenglycol	2.8 mL	–
Acrylamide, Bis 40%T, 6%C	3.2 mL	–
Acrylamide, Bis 40%T, 3%C	–	0.5 mL
Gel buffer	2.0 mL	1.0 mL
TEMED (100 %)	4 µL	2 µL
with H_2O_{dist} fill up	→ 8 mL	→ 4 mL
APS (40 %)	8 µL	4 µL

This is an example for 0.75 mm thin gels and can easily be recalculated for other gel thicknesses and sizes.

- First pipette 3.4 mL resolving gel solution into the cassette. Then carefully apply 1.2 ml of stacking gel solution like an overlay. Insert the comb without trapping air bubbles.

Because of the difference in densities of the solutions they do not mix, a sharp interface is obtained.

Running condition (10 °C):
Two gels 0.75 mm: 200 V   70 mA   18 W   2 h 15 min

Fig. 9 shows a typical separation result in a vertical peptide gel.

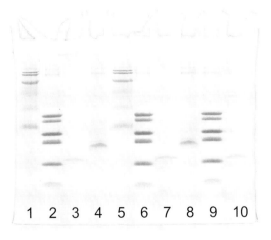

1 2 3 4 5 6 7 8 9 10

Fig. 9: SDS PAGE of small peptides. Samples: lanes 1,5 low molecular weight markers; lanes 2, 6, 9 peptide markers 2.5 – 17 kDa; lanes 3, 7, 10 insulin; lanes 4, 8 aprotinin. Coomassie Brilliant blue staining.

7
Two-dimensional electrophoresis

see page 275 ff.

No stacking gel is required.

In special cases a peptide gel might be useful.

The procedures of 2-D electrophoresis are thoroughly described in Method 11.

When the second dimension is run on a vertical gel, it is suggested to prepare either a homogeneous gel with 12.5 %T or a gradient gel with 12 – 14 %T. Although larger gels are recommended for 2D electrophoresis, here are a few hints for minigels:

At least 1 mm thick gel must be used, in order to accomodate the IPG strips. The gel must be cast up to 0.5 cm below the edge of the ceramics plate. For a single gradient gel: pipet 3.3 mL of each solution into the cylinders of the gradient maker. For multiple gradient gels: pipet 35 mL of each solution into the cylinders of the gradient maker This space is needed for embedding the IPG strip.

For a straight and well polymerized gel edge overlay each cassette with 300 µL of 60 % isopropanol.

Westermeier R, Naven T. Proteomics in Practice. A laboratory manual of proteome analysis. WILEY-VCH, Weinheim (2002).

The casting of large vertical gels for 2-D electrophoresis is thoroughly described in "Proteomics in Practice" (Westermeier and Naven, 2002).

8
DNA electrophoresis

For the separation of DNA fragments homogeneous polyacrylamide gels with 6 or up to 15 % T, 5 % C and mostly the continuous 90 mmol/L Tris-Borate-EDTA buffer are employed.

Tris-Borate-EDTA stock solution (4 × conc):
0.36 mol/L Tris / 0.36 mol/L boric acid / 10 mmol/L EDTA-Na$_2$
(4 × TBE)
43.6 g Tris + 22.25 g boric acid + 3.72 g EDTA-Na$_2$;
make up to 1L with distilled water.

For casting the gels the set-up described above is used, just without stacking gels. Because of the homogeneous buffer system, the separation is run at a constant voltage of 100 V.

These gels can be stained with silver staining or Ethidium bromide.

9
Long shelflife gels

When a long shelflife is required, e.g. for multicast gels, the Tris-acetate / Tris-tricine buffer system can be employed:

Fig. 1 on page 232 shows a separation, which has been performed with this buffer system.

Gel buffer 0.448 mol/L Tris-acetate pH 6.4 (4 × conc):
5.43 g Tris + 0.4 g SDS, dissolve in 80 mL H_2O_{dist}; titrate to pH 6.4 with acetic acid; make up to 100 mL with H_2O_{dist}.

Anode buffer (10 × conc):
7.58 g Tris + 2.5 g SDS, dissolve in 200 mL H_2O_{dist}; titrate to pH 8.4 with acetic acid; make up to 250 mL with H_2O_{dist}.

With this buffer, the storage time of the gel is not limited. Because the pH value of the gel is lower than pH 7, the matrix does not hydrolyse.

Cathode buffer (10 × conc):
4.84 g Tris + 71.7 g tricine + 2.5 g SDS, make up to 250 mL with H_2O_{dist}.

10
Detection of bands

When 0.75 mm thin gels are used, the procedures described in methods 7 and 11 for proteins and method 12 for DNA can be applied without any modifications.

See Fig. 1

The SDS protein gels can be stained as described in Methods 7 and 11.

For some *low molecular weight* peptides it is necessary to prevent diffusion out of or within the gel by an efficient fixing step prior to staining: 60 min in 0.2 % (w/v) glutardialdehyde + 0.2 mol/L sodium acetate in 30 % ethanol.

With this procedure the peptides are crosslinked in the polyacrylamide gel.

For *preservation* the gels are dried either on a filter paper with a gel dryer, which employes heat and vacuum, or between two sheets of wet cellophane, which are clamped in between two plastic frames (see page 298).

Method 9:
Semi-dry blotting of proteins

The principles as well as a few possibilities of application are described in part I [1]). This chapter concerns the transfer technique as well as selected staining methods for blotting from SDS gradient pore gels according to method 7 [2]) and 8 [3]), and from IEF gels according to method 6 [4]). For the electrotransfer, the cooling plate is taken out of the chamber and the graphite plates are used instead (Fig. 1).

[1] see pages 67 and following
[2] see pages 211 and following
[3] see pages 231 and following
[4] see pages 197 and following

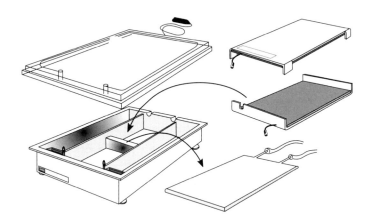

Fig. 1: Installing the graphite anode in place of the cooling plate.

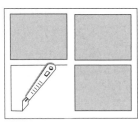

Fig. 2: Mask

A few practical tips will be given first:

- The filter paper and the blotting membrane must be cut to the size of the gel so that the current does not flow around the sides of the actual blot sandwich. Small blot sandwiches can be placed beside one another.
- The filter paper does not have to be cut if a plastic frame is made in which openings for the gels and blotting membrane are cut out. This frame is placed on the stack of anodic filter paper before the blot is built up further (Fig. 2).

The largest gel size is 20 × 27 cm.

When a gel of the size described in method 7 should be blotted, it is easier to cut the gel into two halves before blotting.

Electrophoresis in Practice, Fourth Edition. Reiner Westermeier
Copyright © 2005 WILEY-VCH Verlag GmbH & Co. KGaA, Weinheim
ISBN: 3-527-31181-5

Unfolded SDS-polypeptide micelles behave differently from focused proteins in globular form.

- Only gels of the same kind (IEF, SDS) can be blotted together, because the transfer time depends on the pore size and the state of the proteins.

If the biological activity of enzymes must be preserved, the methanol should be omitted.

- The transfer buffer contains 20% methanol, so that the gels do not swell during the transfer and the binding capacity of the nitrocellulose is increased.

A thin wire is pulled between the gel and support film.

- When gels are bound to a plastic film, the gel and the support film should first be separated with the Film Remover.

Not recommeded, because when several transfer units are blotted, transfer efficiency is lost in the direction of the cathode.

- When several gel-blot layers are blotted over one another, a dialysis membrane (Cellophane) is placed between each layer so that proteins which might migrate through, do not reach the next transfer unit.

This method is the easiest. But the transfer is not as regular, and the bands not as sharp as in discontinuous buffer systems.

- For continuous buffer systems, the same buffer with a pH of 9.5 is used for both the anodic and cathodic sides of the blots. To improve the transfer of hydrophobic proteins and to charge focused proteins (proteins are not charged at their pI), the buffer also contains SDS.

This method is especially valuable for native blotting. The small amounts of SDS (0.01%) in the cathode buffer do not denature the proteins during the short contact time.

In discontinuous buffer systems the speed of migration of the proteins changes during the transfer because of the different ionic strengths of both the anode buffers (0.3 mol/L and 0.025 mol/L). This means that fewer proteins are transferred. The slow terminating ion in the cathode buffer compensates for the differences in speed of migration of the leading ions (here the proteins): a more regular transfer is obtained. In this case, SDS is only added in the cathode buffer.

1
Transfer buffers

Continuous buffer system:

39 mmol/L glycine	2.93 g,
48 mmol/L Tris	5.81 g,
0.0375% (w/v) SDS	0.375 g,
20% (v/v) methanol	200 mL,

make up to 1 L with distilled water.

Discontinuous buffer system
Anode solution I:

acc. to Kyhse-Andersen (1984)

0.3 mol/L Tris   36.3 g,
20% (v/v) methanol   200 mL,
make up to 1 L with distilled water.

Anode solution II:

25 mmol/L Tris   3.03 g,
20 %(v/v) methanol   200 mL,
make up to 1 L with distilled water.

Cathode solution

40 mmol/L 6-aminohexanoic acid   5.2 g,
0.01% (w/v) SDS   0.1 g,
20% (v/v) methanol   200 mL,
make up to 1L with distilled water.

The same as ε-aminocaproic acid.

Transfers from agarose gels
Use the discontinuous buffer system *without* methanol.

2
Technical procedure

The following explanations refer to the discontinuous buffer system. If a continuous buffer system is used, the anode solutions I and II and the cathode buffer are identical.

    To avoid contaminating the buffer, blotting membranes and filter papers, rubber gloves should be worn to carry out the following operations.

- Wet the graphite anode plate (with the red cable) with distilled water, remove the excess water with absorbent paper.
- Cut the necessary filter papers (6 for the anode, 6 for the cathode, 6 per transfer unit) and the blotting membrane to the size of the gel.
- Pour 200 mL of anode solution I into the staining tray;
- slowly impregnate 6 filter papers and place them on the graphite plate (anode) (Fig. 3);
- Pour the anode solution I back into the bottle;
- Pour 200 mL of the anode solution II into the staining tray;
- Stop the electrophoresis or the isoelectric focusing run;

This enables a regular current to be obtained.

Or use a plastic frame as suggested before.

Avoid air bubbles.

The buffer can be used several times.

See methods 7 and 8

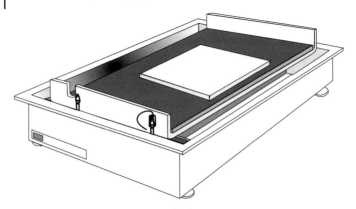

Fig. 3: Assembling the blotting sandwich on the graphite anode plate.

SDS gels are soaked in methanol to prevent them from swelling.

- wipe the kerosene off the bottom of the film;

It is easier to remove IEF gels from the blotting membrane after blotting.

- equilibrate the gels in the anode buffer I:
  SDS-PAGE: 5 min
  IEF: 2 min

Gels on support films are left to "swim" on the top of the buffer with the gel surface on the bottom.

With the refrigeration tubing attached.

- remove the cooling plate from the electrophoresis chamber;
- place the anode plate in the electrophoresis chamber, plug in the cables.

Folds often form when a large support film is removed and it is difficult to get rid of them.

■ Tip!
The handling of a whole gel (25 × 10 cm) during removal is much easier if the gel is cut in two after equilibration, and both halves are put together again after pulling the FilmRemover wire through. A single blotting membrane can nevertheless be used for the whole gel.

The holding mechanism is now in the elevated position.

- Place gels bound to support films with the film side down on the Film Remover; so that a short side is in contact with both the gel clamps;

Both teeth are now pressing on the edge of the film and hold it in place.

- Press on the lever;

The wire now has the mechanical tension necessary to separate the gel completely from the support film.

- Place the wire over the edge of the gel beside the clamps, hook it to the handle on the other side and push the lever down (Fig. 4);

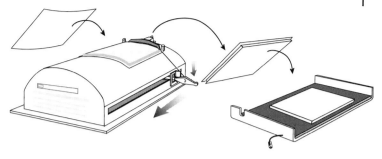

Fig. 4: Removing the support film and transferring the support film-gel-blotting membrane sandwich to the stack of anode paper.

- grasp the handles with both hands and smoothly pull the wire towards you.

While doing this, press the lever down with a finger from the right hand so that it does not spring up.

Film bound gels:

- Briefly soak the blotting membrane in the anode buffer II and place it on the gel;
- slowly soak 3 filter papers in the buffer and place them on the prepared stack of filter papers;
- press on the top handle and lift the whole sandwich with the support film, turn in over and place it on the stack of filter papers.

Gels without support film:

For gels which are not bound to support film, proceed in the opposite way:

The equilibration step is usually not carried out for gels which are not bound to a support film.

- slowly soak 3 filter papers in buffer and place them on the prepared stack of filter papers;
- briefly soak the blotting membrane in the anode buffer II and place it on the stack of filter papers;
- place the gel on the blotting membrane;
- pour off the anode buffer II;

This buffer should only be used once (kerosene).

- rinse out the staining tray with distilled water and dry it with paper;
- pour 200 mL of cathode solution in the staining tray;
- pull the film away slowly, starting at one corner;
- soak 9 filter papers in the cathode buffer and place them on top.

Reusable cathode buffer should not be contaminated with traces of anode buffer.

*See the **tip** above.*

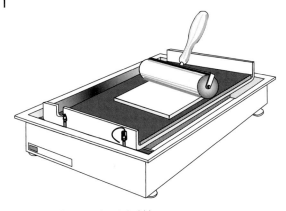

Fig. 5: Rolling out the air bubbles.

When building a blot sandwich as described here, it is difficult to completely avoid air bubbles. They must therefore be pressed out with a roller (Fig. 5):

If not enough pressure is applied, air bubbles will stay in the sandwich and no transfer will take place at these points. On the other hand if too much pressure is applied, the sandwich will be too dry and irregular.

- Start in the centre and roll out in all four directions,

press in such a way that the buffer in the sandwich oozes out but is not completely pressed out. When the roller is removed, the buffer should be "drawn" back in;

See anode plate.

- wet the graphite cathode (black cable) with distilled water; remove the excess water with a paper towell;
- place the cathode plate on the stack, plug in the cable;

Make sure that the polarity is correct.

Higher currents result in a rise of the temperature of the gel and are not recommended.

- place the safety lid on the electrophoresis chamber and connect the cable to the power supply;
- blot at a constant current: 0.8 mA/cm^2

Transfer conditions:
0.5 mm thick gel (250 cm2) SDS pore gradient $T = 8$ to 20%:

	I	U	P	Room temperature
Set:	200 mA	10 V	5 W	20 °C
Read:	200 mA	3 V	1 W	

The blot does not warm up under these moderate conditions.

If thicker or more concentrated gels are blotted, the blotting time *These gels have enough* can be increased up to two hours; it is then recommended to press *mechanical stability that they* down the cathode plate with a 1 to 2 kg weight so that no electrolytic *are not crushed.* gas pockets form.

Blotting native or focusing gels is quicker, because the proteins are *guideline: 30 min* in globular form and, for IEF, the gels have larger pores.

- Switch off the power supply, unplug the cable;
- remove the safety lid and the cathode plate;
- take the blot sandwich apart;
- stain the gel with Coomassie Blue to check. *see methods 7 and 8.*
- Before visualization, either dry the gel overnight or for 3 to 4 h *The proteins bind more firmly* at 60 °C in a heating cabinet. *to the film during drying and are not washed out during*

■ Caution: *staining and specific detection.*
This treatment should not be used before detection of biological activity (zymogram techniques) since most enzymes lose their activity.

3
Staining of blotting membranes

Amido Black staining:
Dissolve 10.1 g of Amido Black in 100 mL of methanol-glacial acetic acid-water (40:10:50 v/v);

- stain for 3 to 4 min in 0.1% Amido Black solution;
- destain in methanol-acetic acid-water (25:10:65 v/v);
- air dry the nitrocellulose.

Reversible staining:
If a specific immuno- or glycoprotein identification test is to be carried out after the general detection, it is recommended that staining with Ponceau S (Salinovich and Montelaro, 1986) or Fast Green FCF is used.

Fast Green staining

- dissolve 0.1% (w/v) of Fast Green in 1% acetic acid;

Mild staining for proteins!

- stain for 5 min;
- destain the background with distilled water for 5 min; *This only works for nitrocellu-*
- complete destaining of the bands is achieved by incubating *lose, in addition alkaline treat-* the film for 5 min in 0.2 mol/L NaOH. *ment increases antigen* Blocking and immunological or lectin detection can now be *reactivity (See page 75: Suther-* carried out. *land and Skerritt, 1986).*

Also a mild staining!

For the sensitivity see Fig. 6 as a comparison with a gel stained with Coomassie Blue. It comes close to the sensitivity of silver staining of gels.

Indian Ink staining (Hancock and Tsang, 1983)

Unfortunately the original Indian Ink ("Fount India") is no longer produced. However, according to Dr. Christian Schmidt, University of Rostock, standard script-ink can be used alternatively, diluted with TBS-Tween instead of PBS-Tween.

- *Soaking:* 5 min in 0.2 mol/L NaOH;
- *Washing:* 4 × 10 min with PBS-Tween (or TBS-Tween)
- (250 mL per wash, agitate).

PBS-Tween: 48.8 g of NaCl + 14.5 g Na_2HPO_4 + 1.17 g NaH_2PO_4 + 2.5 mL Tween 20, fill up to 5 L with distilled water.
TBS-Tween: 2.92 g NaCl + 6.06 g Tris + 0.5 mL Tween 20, fill up to 5 L with distilled water.

Rocking platform! Filter the staining solution first.

- *Staining:* 2 h or overnight with 250 mL of PBS Tween (or TBS Tween) + 2.5 mL of acetic acid + 250 μL of fountain pen ink; scan the image.

Continue with blocking and immunodetection.

- *Remove dye:* 3 × 10 wash with PBS (TBS) Tween.
- *To conserve the image:* Washing: 2 × 2 min with water.
- *Drying:* air-dry.

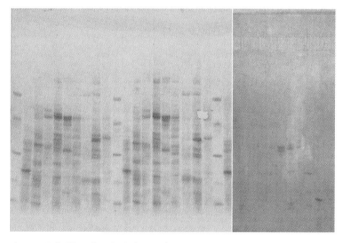

Fig. 6: *Left:* Blot of an SDS electrophoretic separation of legume seed extracts on nitrocellulose; Indian ink stain. *Right:* Coomassie stained gel containing the same protein pattern in the same concentration.

Plastic embedding of the blotting membrane

Pharmacia LKB Development Technique File No 230, Phast-System® (1989).

Nitrocellulose can be made totally and permanently transparent after staining or specific immunological detection methods, so that the result looks like a stained electrophoresis gel.

The membrane is soaked in a monomer solution with the same refractive index as polymerized nitrocellulose. For permanent transparency a photo-initiator, which starts polymerizing in the presence of a UV light source, is added to the monomer solution.

The monomer chosen for this purpose is odorless and relatively safe.

Monomer solution: dissolve 0.5 g of benzoin methylether in 25 mL of TMPTMA. Store at room temperature in a brown flask or in the dark. The solution can be kept for about 2 weeks. For longer storage, keep it in the refrigerator at 4 °C to 8 °C.

Complete dissolution may take several hours. Avoid skin contact: TMPTMA is an irritant for the skin and mucous membranes.

- Dry the nitrocellulose completely;
- cut 2 sheets of PVC to a size larger than that of the blotting membrane;
- pipette 0.3 mL of monomer solution on one of the plastic sheets;
- spread the solution so that it covers an area equal to the size of the membrane;
- stand the blotting membrane on its edge and slowly lower it on to the monomer solution so that the membrane is completely impregnated;
- apply a few drops of solution on to the membrane and place the second plastic sheet on top;
- carefully press out the air bubbles with a roller;
- irradiate both sides for about 15 s with a UV lamp;
- cut the sandwich to the size desired.

Sheets for the overhead projector.

Or more, depending on the size of the blotting membrane.

Method 10:
IEF in immobilized pH gradients

The principles of isoelectric focusing in immobilized pH gradients are described in section I. This technique possesses a number of characteristics, which make it clearly different from IEF in carrier ampholyte gradients:

See pages 59 and following as well as "Strategies for IPG focusing" at the end of this chapter.

Advantages:

- The separations are more reproducible. The pH gradient is absolutely linear and remains so, it is not influenced either by salt or buffer ions or by the proteins of the sample.

  There are no wavy lines, only absolutely straight iso-pH lines.

- The pH gradient is absolutely stable with time, it cannot drift.

  Field strength and time are not limited.

- Basic proteins are better focused in immobilized pH gradients because the gradient remains stable.

  Most problems of carrier ampholyte IEF occur in this range.

- Immobilized pH gradients are more reproducible, because they are produced by at most six different substances which are chemically well defined.

  In contrast to several 100 carrier ampholyte homologues.

- Since immobilized pH gradients can be custom-made and with very narrow pH ranges, very high resolution can be achieved.

  $< \Delta pI = 0.001 pH$

- Since only tertiary amines are used as buffering groups and no carrier ampholytes are present, low molecular weight proteins can be detected directly in the gel with ninhydrin or dansyl chloride.

  Carrier ampholytes are also stained by these substances.

- Immobilized pH gradients are a part of the gel; therefore there are no edge effects and the separations can be carried out in narrow, individual gel strips.

  This is especially practical for the first dimension of two-dimensional electrophoresis.

Disadvantages:

- The gel casting technique for IPG gels is more complicated.

  Pipetting the Immobiline, pouring the gradients.

- Errors can occur during the preparation of the gels.

  E.g. errors during pipetting.

Electrophoresis in Practice, Fourth Edition. Reiner Westermeier
Copyright © 2005 WILEY-VCH Verlag GmbH & Co. KGaA, Weinheim
ISBN: 3-527-31181-5

No agarose gels	• IPGs can only be made with polyacrylamide.
Because of the lower conductivity.	• The separation takes more time.
Also because of the lower conductivity.	• High voltages are necessary.
Tricks are sometimes necessary here.	• Some proteins do not enter the gel easily.

1
Sample preparation

The best results are obtained when concentrated samples are diluted 1:3 or more before application. The loading capacity of Immobiline gels is higher than that of Ampholine gels. The limit is set by the concentration of the sample when the proteins penetrate the gel. When a large amount of proteins is to be applied, it is recommended to dilute the sample and apply it in several steps. In many cases it has proved useful to mix the sample with about 2% of Triton X-100 or ethylenglycol.

Vuillard L, Marret N, Rabilloud T. Electrophoresis 16 (1995) 295–297.

For membrane proteins and proteins from lyophilized platelets, a significant increase in the yield of protein extraction by the addition of nondetergent sulfobetaines has been reported by Vuillard *et al.* (1995).

Strahler JR, Hanash SM, Somerlot L, Weser J, Postel W, Görg A. Electrophoresis 8 (1987) 165–173.

When preparing cell lysates 8 mmol/L of PMSF (at 5×10^8 cell equivalents per mL of lysate solution) should be added as protease inhibitor (Strahler *et al.* 1987).

Since Immobiline gels are extensively washed during their preparation, there are no more toxic monomers and catalysts.

2
Stock solutions

Immobiline® II 0.2 molar stock solutions:

(a) = acid,
(b) = basic

pK 3.6 (a) 10 mL
pK 4.6 (a) 10 mL
pK 6.2 (b) 10 mL
pK 7.0 (b) 10 mL
pK 8.5 (b) 10 mL
pK 9.3 (b) 10 mL

The solutions are stabilized against autopolymerization and hydrolysis and stay stable for at least 12 months when stored in the refrigerator (4 to 8 °C).

Immobilines® II should *not* be frozen!

Acid Immobiline® II (a) dissolved in double-distilled water and stabilized with 5 ppm of polymerization inhibitor (hydrochinone monoethyl-ether), basic Immobiline® II (b) dissolved in n-propanol.

Acrylamide, Bis (T = 30%, C = 3%):

29.1 g of acrylamide + 0.9 g of Bis, make up to 100 mL with distilled water, or reconstitute PrePAG Mix (29.1:0.9) with 100 mL of distilled water.

Dispose of the remains ecologically: polymerize with an excess of APS.

> ■ Caution!
> **Acrylamide and Bis are toxic in the monomeric form. Avoid skin contact, do not pipette by mouth. The solution stays stable for a week when stored in the dark at 4 °C (refrigerator). It can still be used for several weeks for SDS gels.**

The quality of the solutions is not as important for SDS gels as for IEF gels.

Ammonium persulfate solution (APS) 40% (w/v):

dissolve 400 mg of APS in 1 mL of distilled water.

It remains stable for one week when kept in the refrigerator (4 °C).

4 mol/L HCl: 33.0 mL of HCl, make up to 100 mL with distilled water.

2 mmol/L acetic acid: 11.8 µL of acetic acid, make up to 100 mL with distilled water.

2 mmol/L Tris: 24.2 mg of Tris, make up to 100 mL with distilled water.

3
Immobiline recipes

To cast the pH gradient two solutions are prepared, an acid, dense (with glycerol) one and a basic, light one. The acid solution is the acid end point of the gel and the basic one, the basic end point. The standard gel thickness is 0.5 mm; 7.5 mL of each starting solution are necessary for a gel. The recipes in table 1 and 2 are calculated for a optimum ionic strength of about 5 mmol/L, the pH and pK values are given at 10 °C.

From the beginning, it was decided to choose the acid solution as the denser one for Immobiline gels. This means that the acid end of the gel is always at the bottom.

Custom-made pH gradients

If custom-made pH gradients are needed to optimize the resolution (e.g. for preparative uses) the Immobiline volumes of both starting solutions can be calculated with a computer, using the software by Altland (1990) or by Giaffreda *et al.* (1993).

Sometimes more reliable than the computer software.

The recipes can also be deduced graphically (Fig. 1):

For example: desired: pH 5.3 to 5.6, chosen: pH 5.0 to 6.0
Tab. 1

- Choose the gradient which corresponds best to the desired gradient from Tab. 1 or Tab. 2.
- Report the volumes of the Immobiline acid and basic starting solutions (μL) on a scale of the chosen gradient on graph paper.

Linear gradient!

- Connect the quantities (μL) of the related Immobilines with lines.
- Mark the acid and basic end points of the desired gradient on the pH gradient scale and draw vertical lines from this point.

Sometimes several recipes are available for one gradient. Various formulations are then possible for a gradient, yet the resulting gradients will be identical.

- Read the corresponding Immobiline volumes at the intersection of the vertical lines and the joining lines. This is the recipe for the starting solutions of the desired gradient (Fig. 1).

The graphic interpolation is sometimes also used for empirical pH gradient optimization:

A sample is separated in a chosen Immobiline pH gradient. If a better resolution of a definite group of bands within the gradient is desired, draw the pH scale of the gradient used to the length of the separation distance on graph paper. Place the stained gel on the draw-

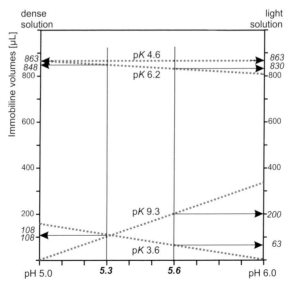

Resulting Immobiline volumes for pH 5.3 - 5.6 [μL]:

pH	5.3	5.6
pK 3.6	108	63
pK 4.6	863	863
pK 6.2	848	830
pK 9.3	108	200

Fig. 1: Graphic interpolation of a custom-made immobilized pH gradient pH 5.3 – 5.6.

Tab. 1: Narrow pH gradients: Immobiline quantities for 15 mL of the starting solutions (2 gels) 0.2 mol/L Immobiline in µL.

Acidic, dense solutions (D.S.) Immobiline pK						pH gradient	Basic, light solution (L.S.) Immobiline pK					
3.6	4.6	6.2	7.0	8.5	9.3		3.6	4.6	6.2	7.0	8.5	9.3
—	904	—	—	—	129	3.8— 4.8	—	686	—	—	—	477
—	817	—	—	—	141	3.9— 4.9	—	707	—	—	—	525
—	755	—	—	—	157	4.0— 5.0	—	745	—	—	—	584
—	713	—	—	—	177	4.1— 5.1	—	803	—	—	—	659
—	689	—	—	—	203	4.2— 5.2	—	884	—	—	—	753
—	682	—	—	—	235	4.3— 5.3	—	992	—	—	—	871
—	691	—	—	—	275	4.4— 5.4	—	1133	—	—	—	1021
—	716	—	—	—	325	4.5— 5.5	—	1314	—	—	—	1208
562	600	863	—	—	—	4.6— 5.6	—	863	863	—	—	105
458	675	863	—	—	—	4.7— 5.7	—	863	863	—	—	150
352	750	863	—	—	—	4.8— 5.8	—	863	863	—	—	202
218	863	863	—	—	—	4.9— 5.9	—	863	863	—	—	248
158	863	863	—	—	—	5.0— 6.0	—	863	803	—	—	338
113	863	863	—	—	—	5.1— 6.1	—	863	713	—	—	443
1251	—	1355	—	—	—	5.2— 6.2	337	—	724	—	—	—
1055	—	1165	—	—	—	5.3— 6.3	284	—	694	—	—	—
899	—	1017	—	—	—	5.4— 6.4	242	—	682	—	—	—
775	—	903	—	—	—	5.5— 6.5	209	—	686	—	—	—
676	—	817	—	—	—	5.6— 6.6	182	—	707	—	—	—
598	—	755	—	—	—	5.7— 6.7	161	—	745	—	—	—
536	—	713	—	—	—	5.8— 6.8	144	—	803	—	—	—
486	—	689	—	—	—	5.9— 6.9	131	—	884	—	—	—
447	—	682	—	—	—	6.0— 7.0	120	—	992	—	—	—
416	—	691	—	—	—	6.1— 7.1	112	—	1133	—	—	—
972	—	—	1086	—	—	6.2— 7.2	262	—	686	—	—	—
833	—	—	956	—	—	6.3— 7.3	224	—	682	—	—	—
722	—	—	857	—	—	6.4— 7.4	195	—	694	—	—	—
635	—	—	783	—	—	6.5— 7.5	171	—	724	—	—	—
565	—	—	732	—	—	6.6— 7.6	152	—	771	—	—	—
509	—	—	699	—	—	6.7— 7.7	137	—	840	—	—	—
465	—	—	683	—	—	6.8— 7.8	125	—	934	—	—	—
430	—	—	684	—	—	6.9— 7.9	116	—	1058	—	—	—
403	—	—	701	—	—	7.0— 8.0	108	—	1217	—	—	—
381	—	—	736	—	—	7.1— 8.1	103	—	1422	—	—	—
1028	—	—	750	750	—	7.2— 8.2	548	—	—	750	750	—
983	—	—	750	750	—	7.3— 8.3	503	—	—	750	750	—
938	—	—	750	750	—	7.4— 8.4	458	—	—	750	750	—
1230	—	—	—	1334	—	7.5— 8.5	331	—	—	—	720	—
1037	—	—	—	1149	—	7.6— 8.6	279	—	—	—	692	—
885	—	—	—	1004	—	7.7— 8.7	238	—	—	—	682	—
764	—	—	—	893	—	7.8— 8.8	206	—	—	—	687	—
667	—	—	—	810	—	7.9— 8.9	180	—	—	—	710	—
591	—	—	—	750	—	8.0— 9.0	159	—	—	—	750	—
530	—	—	—	710	—	8.1— 9.1	143	—	—	—	810	—
482	—	—	—	687	—	8.2— 9.2	130	—	—	—	893	—
443	—	—	—	682	—	8.3— 9.3	119	—	—	—	1004	—
413	—	—	—	692	—	8.4— 9.4	111	—	—	—	1149	—
389	—	—	—	720	—	8.5— 9.5	105	—	—	—	1334	—
1208	—	—	—	—	1314	8.6— 9.6	325	—	—	—	—	716
1021	—	—	—	—	1133	8.7— 9.7	275	—	—	—	—	691
871	—	—	—	—	992	8.8— 9.8	235	—	—	—	—	682
753	—	—	—	—	884	8.9— 9.9	203	—	—	—	—	689
659	—	—	—	—	803	9.0—10.0	177	—	—	—	—	713
584	—	—	—	—	745	9.1—10.1	157	—	—	—	—	755
525	—	—	—	—	707	9.2—10.2	141	—	—	—	—	817
478	—	—	—	—	686	9.3—10.3	129	—	—	—	—	903
440	—	—	—	—	682	9.4—10.4	119	—	—	—	—	1017
410	—	—	—	—	694	9.5—10.5	111	—	—	—	—	1165

Tab. 2: Wide pH gradients: Immobiline quantities for 15 mL of the starting solutions (2 gels), 0.2 mol/L Immobiline in µL.

| Acidic, dense solution (D.S.) Immobiline pK | | | | | | pH gradient | Basic, light solution (l.S.) Immobiline pK | | | | | |
3.6	4.6	6.2	7.0	8.5	9.3		3.6	4.6	6.2	7.0	8.5	9.3
299	223	157	—	—	—	3.5— 5.0	212	310	465	—	—	—
569	99	439	—	—	—	4.0— 6.0	390	521	276	—	—	722
415	240	499	—	—	—	4.5— 6.5	—	570	244	235	—	297
69	428	414	—	—	—	5.0— 7.0	—	474	270	219	—	320
—	450	354	113	—	—	5.5— 7.5	347	—	236	287	284	—
435	—	323	208	44	—	6.0— 8.0	286	—	174	325	329	—
771	—	276	185	538	—	6.5— 8.5	192	—	153	278	362	—
1349	—	—	272	372	845	7.0— 9.0 ·	484	—	—	232	189	546
668	—	—	445	226	348	7.5— 9.5	207	—	—	925	139	346
399	—	—	364	355	94	8.0—10.0	91	—	—	329	366	289
578	110	450	—	—	—	4.0— 7.0	302	738	151	269	—	876
702	254	416	133	346	—	5.0— 8.0	175	123	131	345	346	—
779	—	402	93	364	80	6.0— 9.0	241	—	161	449	237	225
542	—	—	378	351	—	7.0—10.0	90	—	—	324	350	280
588	254	235	117	170	—	4.0— 8.0	—	554	360	142	334	288
830	582	218	138	795	122	5.0— 9.0	—	249	263	212	292	230
941	—	273	243	260	282	6.0—10.0	100	—	333	361	239	326
829	235	232	22	250	221	4.0— 9.0	147	424	360	296	71	663
563	463	298	273	227	127	5.0—10.0	21	59	34	420	310	273
1102	—	455	89	334	—	4.0—10.0	—	114	50	488	157	357

ing so that the trace and the scale are superimposed (do not forget the orientation of the gel: acid end – anode and basic end – cathode!). Draw the end points of the new gradient desired (to the top and the bottom of the group of bands) and continue as described above.

4
Preparing the casting cassette

The "spacer" is the glass plate with the 0.5 mm thick U-shaped silicone rubber gasket.

Before use, treat the inner side of the spacer with Repel Silane so that the gel can be removed after separation.

Making the surface hydrophobic. This treatment only needs to be carried out once, test with a drop of water.

Under the fume hood drop a few mL of Repel Silane on the surface of the plate and spread it over the whole surface with tissue (wear rubber gloves). When the solution has dried, wash the plate under running water and then rinse it with distilled water (to remove the chloride ions which result from the evaporation of the solvent).

It is a good idea to mark the **hydrophobic** *side of the film with a black marker: gradient pH from ... to ... acid (+) and basic (–) side.*

The gel is covalently polymerized on a GelBond PAG film. Place a clean glass plate on absorbent paper and wet it with a few mL of water. Place a GelBond PAG film with the untreated, hydrophobic side down on the glass plate and press it on with a roller (Fig. 2).

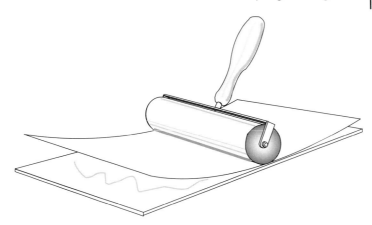

Fig. 2:   Applying the support film with a roller.

A thin layer of water then forms between the glass plate and the film and holds them together by adhesion. The excess water, which runs out is absorbed by the tissue.

To facilitate pouring the gel solution, let the film overlap the length of the glass plate by about 1 mm.

Place the spacer over the glass plate and clamp the cassette together (Fig. 3). Cool the cassette in the refrigerator at 4 °C for about 10 min before filling it, this delays the onset of polymerization. This last step is essential since it takes 5 to 10 min for the poured gradient in the 0.5 mm thick layers to settle horizontally.

In a warm laboratory in the summer, also cool down the starting solution to 4 °C.

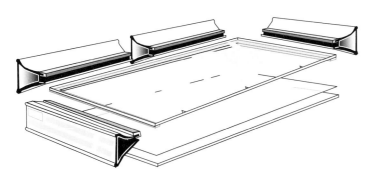

Fig. 3:   Assembling the gel cassette.

5
Preparing the pH gradient gels

Other gradients: use Tab. 1 or Tab. 2.

Tab. 3: Recipes for starting solutions for two 0.5 mm thick IPG gels pH 4 to 10, $T = 4\%$, $C = 3\%$

Solutions	pH 4.0	pH 10.0
Glycerol (85 %)	4.3 mL	0.8 mL
Immobiline pK 3.6	1,102 µL	–
Immobiline pK 4.6	–	114 µL
Immobiline pK 6.2	455 µL	50 µL
Immobiline pK 7.0	89 µL	488 µL
Immobiline pK 8.5	334 µL	157 µL
Immobiline pK 9.3	–	357 µL
Acrylamide, Bis 30 %T, 3%C	2 mL	2 mL
TEMED	7.5 µL	7.5 µL
with distilled water	→ 15 mL	→ 15 mL
mix carefully and measure the pH		
4 mol/L HCl titrate→ pH 7*)	15 µL	20 µL
TEMED (100 %) titrate→ pH 7*)		
APS (40 %)	(15 µL)	(15 µL)

experimental value
experimental value
add later: 7.5 µL each

NaOH can also be used instead of TEMED, but the use of TEMED is easier and not harmful. These substances used for the titration do not co-polymerize with the gel and are removed by washing.

As can be seen from the recipes, both starting solutions are brought to an identical optimum pH value of 7 with HCl or TEMED. This is especially necessary for broad and relatively high or low pH ranges. The effectiveness of the polymerization depends on the pH; this means that when pouring a pH gradient, different or extreme pH values will result in varying rates of polymerization of the Immobiline within the gel layer. The pH gradients occurring in the polymerized gels would then not correspond to the ones calculated.

The quantities of HCl and TEMED listed in Tab. 3 are only valid for the recipes in this example. For other gradients, pH 7 should be cautiously approached in 5 µL-steps. The precision of pH paper is sufficient here.

Only add APS in the mixing chamber so the gel has more time to settle horizontally.

Pouring the gradient

a) Setting up the casting apparatus

It is not recommended to use sucrose for the density gradient of thin gels because the solution would be too viscous.

A gradient mixer is used to prepare the gradient. It consists of two communicating chambers (Fig. 4).The dense, acid solution and a stirrer bar are placed in the front cylinder, the *mixing chamber*. The rear cylinder, the *reservoir*, contains the light basic solution. 25 % glycerol is added to the dense solution and 5 % to the light one so that it is easier to overlay the gel solution in the cassette with water before polymerization.

The gel solutions flow to the cassette through the tubing under the influence of gravity, no pump is necessary.

Position the bottom of the mixing chamber 5 cm above the top of the cassette with a laboratory platform to reach the optimum flow rate (cut the tubing to the correct length).

For reproducible gradients the outlet of the gradient mixer must always be on the same level above the edge of the gel cassette.

Because of the differences in the density of the solutions, the dense solution would flow back into the gradient mixer when the connecting tubing is opened, this can be prevented with the *compensating bar.*

It also compensates the volume of the stirring bar and helps to mix APS into the light solution.

Add the light and the dense solutions directly into the gradient mixer in the following order:

- pour 7.5 mL of light solution into the reservoir;
- briefly open the valve to fill the connecting channel;
- pour 7.5 mL of dense solution into the mixing chamber.

When everything is ready, remove the casting cassette from the refrigerator and place it close to the gradient mixer. The glass plate with the film should be facing the operator.

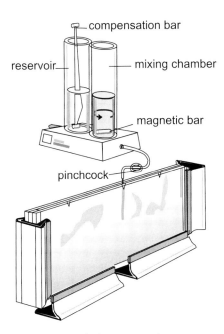

Fig. 4: Pouring the linear pH gradient.

b) Filling the cassette:

The linear gradient is established during pouring by continuous overlayering of the gel solution which always becomes less dense.

- Add APS (40 %) to the starting solutions:
  to the dense solution (7.5 mL) 7.5 μL
  to the light solution (7.5 mL) 7.5 μL
  mix it well by vigorously stirring with the stirring bar in the mixing chamber and the compensating stick in the reservoir;
- insert the tubing in the middle notch of the glass plate;
- set the stirrer at a moderate speed;
- open the valve connecting the chambers;
- open the pinchcock.

Immediately rinse the gradient mixer with double-distilled water.

After a few minutes, when the liquid level has reached a height of about 1 cm below the edge of the cassette, the gradient mixer should be empty.

Water works best; nothing else, e.g. butanol, has proved comparable.

■ Important:
Overlay the gel with about 0.5 mL of distilled water, so that the edge of the basic part becomes even and to prevent oxygen from diffusing into the gel;

- let the gel polymerize for 10 min at room temperature, so the density gradient can settle;

make sure the cassette is horizontal.

- polymerize the gel for 1 h in a heating cabinet at 37 °C;
- cool the cassette under running water and remove the gel from the cassette.

c) Washing the gel

The gels swell during washing. Sometimes in a specific pH interval they acquire a characteristic surface structure (snake skin). But the surface becomes smooth again once the gel is dry.

Immobiline gels exhibit low conductivity because the buffering groups of the pH gradient are firmly bound in the gel and cannot move freely like carrier ampholytes. Ions, such as TEMED and APS, which are not co-polymerized cannot be transported electrophoretically out of the gel during IEF. Therefore Immobiline gels must be washed with double-distilled water after polymerization.

Using a cassette has the additional advantage that one has a quantitative control over the additive concentration.

The gels are then dried and rehydrated with volume control before use. If the gels were used directly after washing, the excess water would seep out of the gel during IEF ("sweating"). They are rehydrated in a cassette or a tray (Fig. 5) .

On a laboratory shaker.

- Wash the gel in distilled water:
  4 × 15 min in 300 mL of distilled water each time.

So that it does not roll up during drying.

In a dust free cabinet if possible.

- Equilibrate the gel for 15 min in 1.5% (v/v) glycerol;

- dry the gel at room temperature with a fan or a ventilator;

- as soon as possible, when the gel is dry, cover it with Mylar film and store in the freezer in plastic bags (Rossmann and Altland, 1987).

Rossmann U, Altland K. Electrophoresis 8 (1987) 584–585.

d) Storage
The dry gels should be stored at −20 °C, so that their swelling properties are retained.

e) Rehydration
The dry gels are reconstituted in a reswelling cassette or in a rehydra-

Loessner MJ, Scherer S. Electrophoresis 13 (1992) 461–463.

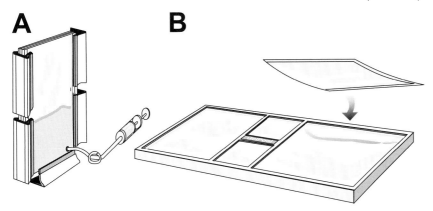

Fig. 5: Rehydration of an IPG gel in a vertical cassette (A) or alternatively in the rehydration tray GelPool (B).

tion tray (Loessner and Scherer, 1992) in the calculated volume of rehydration solution (Fig. 5), so that the gel reswells evenly and to ensure a quantitative control over the additive concentration. A few examples for the rehydration of Immobiline gels are shown in Tab. 4.

Tab. 4: Rehydration of Immobiline gels.

Rehydration solution	Time	Samples
Dist. water	1 h	Water soluble proteins, enzymes
2 mmol/L acetic acid (anodal sample appl.: – prevents lateral band spreading)	1 h	Water soluble proteins, enzymes
2 mmol/L Tris (cathod. sample appl.: – prevents lateral band spreading)	1 h	Water soluble proteins, enzymes

Bjellqvist B, Linderholm M, Östergren K, Strahler JR. Electrophoresis 9 (1988) 453–462.

	Rehydration solution	Time	Samples
LKB Application Note 345 (1984).	25 % (v/v) glycerol	1 h	Serum for determin. of PI, GC, TF
Altland K, Banzhoff A, Hackler R, Rossmann U. Electrophoresis 5 (1984) 379–381.	8 mol/L urea	2 h	Poorly soluble proteins, protein complexes. Plasma: prealbumin
Rossmann and Altland (1987).	8 mol/L urea, 0,5 % (g/v) Ampholine, 2 % (v/v) 2-mercaptoethanol or 50 mmol/L DTT	overnight	Dried blood, plasma, erythrocyte lysates, globins
Baumstark M, Berg A, Halle M, Keul J. Electrophoresis 9 (1988) 576–579.	8 mol/L urea, 0.5 % (w/v) Ampholine, 30 % (v/v) glycerol	overnight	Serum VLDL for the determination of apolipoprotein E
Günther S., Postel, W., Weser, J, Görg A. In: Dunn MJ, Ed. Electrophoresis'86. VCH Weinheim (1986) 485–488.	5 mol/L urea, 20 % (v/v) glycerol	2 h	Alcohol soluble proteins
LKB Appl. Notes 345 and 373 (1984).	0.5 % (w/v) Ampholine	1 h	PGM, time sensitive enzymes
Rimpilainen M, Righetti PG. Electrophoresis 6 (1985) 419–422.	4 % (w/v) Ampholine, 2 % (v/v) Nonidet NP-40	overnight	Membrane proteins
Sinha PK, Bianchi-Bosisio A, Meyer-Sabellek W, Righetti PG. Clin Chem. 32 (1986) 1264–1268.	4 % (w/v) Ampholine,	overnight	Alkaline phosphatase
Görg A, Postel W, Weser J, Günther S, Strahler JR, Hanash SM, Somerlot L. Electrophoresis 8 (1987) 45–51.	up to 9 mol/L urea, 0.5 % (v/v) Nonidet NP-40	overnight	Hydrophobic proteins, complex protein mixtures, 2D-electrophoresis
Hanash SM, Strahler JR, Somerlot L, Postel W, Görg A. Electrophoresis 8 (1987) 229–234.	8 mol/L urea, 2 % (v/v) Nonidet NP-40, 10 mmol/L DTE	16–18 h	2-D electrophoresis of myeloblast lysates

If merely a few samples are to be separated, only part of the IPG gel needs to be rehydrated. The unused part should be stored at –20 °C tightly packed. For 2D electrophoresis the gel is cut into individual strips with a paper cutter before rehydration.

The rehydration cassette can also be used to cast immobilized pH gradient gels for long separation distances (up to 25 cm).

To optimize additive concentration or to investigate conformational changes and ligand binding properties of certain proteins with regards to additive concentrations, an *additive gradient* perpendicular to the immobilized pH gradient (s. Fig. 6) can be established (Altland et al. 1984).

To prevent the surface of the gel from sticking to the glass (with the U-shaped gasket), the glass plate should first be coated with Repel Silane.

After rehydration, the gel surface should be dried, that is the whole solution should be soaked up by the gel. If the reswelling solution contains a high concentration of a non-ionic detergent (e.g. 1% Triton X-100), the surface may be slightly sticky after rehydration, in which case, the surface should be dried with precision wipe.

It is important to place Immobiline gels properly – the acidic end at the anode (Fig. 7).

The *basic* end of the gel is easy to recognize by:
1) the wavy edge of the gel (the anodic edge is straight);
2) the wider film edge (ca 1.5 cm).

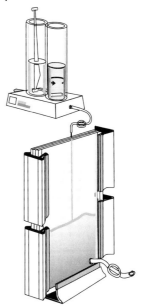

Fig. 6: Rehydrating in an additive gradient using the reswelling cassette.

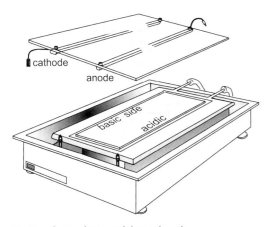

Fig. 7: Placing the Immobiline gel on the cooling plate of the separation chamber.

Use kerosene between the cooling plate and the support film. If other fluids are used (e.g. distilled water, 1% Triton X-100) discharges or sparks can occur at the edge of the support film, since very high voltages are used with Immobiline gels.

6
Isoelectric focusing

Sample application
No prefocusing is done with Immobiline gels, the samples can be directly applied in the form of drops (10 to 20 μL on the surface). It has proved best if the gel/sample contact surface is as small as possible especially if the gel and/or sample contain nonionic detergents (Strahler *et al.* 1987).

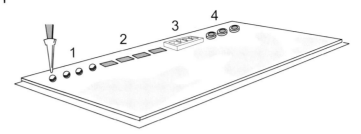

Fig. 8: Possibilities for sample application for IPG. – (1) droplets (2) pieces of cellulose (3) sample application mask (4) cut-off silicone tubing.

This has reduced lateral band spreading significantly.

For separating hydrophobic proteins in the presence of 8 mol/L urea and 0.5 % NP-40 application in preformed sample wells or into holes, which have been punched into the gel (penetrating the support film) have proved to be suitable (Loessner and Scherer, 1992).

The low conductivity of IPGs should be taken into consideration (see the end of this chapter).

Large sample volumes can also be applied with cut-out 2 mm thick silicone rubber frames (Hanash *et al.* 1987), cut-out sample application strips or silicone tubing, which are placed on the gel (Fig. 8). For IPG as well it is recommended to carry out a step test to find the optimum sample application point (s. method 6).

Pflug W, Laczko B. Electrophoresis 8 (1987) 247–248.

When many samples are focused together, silicone rubber application masks with funnel shaped holes and grooves on the bottom are particularly efficient (Pflug and Laczko, 1987).

Electrode solutions

For immobilized pH gradients both electrode strips only have to be soaked in distilled water. This saves work, rules out mistakes, increases the absorption capacity of the strips for salt ions and reduces the field strength at the beginning (better sample entry!).

Focusing conditions

See also: "strategies for IPG focusing" at the end of this chapter.

Normally focusing is carried out at 10 °C, lower (sensitive enzymes) or higher (e.g. 8 mol/L urea) temperatures being used for special applications. The pH values indicated on the packages are only valid at 10 °C. Many additives, e.g. urea, shift the pH value.

The separation times depend on the steepness of the pH gradient, the additives, molecular sizes and power supply settings.

Make sure that the anodic cable is plugged in in front.

The *current* is set so that the field strength in the gel is low at first (40 V/cm); if the field strength is too high at the beginning some of the proteins will precipitate.

Example:

Whole IPG gel, pH 4.0 to 7.0, reswollen in distilled water, 10 °C:
Maximal value for the power supply: 3000 V, 1.0 mA, 5.0 W
Focusing time: 5 h

- After IEF, switch off the power supply and open the safety lid.
- The proteins can now be stained or blotted. Should problems occur, consult the trouble-shooting guide in the appendix.

Consult the strategy graph at the end of the instructions.

Measuring the pH gradient:

The pH values cannot be measured with a pH meter because of the low conductivity. However, the pH gradient can easily be determined with marker proteins.

7
Staining

Colloidal Coomassie staining

The result is quite quickly visible with this method. Few steps are necessary, the staining solutions are stable and there is no background staining. Oligopeptides (10 to 15 amino acids) which are not properly fixed by other methods can be revealed here. In addition, the solution is almost odorless (Diezel *et al.* 1972; Blakesley and Boezi, 1977).

This method is especially recommended for Immobiline gels, because all other Coomassie Blue stains result in strongly colored backgrounds

Preparation of the staining solution: Dissolve 2 g of Coomassie G-250 in 1 L of distilled water and add 1 L of sulfuric acid (1 mol/L or 55.5 mL of conc H_2SO_4 per L) while stirring. After stirring for 3 h, filter (paper filter) and add 220 mL of sodium hydroxide (10 mol/L or 88 g in 220 mL). Finally add 310 mL of 100% TCA (w/v) and mix well, the solution will turn green.

Fixing and staining: 3 h at 50 °C or overnight at room temperature in the colloidal solution;

Washing out the acid: soak in water for 1 or 2 h, the green color of the curves will become blue and more intense.

Acid violet 17 staining

Immobiline gels containing 8 mol/L urea and nonionic or zwitterionic detergents normally show a high background with staining. The results are much better, when the gels are stained with Acid violet 17 (Patestos *et al.* 1988).

Patestos NP, Fauth M, Radola BJ. Electrophoresis 9 (1988) 488–496.

3% phosphoric acid: 21 mL 85 % H_3PO_4 in 1 L H_2O_{dist}.

11% phosphoric acid: 76.1 mL 85 % H_3PO_4 in 1 L H_2O_{dist}.

1 % Acid violet stock solution: 1 g Acid violet 17 in 100 mL H_2O_{dist}. Heat to 50–60 °C with magnetic stirrer.

0.1 % Acid violet staining solution: 10 mL of 1 % acid violet stock solution plus 90 mL 11% phosphoric acid.

Staining procedure

- *fix* for 20 min in 20% TCA,
- *wash* for 1 min in 3% phosphoric acid,
- *stain* for 10 min in 0.1 % Acid violet 17 solution.
- *destain* 3 × in 3% phosphoric acid until background is clear,
- *wash* 3 × 1 min with H_2O_{dist}.,
- *impregnate* with 5 % glycerol,
- air dry.

Silver staining

The silver staining techniques described in method 6 on page 206 and the following can be applied.

Practical tip

In the next paragraph (strategies for IPG focusing) it is recommended, in specific situations, to cut out strips of gel between the sample tracks. The rehydrated gel is placed on the line template (see appendix). Two glass plates with U-shaped gaskets are placed along the long edges of the support film and finally a simple glass plate is laid on top. Strips of gel are scraped out with a 4 or 5 mm thick spatula.

8
Strategies for IPG focusing

Sample application: concentration test

Burning line between sample application points

Cut out 5mm wide gel strips between the tracks (Fig. 10) and apply the sample.

The sample has a higher conductivity than the gel

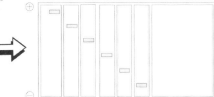

Lateral band broadening

That also helps

Anodal   Cathodal
sample application

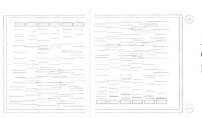

Also a conductivity problem

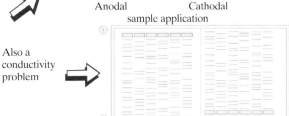

Rehydrate the gel in
2 mmol/L acetic acid   2 mmol/L Tris

The proteins smear on the surface

Part of the proteins tend to aggregate because of the low conductivity and do not enter the gel

Rehydrate in urea

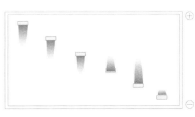

If >6 mol/L urea:
add 0.1 to 0.5% non-ionic detergent

Oxidation sensitive proteins e.g. hemoglobin

The gel is oxidized twice:

1. during polymerization (by APS)
2. during drying.

The proteins are oxidized during the run.

1. fist wash the gel for 30 min in 0.1 mol/L ascorbic acid, titrate with 1 mol/L Tris to pH 4.5 or in 2 mmol/L DTT solution and then in double-distilled water
or

2. Rehydrate the gel in 10 mmol/l DTT or in 2% 2-mercaptoethanol.

Method 11:
High resolution 2-D electrophoresis

With the advent of the proteome analysis approach 2-D electrophoresis has gained high importance:

- It is the core technique to find proteins of interest among thousands of other gene products in cell lysates and tissue extracts.
- Because of its very high resolving power it affords a high level of purification of a protein for its identification and further characterization with mass spectrometry.
- The separated proteins are conserved in the gel matrix for further analysis at any desired time.

The goal is to achieve a clear and reproducible spot pattern, which represents the actual condition in a cell at the moment of taking the sample. However, 2-D electrophoresis is a multistep procedure with many "opportunities" to lose or modify proteins. A general overview over the technique's state of the art can be found in the "Proteome analysis" chapter in this book (pages 91–119). The practical aspects are described in the following chapter.

Already at this point it should be mentioned, that inappropriate sample preparation and overloading cause more than 95 % of faulty gels.

The following points should be considered for choosing the optimal separation technique:

Gel formates:

- For the complex protein mixtures usually studied with 2-D electrophoresis, high resolution and high purity of spots can solely be achieved by adequate spatial resolution: thus *large* gel sizes are needed in the most cases.
- Less complex protein subsets analyses, or protein identification with antibodies can be performed in medium sized to miniformate gels.
- Optimization of sample preparation is usually checked with miniformate gels for quick and reliable results.

Prepurification can cause uncontrollable protein losses.

Also the correct sample load can be checked with small gels.

Electrophoresis in Practice, Fourth Edition. Reiner Westermeier
Copyright © 2005 WILEY-VCH Verlag GmbH & Co. KGaA, Weinheim
ISBN: 3-527-31181-5

Horizontal or vertical second dimension:
Horizontal gels

Görg A, Boguth G, Ober-maier C, Posch A, Weiss W. Electrophoresis 16 (1995) 1079–1086.

- When just one gel has to be run, the experimental set-up is very easy to handle and versatile for different gel sizes.
- Due to the very efficient cooling and possibility of using thin gels (0.5 mm), running time is very short. This results in very sharp spots and high sensitivity of the detection techniques (Görg *et al.* 1995).

Vertical gels

Anderson NG, Anderson NL. Anal Biochem. 85 (1978) 331–340.

Eckerskorn C, Lottspeich F. Chromatographia. 28 (1989) 92–94.

- Ten or more gels can be run in one apparatus in parallel for reproducible separation conditions (like the IsoDalt system by Anderson and Anderson, 1978).
- The use of gels thicker than 1 mm is necessary for semi-preparative separations needed for *de novo* sequencing of proteins (Eckerskorn and Lottspeich, 1989).

The techniques of gel preparation for immobilized pH gradient and SDS gels are described in methods 10, 7 and 8. The following instruction contains some optimized recipes for high resolution 2-D electrophoresis.

1
Sample preparation

Some important facts on sample preparation are listed in chapter 6. The following are recipes for practical experiments.

Washing cells: Instead of PBS, which contaminates the cells with too much salt, Tris-buffered sucrose should be employed (10 mmol/L Tris, 250 mmol/L sucrose pH 7): 1.21 g Tris, 85.6 g sucrose, fill up to 1 L with distilled water and titrate to pH 7 with 4 mol/L HCl.

Cell disruption: The methods are dependent on the type of sample. Still the most employed procedure is to freeze the cells or the tissue with liquid nitrogen and grind them with mortar and pestle. Cell proteins can be extracted by sonication.

Standard solubilization mixture "lysis buffer" (10 mL):

9 mol/L urea 5.4 g
2% (w/v) CHAPS 200 mg
1% (w/v) DTT 100 mg
1 mmol/L EDTA*) 4 mg
4 mmol AEBSF*)**) 10 mg
0.8% (w/v) Pharmalyte pH 3 to 10***) 200 μL
0.002 % Bromophenol blue 20 μl****)
make up to 10 mL with distilled water

Prepare the solution freshly, shake to dissolve the urea, do not warm it higher than 30 °C to avoid carbamylation.
*) Protease inhibitors.
**) = Pefabloc® SC
***) or IPG buffer of the corresponding pH range.
****) 1% (w/v) bromophenol blue solution.
containing 0.6 % Tris base

Protein loads:	Radiolabelled:	1 ng or less
	Coomassie Blue stain:	500 μg – 2 mg
	Silver stain:	20 μg – 100 μg

The maximum applicable sample loads depend on the length of the first dimension gel and the pH gradient.

To separate protein solutions such as serum, plasma etc. the solubilizing mixture is diluted accordingly.

Due to the presence of urea, detergent and carrier ampholytes, the standard protein assays lead to a strong underestimation of protein content in these solutions. The most reliable results are achieved with dedicated quantification kits, which are based on precipitation of the proteins and subsequent colorimetric measurement.

The precipitation chemistry should be the same like for sample clean-up.

Example 1: yeast cell lysate

- mix 600 mg of lyophilized yeast (Saccharomyces cerevisiae) with 2.5 mL lysis buffer;
- sonicate for 10 min at 0 °C;
- centrifuge for 10 min at 10 °C with 42,000 g.

Sample concentration for silver staining

- mix 10 μL of the supernatant
  – with 340 μL rehydration solution and soak 18 cm dry IPG strip in this solution.
  – with 30 μL lysis buffer and apply 20 μL to the acidic end of the IPG strip

Rehydration-loading, see page 288
Cup-loading, see page 289.

Sample concentration for Coomassie Blue staining

- mix 200 μL of the supernatant with 150 μL rehydration solution and soak 18 cm dry IPG strip in this solution.
  Or apply 100 μL to the acidic end of the IPG strip.

Rehydration-loading
Cup-loading

Example 2: E. coli extract

- mix 400 mg of lyophilized Escherichia coli with 10 mL lysis buffer;
- sonicate for 10 min at 0 °C;
- centrifuge for 10 min at 10 °C with 42,000 g.

Sample concentration for silver staining

Rehydration-loading

- mix 20 μL of the supernatant with 330 μL rehydration solution and soak 18 cm dry IPG strip in this solution.

Cup-loading

Or apply 20 μL to the acidic end of the IPG strip.

Sample concentration for Coomassie Blue staining

Rehydration-loading

- mix 100 μL of the supernatant with 250 μL rehydration solution and soak 18 cm dry IPG strip in this solution.

Cup-loading

Or apply 100 μL to the acidic end of the IPG strip.

However, also some acidic proteins get lost with this procedure.

A *crude extract,* however, shows contaminations with phospholipids and nucleic acids, which are visualized with silver staining as horizontal streaks in the acidic part of the gel. These disturbances are gone after TCA-acetone precipitation and resolubilization with lysis buffer (see procedure below). Fig. 1 shows the silver stained 2-D maps of E. coli lysate with and without precipitation.

The precipitation method described below also removes nucleic acids.

Nucleic acids can be removed with DNAse and RNAse treatment; the easiest technique is sonication, which breaks nucleic acids into little fragments.

... and salts.

Desalting with micro-dialysis tubes shows less protein losses than with gel filtration. In special cases desalting by a low voltage prefocusing is employed (Görg *et al.* 1995).

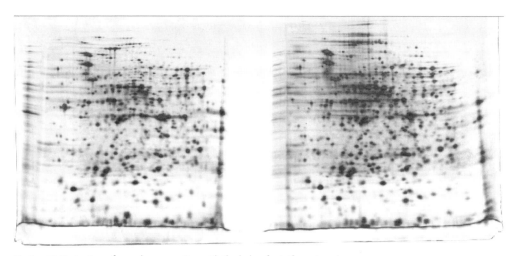

Fig. 1: Optimization of sample preparation with the help of miniformate gels. E. Coli extract after precipitation and resuspension and crude extract were first separated in 7 cm IPG strips pH 4-7, and then together on one 16 × 8 cm vertical SDS gel. Silver staining. From Tom Berkelman, Amersham Biosciences San Francisco with kind permission.

Precipitation of low concentrated proteins (e.g. plant tissue)
based on the procedure by Damerval et al. (1986):

Damerval C, DeVienne D, Zivy M, Thiellement H. Electrophoresis 7 (1986) 53–54.

- *freeze* plant material in liquid nitrogen and grind it in a pre-frozen mortar;
- *mix* the powder with 10 % TCA in acetone (–20°C) containing 1 % 2-mercaptoethanol; *or 0.2 % DTT*
- *precipitate* overnight in a freezer;
- *centrifuge* at +20 °C with 13,000 rpm for 10 min;
- *wash* with 90 % acetone / 10 % water (–20°C) containing 1 % 2-mercaptoethanol and centrifuge again; *or 0.2 % DTT*
- *resuspend* the pellet in lysis buffer, carefully sonicate; *Avoid heating!*
- *centrifuge* at +20 °C with 13,000 rpm for 10 min.

A novel clean-up kit based on acidic precipitation with detergent co-precipitants and organic washing solutions allow a sample clean-up within 1 ½ hours (Stasyk *et al.* 2001).

For *membrane proteins* and other *very hydrophobic* proteins a combination urea / thiourea in the solubilization solution can be very helpful to get more proteins into solution (Rabilloud, 1998). The modified lysis buffer is then composed as follows: 7 mol/L urea, 2 mol/L thiourea, 4 % CHAPS, 1 % DTT, 4 mM AEBSF, 0.8 % Pharmalytes pH 3–10.

Rabilloud T. Electrophoresis 19 (1998) 758–760.

Also *SDS* is used in some cases (Sanchez *et al.* 1995): Mix 10 µL human plasma with 6.25 µL 10% SDS, 2.3% DTT. Heat for 5 min at 95 °C. Dilute with 500 µL 8 mol/L urea, 4% CHAPS, 40 mmol/L Tris base, 65 mmol/L DTT.

Sanchez J-C, Appel RD, Golaz O, Pasquali C, Ravier F, Bairoch A, Hochstrasser DF. Electrophoresis 16 (1995) 1131–1151.

2
Stock solutions

Acrylamide, Bis solution (30%T, 2%C):
29.4 g of acrylamide + 0.6 g of Bis, make up to 100 mL with distilled water.

Acrylamide, Bis solution (30%T, 3%C):
29.1 g of acrylamide + 0.9 g of Bis, make up to 100 mL with distilled water.

C = 2% in gradient gel solutions prevents the resolving gel from peeling off the support film and cracking during drying.

■ Caution!
Acrylamide and Bisacrylamide are toxic in the monomeric form. Avoid skin contact and dispose of the remains ecologically (polymerize the remains with an excess of ammonium persulfate).

This solution is used for dilute gels (IPG) and plateaus, since they would be unstable at lower crosslinking.

Gel buffer pH 8.8 (4 × conc):
18.18 g of Tris + 0.4 g of SDS , make up to 80 mL with distilled water.
Titrate to pH 8.8 with 4 mol/L HCl; make up to 100 mL.

Ammonium persulfate solution (APS):
dissolve 400 mg of APS in 1 mL of distilled water.

Can be stored for one week in the refrigerator at +4 °C.

Cathode buffer (10 × conc):
30.28 g of Tris + 144 g of glycine + 10 g of SDS,make up to 1 L with distilled water.

Do not titrate with HCl!

Anode buffer (10 × conc):
30.28 g of Tris + 200 mL of distilled water.
Titrate to pH = 8.4 with 4 mol/L HCl;
make up to 1 L with distilled water.

Economy measure: the cathode buffer can also be used here.

Equilibration stock solution:

Urea and glycerol slow down electroendosmosis; EDTA inhibits the oxidation of DTT.

2% SDS	2.0 g
6 mol/L urea	36 g
0.1 mmol/L EDTA	4 mg
0.01% Bromophenol Blue	10 mg
50 mmol/L Tris HCl pH 8.8	3.5 mL
30% glycerol (v/v)	35 mL of an 85% solution
make up to 100 mL with distilled water	

Agarose sealing solution:

Not needed for horizontal SDS gel! Only for vertical second dimension!

0.5 % Agarose	0.5 g
0.01 % Bromophenol Blue	10 mg
SDS cathode buffer (1 ×conc)	100 mL

Heat on a heating stirrer until agarose is completely dissolved. Pipet 2 mL aliquots into reaction cups and store them at room temperature. Alternatively use a microwave oven.

3
Preparing the gels

IPG strips

As demonstrated here, casting IPG strips in the lab is a relatively complicated procedure.

For high resolution 2-D electrophoresis, the IPG gel strips are reconstituted with 8 mol/L urea, 0.5 % CHAPS (w/v), and 10 mmol/L DTT.

Because of the guaranteed reproducibility of industrial produced ready-made Immobiline pH gradient (IPG) strips, those are usually employed for 2-D electrophoresis.

Nevertheless a procedure is described here, how to prepare these immobilized pH gradient strips (see Tab. 1) in the laboratory. Perhaps it is also useful for a training course. Immobiline recipes for more gradients are listed in method 10.

Gels for short separation distances of 7 or 11 cm are cast in the standard cassette with 4.8 respectively 7.5 mL of the solutions.

The reswelling cassette is used for casting long gels (Fig. 2). To make an 18 cm gel, 5.2 mL of each solution is pipetted, for 24 cm strips 7 mL of each.

Tab. 1: Composition of the polymerization solutions for IPG gels

IPG gradient	pH 4 to 10		pH 4 to 7		pH 7 to 10	
stock solutions	dense	light	dense	light	dense	light
Immob. pK 3.6	551 μL	–	289 μL	151 μL	271 μL	45 μL
Immob. pK 4.6	–	57 μL	55 μL	369 μL	–	–
Immob. pK 6.2	227 μL	25 μL	225 μL	75 μL	–	–
Immob. pK 7.0	45 μL	244 μL	–	135 μL	189 μL	162 μL
Immob. pK 8.5	167 μL	78 μL	–	–	175 μL	175 μL
Immob. pK 9.3	–	179 μL	–	438 μL	–	140 μL
Acrylamide, Bis (30%T, 3%C)	1.0 mL	1.0 mL	1.0 mL	1.0 mL	1.0 mL	1.0 mL
Glycerol (85%)	2.0 mL	0.3 mL	2.0 mL	0.3 mL	2.0 mL	0.3 mL
TEMED (100%)	3.5 μL	3.5 μL	3.5 μL	3.5 μL	3.5 μL	3.5 μL
with $H_2O_{dist} \longrightarrow$	7.5 mL	7.5 mL	7.5 mL	7.5 mL	7.5 mL	7.5 mL

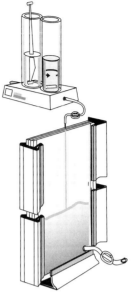

Fig. 2: Casting IPG gels for long separation distances.

Carefully mix the solutions, measure the pH, then titrate to pH 7 with TEMED or 4 mol/L HCl respectively.

for an optimal and regular polymerization, s. Method 10.

Tab. 2: Pipetting volumes for the gradient mixer

IPG strip	7 cm		11 cm		18 cm		24 cm	
	mix.ch. dense	reservoir light	mix.ch. dense	reservoir light	mix.ch. dense	reservoir light	mix.ch. dense	reservoir light
Polym. sol.	4.8 mL	4.8 mL	7.5 mL	7.5 mL	5.2 mL	5.2 mL	7 mL	7 mL
APS (40%)	5 μL	5 μL	7 μL	7 μL	5 μL	5 μL	7 μL	7 μL

Before casting:
Cool the cassette in the refrigerator to delay the start of the polymerization.

After casting:
Overlay the gel with distilled water to prevent polymerization inhibition by oxygen, leave the cassette for 10 min, so that the gradient can settle horizontally.
Polymerization: 1 h at 50 °C (heating cabinet).

To facilitate this process add a small volume of glycerol also to the light solution (see table 1).

Washing and drying

- Remove the gel from the cassette;
- wash 3 × 20 min in 300 mL of distilled water on a laboratory shaker;
- wash 1 × in 300 mL of 2% glycerol for 30 min;
- dry at room temperature in a dust-free cabinett.
- Store wrapped in film at –20 °C (frozen).

The dried IPG gels are cut into 3 mm wide individual strips with a paper cutter (see Fig. 3). The strips can be marked on the hydrophobic back with a waterproof marker pen for the identification of the gradient direction and the sample.

dried gel with
immobilized pH gradient

pH
10.0

pH
4.0

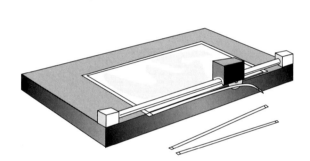

Fig. 3: Precise cutting of the dry IPG strips with a special paper cutter "Roll and cut"

Rehydration

- There are two possibilities:

a) The strip is rehydrated in rehydration solution, the sample is applied to the strip in the IEF system with a loading cup: "Cup-loading".

Rabilloud T, Valette, Lawrence JJ. Electrophoresis 15 (1994) 1552–1558.

b) The sample is included in the rehydration solution as suggested by Rabilloud *et al.* (1994): "Rehydration-loading".

- In both cases it is important to apply the exact volumes:

  7 cm strip: 125 μL
  11 cm strip: 200 μL
  18 cm strip: 350 μL
  24 cm strip: 450 μL

- Rehydration solution:

8 mol/L urea	24 g
0.5% CHAPS	250 mg
0.2 % (w/v) DTT	100 mg
0.25% (w/v) Pharmalytes pH 3–10	310 µL
0.01% Bromophenol Blue	5 mg

make up to 50 mL with distilled water.

Fig. 4 shows a rehydration tray, which can be used for different gel strip lengths up to 24 cm.

With novel equipment, rehydration-loading and IEF are combined (see page 288).

Strips and solutions must be covered with 1 to 2 mL paraffin oil during reswelling to avoid crystallization and oxygen contact.

- Rehydration time:
  > 6 h for rehydration without sample, > 16 h or overnight for rehydration loading.

Large protein molecules need a long time to diffuse into the strip.

- After rehydration:
  Rinse the surface of the strips with distilled water using a squeeze bottle and then place them for a few seconds on their edges on a damp filter paper to drain excess liquid.

so that the urea on the surface does not crystallize out.

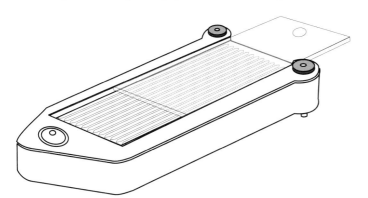

Fig. 4: Rehydration tray for reswelling the IPG strips in grooves: either in rehydration solution (for cup-loading) or in the sample solution (rehydration-loading).

SDS polyacrylamide gels

In *vertical* instruments usually homogeneous gels with 12 or 13 %T, 2.5 or 3 % C are used without stacking gel (see page 101). Only in special cases the matrix concentration is modified to increase the resolution in certain molecular size ranges. As demonstrated by Langen *et al.* (1997), it is impossible to predict the correct acrylamide concentration according to a mathematical function; also here a try and error procedure must be used.

Langen H, Röder D, Juranville J-F, Fountoulakis M. Electrophoresis18 (1997) 2085–2090.

There are ready-made gels available for large and small formates. However, as mentioned in the previous paragraph, sometimes gels with special acrylamide concentrations must be prepared. For a high-resolution high-throughput system as described on page 110, up to 14 gel cassettes are simultaneously filled inside one box. 1 L polymerization solution is needed for a set of 1 mm thin gels of 25×20 cm size. In chapter 8 several casting procedures for miniformate gels are described.

Very good results are achieved with homogeneous resolving gels (12.5 %T, 2 % C) in the horizontal flatbed technique. Here as well, as in method 7, the casting cassette should be cooled and APS added only shortly before pouring, to delay the onset of polymerization.

The preparation of a *horizontal* gel on film support with a porosity gradient is described here. The acrylamide concentrations for the linear pore gradient gels given have been optimized for the separation of yeast cell lysates. The pore gradient can be adapted to the type of separation, which depends on the composition of the sample, by changing the volumes of acrylamide, Bis solution. The sample application plateau contains $T=6\%$ so that it better resists the eventual electroendosmotic influences of the IPG strips (see Tab. 3).

A "very dense" plateau does not mix with the gradient.

Tab. 3: Composition of the three gel solutions for a gradient 12 to 15%*T* and a plateau with 6%*T*

Stock solutions		Gel solutions	
	very dense 6%T, 3%C	dense 12%T, 2%C	light 15%T, 2%C
Glycerol (85 %)	6.5 mL	4.3 mL	–
Acrylamide, Bis 30%T, 3%C	3.0 mL	–	–
Acrylamide, Bis 30%T, 3%C	–	6.0 mL	7,5 mL
Gel buffer	3.75 mL	3.75 mL	3.75 mL
Orange G (1 % w/v) in H_2O_{dist}	100 µL	–	–
TEMED (100%)	10 µL	10 µL	10 µL
Bring to final volume with H_2O_{dist}	→15 mL	→15 mL	→15 mL

Casting volumes

See Tab. 4.

The volumes of the solutions depend on the gel size desired. There are two possibilities here:

Casting cassettes which are as large as the cooling plate of the Multiphor II are used for this.

- use the standard gel format as described in method 7.
- use a longer separation distance to obtain a better resolution for highly complex protein mixtures:
  gel dimensions: $250 \times 190 \times 0.5$ mm ("Large scale").

Tab. 4: Pipetting volumes for the gradient mixer

SDS gel	Standard gel			Large scale		
	cassette very dense	mix.ch. dense	reservoir light	cassette very dense	mix.ch. dense	reservoir light
Polym. sol.	3.5 mL	7.0 mL	7.0 mL	3.5 mL	13.5 mL	13.5 mL
APS (40%)	4 µL	7 µL	7 µL	4 µL	14 µL	14 µL

Before casting:

Cool the cassette in the refrigerator to delay the start of the polymerization.

After casting:

Overlay the gel with distilled water to prevent polymerization inhibition by oxygen, leave the cassette for 10 min, so that the gradient can settle horizontally.

- add the plateau (super dense) with a pipette or a 20 mL syringe. The gradient is added directly, without prior polymerization, onto the plateau. *See method 7, SDS-PAGE.*
- overlay with 60 % v/v isopropanol-water.

4
Separation conditions

First dimension (IPG-IEF)

IPG focusing in individual strips can be performed with conventional equipment directly on the cooling plate (Fig. 5). To improve sample application and facilitate placing the IPG strips as well as more efficient protection from CO_2 – especially in the basic range of the gradient – it is recommended to use the IPG strip kit (Fig. 6). *In the IPG strip tray the gels can also be run under a layer of silicon or paraffin oil.*

In general, separation conditions have to be adjusted to the needs of the sample analysed. Horizontal streaking in the 2-D pattern, for instance, can have many different reasons: too short focusing time, too long focusing time (some proteins became unstable at their pIs), oxidation of DTT (more DTT has to be added to the cathodal electrode strip), too much salt, nucleic acids, phospholipids etc.

In this chapter, running condition parameters for a few examples are given in Tab. 5, comprehensive collections of running conditions for different gradients and strip lengths of are found in the the already mentioned handbook by Berkelman and Stenstedt (2002), and the paper by Görg *et al.* (2000). *These publications are also very helpful for trouble-shooting. More hints can be found on this website: IPG.Dalt@lrz.tu-muenchen.de*

a) IPG-IEF with conventional equipment:

Use the correct orientation! See method 10.

- Place reconstituted IPG strips beside one another – with the gel side facing up – on the cooling plate wetted with kerosene. Place 5 mm wide electrode paper strips, soaked in distilled water, over the ends of the IPG strips (Fig. 5). When "rehydration-loading" was performed, start IEF now.

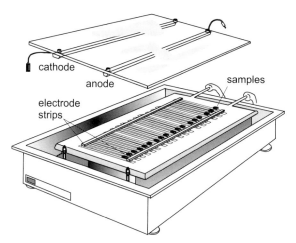

Fig. 5: IEF in individual strips directly on the cooling plate.

For "cup-loading":

Cut the applicator strips into pieces

Place the silicone rubber frames on the anodic (or cathodic) sides of the strips; apply 20 µL of each sample into each well;

Make sure that the long anodic connecting cable is plugged in.

- place the focusing electrodes and connect the cables (Fig. 5).

b) IPG-IEF with IPG strip kit (Fig. 6):

- Place the IPG strip tray on the cooling plate coated with kerosene and plug in the cables;

Do not use kerosene here.
Make sure the orientation is correct! See method 10.

- add 1 mL of paraffin oil to the tank;
- place the reconstituted IPG strips – with the gel side facing up – in the grooves;
- place 5 mm wide, 11 cm long electrode paper strips, soaked in distilled water, over the ends of the IPG strips (Fig. 6);
- insert the electrodes;

Do not use silicon oil, it can take up oxigen.

- pour about 50 mL of paraffine oil onto the strips. When "rehydration-loading" was performed, start IEF now.

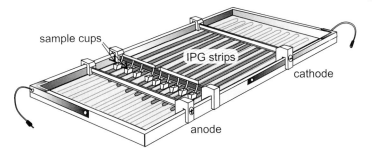

Fig. 6: IEF in individual IPG strips in the IPG
strip kit.

For "cup-loading":

- place the sample applicator holder on the anodic side; or on the cathodal side for some samples; or on both sides according to Langen *et al.* (1997);

- place the sample applicators in the holder in such a way that the cups are pressed onto the surface of the IPG strips; *Good contact between the cups and the gel prevents the sample from leaking out.*

- pipette the samples into the cups. *Sample volumes up to 100 μL can be used.*

Separation conditions

- set the temperature at 20 °C;
- set the running conditions, a few examples are given in Tab. 5. *IEF with IPG strips is done with voltage control.*

Tab. 5: Power supply settings for IPG-IEF on Multiphor II

					Current: 50 μA per strip, Power: 5W				
	rehydration-loading					**cup-loading**			
pH	7 cm 4–10	7 cm 4–7, 7–10	18cm 4–10	18cm 4–7, 7–10	pH	7 cm 4–10	7 cm 4–7, 7–10	18cm 4–10	18cm 4–7, 7–10
					Phase 1 150 V	30 min	30 min	30 min	30 min
Phase 1 500 V	1 h	1 h	1 h	1 h	Phase 2 300 V	1 h	1 h	1 h	1 h
Phase 2 3.5 kV	1.5 h	2.5 h	7 h	over night	Phase 3 3.5 kV	1.5 h	2.5 h	7 h	over night
Phase 3	5 kVh	7 kVh	20 kVh	42 kVh	Phase 4 3.5 kV	5 kVh	7 kVh	20 kVh	42 kVh

For rehydration-loading the start conditions are less critical.

■ **Note:**

For cup-loading, two phases with low voltages independent of the gel length are applied for optimized starting conditions: emigration of salt ions and slow sample entry.

c) IPG-IEF in individual ceramic trays

With individual trays any carry-overs of samples are avoided.

Rehydration-loading and IEF can be combined in an isoelectric focusing specially developed for the IPG technique in 2-D electrophoresis. The samples are pipetted into individual ceramic trays with built-in platinum electrode contacts. The IPG strips are rehydrated and run with the gel facing down (see Fig. 7).

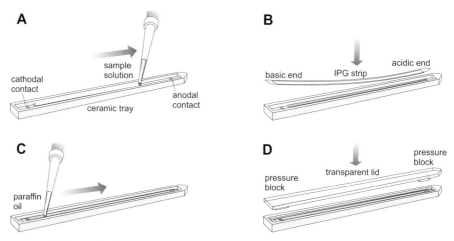

Fig. 7: Rehydration-loading in individual ceramic trays. (A) Pipetting the sample-rehydration solution into the tray; (B) placing the IPG strip with dried gel side down onto the fluid; (C) pipetting a few mL of paraffin oil on the IPG strip support film; (D) closing the tray with the lid.

Görg A, Boguth G, Obermaier C, Harder A, Weiss W. Life Science News 1 (1998) 4–6.

These trays are placed on the cooled electrode contact areas of the power supply (see Fig. 66 on page 130). It has been shown, that rehydration under a low voltage (30 – 60 V) for 10 hours facilitates the entry of high molecular weight proteins larger than 150 kDa into the strips (Görg *et al.* 1998).

Local overheating is very dangerous: Carbamylation can occur, strips can even start to burn.

The ceramic material dissipates very efficiently the Joule heat, which is produced during the run. This is particularly necessary, when a sample load of several mg protein in one strip has to be separated. The ceramic surface is specially treated to avoid protein absorption.

The pressure blocks on the transparent lids are needed to hold the ends of the strip down on the contacts, because electrolysis gas will be produced during the run.

Because the separation is performed in a closed system, up to 8 kV can be applied on the gels. This leads to shorter separation times and sharper spots.

Because of this feature a high thermal conductivity of the tray is very important.

The *main benefit* of this system is that rehydration-loading and isoelectric focusing are combined to one procedure, thus reducing the number of steps in 2-D electrophoresis.

This saves time and reduces possible errors.

Rehydration and IEF separation are carried out at 20 °C:

- Rehydration (without voltage or at 30 – 60 V) for 10 hours.
- *IEF:* 50 µA per strip are applied; voltage and time settings are depending on strip length, gradient and sample load; it is done in several voltage steps up to 8 kV.

More details are found in the handbook by Berkelman and Stenstedt (2002).

Narrow gradient intervals

Very high spacial resolution and increased protein loading capacities are achieved with one-unit or more narrow pH gradients in long strips (Langen and Röder, 1999; Wildgruber *et al.* 2000). These gels allow the detection of thousands more proteins in a sample of complex cell lysate than wide gradients. They require, however, very long running periods, up to ca. 200 kVh.

Langen H, Röder D. Life Science News 3 (1999) 6–8. Wildgruber R, Harder A, Obermaier C, Boguth G, Weiss W, Fey SJ, Larsen PM, Görg A. Electrophoresis 21 (2000) 2610–2616.

d) Ceramic tray for cup-loading:

Some of these narrow gradients, very alkaline and very acidic gradients give better results with cup-loading rather than with rehydration-loading. Also high sample loads require sometimes runs with the gel surface facing up. In these cases rehydration is performed in an external reswelling tray (Fig. 4), the gels are run in special ceramic trays, which allow also cup-loading (Fig. 8).

Filter paper pads are inserted between the electrodes and the IPG strips for removal of accumulated extra-gradient proteins and salt ions.

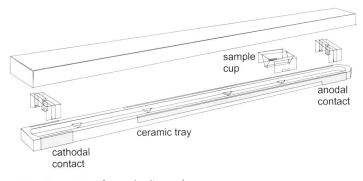

Fig. 8: Ceramic tray for cup-loading and running IPG gel strips facing up.

After IEF

−20 to −40 °C is not enough: proteins become modified.

Either the strips are equilibrated in SDS buffer and run on the second dimension right away, or they are stored at −60 to −80 °C in a deep-freezer.

Patestos NP, Fauth M, Radola BJ. Electrophoresis 9 (1988) 488–496.
Note: *This staining is **not** reversible!*

Sometimes it might be useful to check, whether the separation in the first dimension worked well before all the work with the second dimension run is started. The most sensitive technique for staining these urea and detergent containing strips is Acid violet 17 staining according to Patestos *et al.* (1988):

Acid violet 17 staining of IPG strips:

3% phosphoric acid: 21 mL 85 % H_3PO_4 in 1 L H_2O_{dist}.
11% phosphoric acid: 76.1 mL 85 % H_3PO_4 in 1 L H_2O_{dist}.
1 % Acid violet stock solution: 1 g Acid violet 17 in 100 mL H_2O_{dist}. Heat to 50–60 °C with magnetic stirrer.
0.1 % Acid violet staining solution: 10 mL of 1 % acid violet stock solution plus 90 mL 11% phosphoric acid.

Staining procedure

- *fix* for 20 min in 20% TCA,
- *wash* for 1 min in 3% phosphoric acid,
- *stain* for 10 min in 0.1 % Acid violet 17 solution.
- *destain* for 10 min in in 3% phosphoric acid until the background is clear.
- *wash* 3 × 1 min with H_2O_{dist},
- *impregnate* with 5 % glycerol, air dry.

Equilibration

Westermeier R, Postel W, Weser J, Görg A. J Biochem Biophys Methods. 8 (1983) 321–330.
Görg A, Postel W, Günther S, Weser J. Electrophoresis 6 (1985) 599–604.

Prior to SDS-PAGE the strips have to be equilibrated in SDS buffer. Since the carboxylic groups of the immobilized pH gradient gels pick up charges and cause electroendosmotic effects in SDS gels (Westermeier *et al.* 1983), special precautions must be taken, when IPG strips are employed. Görg *et al.* (1985) have modified the composition of the equilibration buffer by adding glycerol and urea to hold the electroosmotic water flow back (see also page 7), which can cause partial losses of proteins, which are carried towards the cathode with the electroosmotic flow.

on a shaker!

Prevention of silver staining artefacts which can come from dust particles etc.

- Equilibrate for 15 min with 10 mL equilibration stock solution plus 100 mg DTT.
- Equilibrate for another15 min with 10 mL equilibration stock solution plus 480 mg iodoacetamide.
- Slightly bend the equilibrated IPG strips to the shape of a "C" and lay them on their side on dry filter paper for a minute to remove the excess fluid.

The equilibration periods seem rather long, but the charged carboxylic groups of the immobilized pH gradients act like a weak ion exchanger and prevent the diffusion of proteins out of the strip.

The long equilibration time is needed, because SDS is also negatively charged, thus diffusion of SDS into the strips is slow.

Second dimension (SDS electrophoresis)
Vertical gels

Usually1 or 1.5 mm thick SDS gels are used. With 1 mm thick gels, embedding of the IPG strip with agarose should not be needed, because the thickness of the strip increases during equilibration, and together with the 0.2 mm thick support film it will stay in place betweeen the glass plates. But the use of agarose sealing has a few advantages:

The IPG strip is placed on the SDS gel edge to edge. The surfaces of IPG strips with high sample loads are not even: highly abundant proteins form gel ridges. In order to minimize the danger of trapping air bubbles between the two gels, agarose solution is pipetted onto the SDS gel first, and then the strip is inserted (Fig. 9). This measure also helps to level out an uneven SDS gel edge. Sliding the strip into the cassette is facilitated, when it has been shortly dipped into the SDS electrophoresis buffer.

The agarose also seals the lateral edges and the spacers of film-supported ready-made gels, which are placed into a special cassette prior to electrophoresis, see Fig. 67 on page 131.

Molecular weight standards are pipetted on sample applicator pieces; those are placed onto the SDS gel edge before application of the agarose.

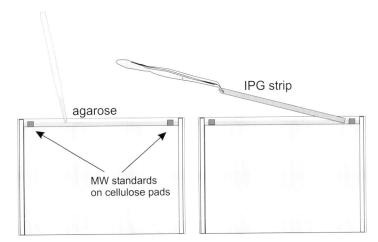

Fig. 9: Pipetting the agarose sealing solution and insertion of the equilibrated IPG strip into the SDS gel cassette. Marker proteins are pipetted on sample applicators and placed on the gel edge before applying the agarose.

Separation time varies from an hour to overnight depending on the gel size and the type of the instrument. A short separation time for large gels is very advantages, because the spots show less diffusion. Spacial resolution and protein concentration in the spots are higher.

Horizontal gels

Water is not suitable, as it can cause electric shorting.

Wet the cooling plate with 2 to 3 mL of kerosene.

Wear gloves.

- Place the gel onto the center of the cooling plate with the film on the bottom; the side with the stacking gel must be oriented towards the cathode (–). The anodal edge should match exactly with line "5" on the scale on the cooling plate.

*Be sure to use **very** clean wicks, SDS would dissolve any traces of contaminating compounds.*

- Lay two of the electrode wicks into the compartments of the PaperPool. If smaller gel portions are used, cut them to size.
- Mix 10 mL of the cathode buffer with 10 mL distilled water and apply it onto the respective strip (Fig. 10).

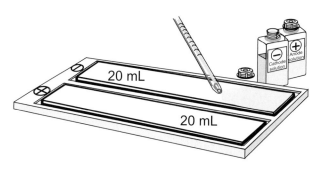

Fig. 10: Soaking the electrode wicks with the respective solution in the PaperPool.

- Mix 10 mL of the anode buffer with 10 mL distilled water and apply it onto the respective strip.

Always apply cathode wick first to avoid contamination of cathode buffer with leading ions.

- Place the cathode wick onto the cathodal edge of the gel; the edge of the wick matching "3.5" on the cooling plate; the anodal wick over the anodal edge, matching "13.5".

The size of the SDS gels is such that two separations in 11 cm long IPG strips can be run in parallel in a gel.

- Place the equilibrated and drained IPG strips, with the gel sides to the bottom, on the surface of the SDS gel 1 cm apart from the cathodal wick (Fig. 11).
- Electrophoresis: 15 °C, max. 30 mA, max. 30 W.

Stain the IPG strips with Acid violet to check whether all the proteins have been transferred.

- 75 min at max. 200 V, then remove the IPG strips (Fig. 12A) and place the cathodal electrode wick over the IPG-SDS gel contact area (Fig. 12B).

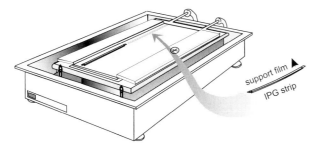

Fig. 11: Placing the equilibrated IPG strip on the SDS gel.

- Continue the separation at max. 800 V till the Bromophenol Blue has reached the anodic edge of the gel:
  Standard gel: 90 min Large scale: 5 h

Gels with the Tris-acetate/tricine buffer system (see page 38) need only half the separation time, because due to the low molarity of the gel buffer (0.12 mol/L Tris-acetate pH 6.6) the conductivity is lower:
First phase: 40 min
Second phase: Standard size: 40 min Large scale: 2 h 40 min

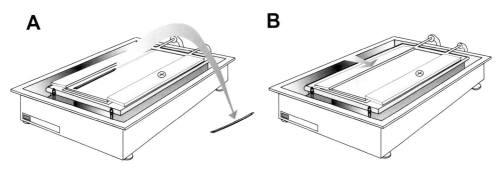

Fig. 12: Removing the IPG strips (A) and shifting the cathodal wicks over the contact surface (B).

5
Staining procedures

The most accurate results are obtained with the 2-D DIGE system: pre-labelling of the proteins with cyanine dyes, because the multiplexing method allows the use of a pooled sample mixture as internal standard.

See page 104 ff.

Colloidal Coomassie Blue staining

Neuhoff V, Stamm R, Eibl H. Electrophoresis 6 (1985) 427–448.

This method has the highest sensitivity of all Coomassie staining methods (ca. 30 ng per band), but takes overnight (Neuhoff *et al.* 1985).

Preparation of the staining solution:

The staining solution can be used several times.

Slowly add 100 g of ammonium sulfate to 980 mL of a 2% H_3PO_4 solution till it has completely dissolved. Then add Coomassie G-250 solution (1 g per 20 mL of water). Do not filter the solution! Shake before use.

Staining

The use of a stirrer is recommended for this step.

- *Fixing* 1 h in 12% (w/v) TCA, for gels on support films it is best to place the gel surface on the bottom (on the grid of the staining tank), so that additives with a higher density (e.g. the glycerol for gradient gels) can diffuse out of the gel.
- *Staining* overnight with 160 mL of staining solution [0.1% (w/v) Coomassie G-250 in 2% H_3PO_4, 10% (w/v) ammonium sulfate see above] plus 40 mL of methanol (add during staining).
- *Wash* for 1 to 3 min in 0.1 mol/L Tris, H_3PO_4 buffer, pH 6.5.
- *Rinse* (max. 1 min) in 25% (v/v) aqueous methanol.
- *Stabilize* the protein-dye complex in 20% aqueous ammonium sulfate.

Fast, cheap and environment friendly.

Fast Coomassie Brilliant Blue staining

0.025 % (w/v) Coomassie R-350 staining solution:

Dissolve 1 PhastGel Blue tablet in 1.6 L of 10% acetic acid.

Staining

- Heat the solution to 90 °C and pour it over the gel, which is in a stainless steel tray.
- Place the tray on a laboratory shaker for 10 min.

Destaining

Placing a paper towel into the destaining solution adsorbs the Coomassie dye.

- On a rocking table in 10% acetic acid for at least 2 h at room temperature. Change solution several times; recycle it by pouring it through a filter filled with activated charcoal.

Staining and destaining solutions can be used repeatedly.

Subsequent silver staining will reveal more protein spots. Silver staining with prior Coomassie Brilliant Blue staining is more sensitive than without prestaining, and detects proteins, which are otherwise missing.

Reversible imidazole-zinc negative staining

This staining procedure for SDS gels according to Hardy *et al.* (1996) produces non stained spots against a white background. A sensitivity down to 15 ng per spot has been reported. Eventually with an EDTA mobilization buffer the zinc-imidazole complex can be dissolved prior to electrophoretic blotting.

Some researchers prefert this method, when subsequent mass spectrometry analysis of some spots is performed because it does not modify the proteins.

- *Fix* the gel in 200 mol/L Imidazole containing 0.1 % SDS: 2.72 g Imidazole + 0.2 g SDS, dissolve in 200 mL distilled water. 15 min with continuous shaking.
- *Rinse* with distilled water.
- *Stain* (negative) with 200 mmol/L Zinc sulphate: 5.74 g $ZnSO_4$, dissolve in 200 mL H_2O_{dist}. Shake for 30 to 60 s until white background has developed.
- *Rinse* with distilled water.
- *Store* in 200 mL of new fixing solution diluted 1 : 10 with distilled water.
- *Mobilize* proteins with 50 mmol/L EDTA, 25 mmol/L Tris; pH 8.3: 0.61 g Tris + 3.72 g EDTA-Na_2, dissolve in 200 mL distilled water, adjust to pH 8.3 with a few grains of Tris when necessary. 6 min 200 mL with vigorous shaking.

Deep Purple™ fluorescent staining

All steps should be performed in dark containers at room temperature on a shaking platform.

Only for gels, which are backed by a non-fluorescent support or non-backed.

- *Fix* the gel in 7.5% (v/v) acetic acid, 10% (v/v) methanol for 1 h (overnight for backed gels).
- *Wash* non-backed gels in 200 mmol/L Na_2CO_3 for 30 min (for backed gels use 35 mmol/L $NaHCO_3$)
- *Stain* in 1:200 stain dilution for 1 h
- *Destain two times* for 15 min in 7.5% (v/v) acetic acid.
- *Store* the gels in 7.5% (v/v) acetic acid at 2–8 °C in the dark (several weeks possible).

Silver staining

Many different modifications of silver staining have been published. A detailed description of some of these procedures has been published by Rabilloud and Charmont (2000). Ammoniacal silver staining procedures are not as easy to use as silver nitrate methods: negative spots and the danger of the development of a silver mirror have often been reported.

The following modification of the silver nitrate technique, based on the method by Heukeshoven und Dernick (1986) is probably the most sensitive and reproducible one, when the quality of the chemi-

Rabilloud T, Chalmont S. In Rabilloud T, Ed. Proteome research: Two-dimensional gel electrophoresis and identification methods. Springer Berlin Heidelberg New York (2000) 107–126.
Heukeshoven J, Dernick R. Electrophoresis 6 (1985) 103–112.

cals and the water is high, and when the timing of the steps is exactly kept. The sensitivity of this method is of about 0.05 to 0.1 ng/mm^2.

Tab. 6: Silver staining acc. to Heukeshoven and Dernick (1986) for a 0.5 and a 1 mm thick gel on support film.

Step	Solution	V [mL]	t [min]	
		per step	0.5 mm	1 mm
Fixing	200 mL ethanol + 50 mL acetic acid with $H_2O_{dist} \rightarrow$ 500 mL 75 mL ethanol*) 17 g Na-acetate	250	2 × 15	2 × 60
Sensitizer	1.25 mL glutardialdehyde (25% w/v) 0.50 g $Na_2S_2O_3 \times 5\ H_2O$ with $H_2O_{dist} \rightarrow$ 250 mL	250	30	60
Washing	H_2O_{dist} 0.625 g AgNO$_3$**)	250	3 × 5	5 × 8
Silver	100 µL formaldehyde (37%) with $H_2O_{dist} \rightarrow$ 250 mL	250	20	60
Developer	6.25 g Na_2CO_3 100 µL formaldehyde (37%) with $H_2O_{dist} \rightarrow$ 250 mL ***)	250	3 to 7	7 to 10
Stopping	3.65 g EDTA-Na$_2 \times 2H_2O$ with $H_2O_{dist} \rightarrow$ 250 mL	250	10	50
Washing	H_2O_{dist}	250	3 × 5	3 × 30
Preserving	12.5 mL glycerol (87% w/v) with $H_2O_{dist} \rightarrow$ 250 mL	250	30	40
Total time			2 h 37 min	7 h
Drying	air-drying (room temperature)			

* First dissolve NaAc in water, then add ethanol. Add the thiosulfate and glutaraldehyde just before use.

** Dissolve AgNO$_3$ in water, add the formaldehyde before use.

*** Sensitivity can be improved by adding 7 µL of a 5 % (w/v) stock solution of $Na_2S_2O_3 \times 5\ H_2O$ to the 250 mL developer.

■ Note:

250 mL per step are enough for an automated gel stainer as decribed on page 128. Manual staining may require larger volumes.

Gels without film:

The staining conditions for 0.5 mm film-supported gels are applied on 1 mm unsupported gels, those for 1mm film-supported gels on 1.5 mm unsupported gels.

If larger volumes are prepared in advance: fixing solution should be made fresh, silver nitrate should be kept as 10 ×concentrated stock solution in a dark bottle, the aldehydes are added before use.

Silver staining of multiple gels

Novel 2-D electrophoresis instruments allow simultaneous runs of up to twelve gels. Silver staining of so many gels can become a bottleneck.

In Fig. 13 equipment is shown, which facilitates simultaneous staining of four large formate gels – with or without film support. The solutions are easily drained from the sieve-tray without mechanical stress for the gels. 1.5 L solution is needed per step. Several sets can be stacked on top of each other.

Fig. 13: Staining of multiple large formate gels on an orbital shaker. One tray set accommodates up to four gels. The tray sets can be stacked on top of each other.

Mass spectrometry analysis of silver stained spots

For subsequent spot analysis with mass spectrometry crosslinking of proteins within the gel must be avoided: omit glutardialdehyde from the sensitizer and formaldehyde from the silver nitrate solution (Shevchenko *et al.* 1996). The silver staining protocol described above

Shevchenko A, Wilm M, Vorm O, Mann M. Anal Chem 68 (1996) 850–858.
Yan JX, Wait R, Berkelman T, Harry RA, Westbrook JA, Wheeler CH, Dunn MJ. Electrophoresis 21 (2000) 3666–3672.

can be modified in the same way, the compatibility of this modification with mass spectrometry has been verified in a paper by Yan *et al.* (2000). It should be noted, that omission of aldehydes from the sensitizer and the silver solution leads to a loss of sensitivity by about 80 %.

Gharahdaghi F, Weinberg CR, Meagher D, Imai BS, Mische SM. Electrophoresis 20 (1999) 601–605.

Gharahdaghi *et al.* (1999) have published a paper, which could show, that sensitivity in mass spectrometry is enhanced, when the silver is removed prior to in-gel digestion of the protein.

It is very important, that staining is performed in closed trays, because contaminations with keratin and other stuff has to be prevented.

Preserving and drying of the gels

Film-supported gels with 2 %C crosslinking are dried at room temperature, the gel will not crack. For 3 %C crosslinked gels the preserving solution should contain 30 % glycerol. In order to prevent sticking of the surface the gels should be placed into a sheet protector.

Gels without film support can be dried between two sheets of cellophane as shown in Fig. 14. Because the gels swell during staining, they sometimes – particularly, when they do not fit into the frames – have to be shrunk back to the original size: they are placed into 30 % ethanol / 10 % glycerol until they have reached the appropriate formate. The cellophane must be prewetted with 10 % glycerol / water without the alcohol.

For *autoradiography* the gels are dried onto a thick filter paper in a hot vacuum dryer.

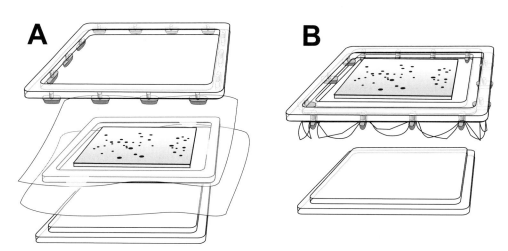

Fig. 14: Drying of a polyacrylamide gel between two sheets of cellophane using two plastic frames. (A) From the bottom: Loading platform – smaller frame – cellophane – gel – cellophane – larger frame. (B) The two frames with the sandwich are removed from the platform, the screws are turned by 90°.

Method 12:
PAGE of double stranded DNA

With the PCR® technique DNA fragments in the size range of 50 to 1,500 bp are amplified. For the analysis of these fragments, the application of thin horizontal polyacrylamide gels on film support with subsequent silver staining show several advantages over conventional agarose gel electrophoresis of DNA fragments:

These gels can be used directly, or after washing, drying and rehydration in the appropriate buffer. The gels can be prepared in the laboratory or purchased ready-made in the dried form.

- Polyacrylamide gels have a higher resolution than agarose gels.

and discontinuous buffer systems can be applied.

- Silver staining has a higher sensitivity (15 pg/band) and is less toxic than ethidium bromide. Silver staining is particularly useful for staining small fragments, because it is independent on the size.

Silver staining does not work well for agarose gels

- Bands are visible without an UV lamp, no photography is necessary for a permanent record.

Polyacrylamide gels can be dried and put in files.

- Due to the high sensitivity of silver staining, autoradioactivity can be replaced in many cases.

Thus no x-ray films are needed, no radioactive waste.

It should , however, not be forgotten, that the migration in native polyacrylamide gels is not only dependent on the size, but also on the sequence. Thus A and T rich DNA fragments migrate slower than they should according to their sizes.

Example: the "spike" of the 100 bp ladder shows up at 800 bp in agarose gels, but at 1,200 bp in native polyacrylamide gels.

A number of DNA typing methods are performed in gels with the short separation distance: Screening of PCR products, separation of RNA, random amplified polymorphic DNA (RAPD) and DNA amplification fingerprinting (DAF), heteroduplex analysis, and a part of minisatellite analysis.

The gel concentrations can vary from 5 to 15 % T.

In some cases, however, a higher resolution and more space for multiple bands can only be achieved by running gels in the long separation distance: differential display reverse transcription (DDRT) and minisatellite samples with shorter repeats, like the VNTR (variable number of tandem repeat) sample D1S80.

The electrode plate of the Multiphor chamber is square, thus it can be turned 90° and used for long distance separations.

Electrophoresis in Practice, Fourth Edition. Reiner Westermeier
Copyright © 2005 WILEY-VCH Verlag GmbH & Co. KGaA, Weinheim
ISBN: 3-527-31181-5

Both buffer systems can be employed for directly used gels or for washed gels.

The best results are obtained with discontinuous 0.5 mm thin gels and discontinuous buffer systems. In the following instruction two buffer systems are described: Tris-acetate / Tris-tricine and Tris-phosphate / Tris-borate-EDTA.

Seymour C, Gronau-Czybulka S. Pharmacia Biotech Europe Information Bulletin (1992).

Washed, dried and rehydrated 0.5 mm thin polyacrylamide gels show very high resolution, when rehydrated in the "PhastGel buffer system" (Seymour and Gronau-Czybulka, 1992); and they are free from acrylamide monomers.

1
Stock solutions

C = 2% in the resolving gel solution prevents the separation gel from peeling off the support film and cracking during drying.

Acrylamide, Bis solution(T = 30%, C = 2%):
29.4 g of acrylamide + 0.6 g Bis, to 100 mL with H_2O_{dist}.

*This solution is used for slightly concentrated **plateaus** with C = 3%, because the slot would become unstable if the degree of polymerization were lower.*

Acrylamide, Bis solution" (T = 30%, C= 3%):
29.1 g of acrylamide + 0.9 g Bis, to 100 mL with H_2O_{dist}.

■ Caution!
Acrylamide and Bis are toxic in the monomeric form. Avoid skin contact and dispose of the remains ecologically (polymerize the remains with an excess of APS).

It can be stored for one week in the refrigerator (4°C).

Ammonium persulfate solution (APS) 40% (w/v):
Dissolve 400 mg of ammonium persulfate in 1 mL H_2O_{dist}.

Buffer System I (Tris-acetate / Tris-tricine)

With this buffer, the storage time of the gel is not limited. Because the pH value of the gel is lower than pH 7, the matrix does not hydrolyse.

Gel buffer 0.448 mol/L Tris-acetate pH 6.4 (4 × conc):
5.43 g Tris, dissolve in 80 mL H_2O_{dist}; titrate to pH 6.4 with acetic acid; make up to 100 mL with H_2O_{dist}.

Anode buffer 0.45 mol/L Tris-acetate pH 8.4:
27.3 g Tris, dissolve in 400 mL H_2O_{dist}; titrate to pH 8.4 with acetic acid; make up to 500 mL with H_2O_{dist}.

Cathode buffer 0.08 mol/L Tris / 0.8 mol/L tricine:
4.85 g Tris + 71.7 g tricine, make up to 500 mL with H_2O_{dist}.

For washed gels, electrode solutions with lower concentrations can be employed (than for nonwashed gels).

Electrode buffer for washed and rehydrated short distance gels 0.2 mol/L Tris-0.2 mol/L tricine – 0.55 % SDS:
24.2 g Tris + 35.84 g tricine + 5.5 g SDS, to 1 L with H_2O_{dist}.

Buffer System II (Tris-phosphate / TBE)

Gel buffer 0.36 mol/L Tris-phosphate pH 8.4 (4 × conc):
4.36 g Tris, dissolve in 80 mL H_2O_{dist}; titrate to pH 8.4 with phosphoric acid; to 100 mL with H_2O_{dist}.

Electrode buffer 450 mmol/L Tris / 75 mmol/L boric acid / 12.5 mmol/L EDTA-Na₂ (5 × TBE):
54.5 g Tris + 23.1 g boric acid + 4.65 g EDTA Na_2, to 1 L with H_2O_{dist}.

The Tris-phosphate buffer is much better for polymerization than TBE, because borate inhibits the copolymerization of the gel and the Gelbond PAGfilm.

Bromophenol blue solution (1%):
100 mg Bromophenol blue, make up to 10 mL with H_2O_{dist}.

Xylencyanol solution (1%):
100 mg Xylencyanol, make up to 10 mL with H_2O_{dist}.

0.2 mol/L EDTA Na₂ solution:
7.44 g EDTA Na_2, make up to 10 mL with H_2O_{dist}.

Sample buffer:
22 mL H_2O_{dist} + 3 mL gel buffer + 60 µL bromophenol blue solution (1%) + 40 µL Xylencyanol solution (1%) + 250 µL 0.2 mol/L EDTA-Na_2 solution.

2
Preparing the gels

Slot former
Samples are applied in small wells which are molded in the surface of the gel during polymerization. To form these slots a mould must be fixed on a glass plate with a 0.5 mm thin U-shaped silicon rubber gasket.

The cleaned and degreased glass plates with 0.5 mm U-shaped spacer is placed on the template (slot former template in the appendix). For long distance gels the "reswelling cassette" is used. A layer of "Dymo" tape (6 mm wide embossing tape, 250 µm thick) is applied, avoiding air bubbles, at the starting point. The slot former is cut out with a scalpel (see Fig. 1). After pressing the individual slot former pieces against the glass plate, the remains of sticky tape are removed with methanol.

"Dymo" tape with a smooth adhesive surface should be used. Small air bubbles can be enclosed when the adhesive surface is structured, these inhibit polymerization and holes appear around the slots.

The casting mold is then made hydrophobic by spreading a few mL of Repel Silane over the whole slot former with a tissue under the fume hood. When the Repel Silane is dry, the chloride ions, which result from the coating are rinsed off with water.

This treatment only needs to be carried out once

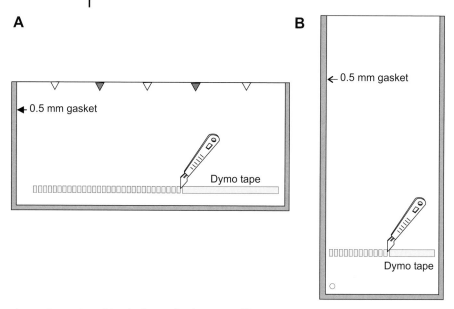

Fig. 1: Preparation of the slot former for short (A) and long (B) separation distances.

Assembling the casting cassette

For mechanical stability and to facilitate handling, the gel is covalently polymerized on a support film. The glass plate is placed on an absorbent tissue and wetted with a few mL of water. The GelBond PAG film is applied with a roller with the untreated hydrophobic side down (Fig. 2). A thin layer of water then forms between the film and the glass plate and holds them together by adhesion. The excess water, which runs out, is soaked up by the tissue. To facilitate pouring in the gel solution, the film should overlap the long edge of the glass plate by about 1 mm.

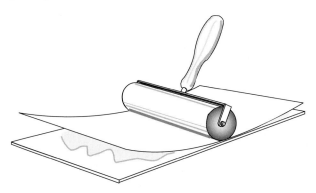

Fig. 2: Applying the support film with a roller.

The finished slot former is placed on the glass plate and the cassette is clamped together (Fig. 3)

The cassette for long gels is clamped along the long side.

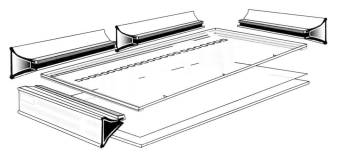

Fig. 3: Assembling the gel cassette for short distances.

Cool the casting cassette in the refrigerator at 4 °C for about 10 min: this delays the onset of polymerization. This step is necessary because the stacking gel with large pores and the resolving gel with small pores are cast in one piece. The polymerization solutions which have different densities take 5 to 10 min to settle.

In the summer in a warm laboratory, the gel solutions should also be brought to 4 °C.

Tab. 1: Composition of the gel solutions for one gel

	Stacking gel 4%T / 3%C	Resolving gel 10%T / 2%C	15%T / 2%C
Glycerol (85 %)	1.0 mL	0.3 mL	0.3 mL
Acrylamide, Bis 30%T, 3%C	0.65 mL	–	–
Acrylamide, Bis 30%T, 2%C	–	5.0 mL	7.5 mL
Gel Buffer	1.25 mL	3.75 mL	3.75 mL
TEMED	3 µL	8 µL	8 µL
With H_2O_{dist} fill up	→ 5 mL	→ 4 mL	→ 4 mL
APS (40 %)*	5 µL	15 µL	15 µL
Pipet into cassette:	3 mL	13 mL	13 mL

*) APS is only added shortly before filling the cassette.

Filling the cooled gel cassette

The cassettes are filled with a 10 mL pipette or a 20 mL syringe (Fig. 4). Draw the solution into the pipette with a pipetting device. The stacking gel plateau is introduced first, and then the resolving gel which contains less glycerol and is less dense. Pour the solutions in slowly. The gel solution is directed into the cassette by to the piece of film sticking out.

Never pipette the toxic monomers by mouth!

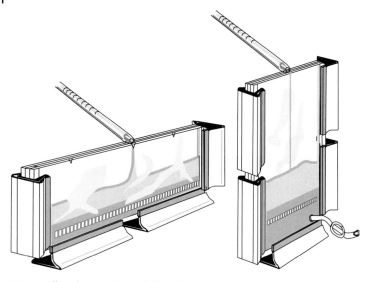

Fig. 4: Filling the cassettes with the polymerization solutions.

If air bubbles are trapped in the solution, they can be dislodged with a long strip of polyester film.

100 μL of 60 % (v/v) isopropanol are then layered in each filling notch. Isopropanol prevents oxygen, which inhibits polymerization, from diffusing into the gel. The gel will then present a well-defined, aesthetic upper edge.

Polymerization

Let the gel stand at room temperature.

Removing the gels from the casting cassette

There is a slow "silent polymerization" after the gel has solidified, which should be completed before the gel is used or washed.

After the gel has polymerized over night, the clamps are removed and the glass plate gently lifted off the film with a spatula. The gel can slowly be pulled away from the spacer by grasping a corner of the film.

The gel is used directly or after it has been washed, dried and rehydrated in the gel buffer.

Washing the gels

The remains of monomers, APS and TEMED are removed from the gel.

The gels are washed three times for 20 min by shaking in double-distilled water. The last washing solution should contain 5% glycerol. Air-dry the gel overnight. Store in the refrigerator.

3
Sample preparation

PCR® products in general

PCR® products, which are well detectable with Ethidium bromide, must be diluted: 1 + 3 parts of sample buffer. 6 μL are applied.

If they are applied nondiluted, the resolution is inadequate.

RAPD and DAF samples

Low stringency PCR conditions are applied to allow the primer to anneal to multiple sites on the DNA. A wide spectrum of fragments are amplified. Practice has shown, that RAPD samples should be more diluted (up to 1 + 9 parts of sample buffer). 6 μL are applied. A typical result is shown in Fig. 17 on page 29.

Typical low stringency PCR conditions for RAPD: 45 cycles 94 °C 1 min; 36 °C 1 min; 72 °C 2 min.

Heteroduplex samples

The concentration for Ethidium bromide staining (50 ng/μL) must be used to achieve heteroduplexes instead of exclusively single strands and homoduplexes.

The PCR® amplicates of the wildtype and the mutant are mixed 1 :1 and heated to 95 °C for 2 minutes without formamide or urea. The samples are immediately cooled in an icebath for 10 minutes for annealing. For the separation they are diluted 1 + 9 parts sample buffer. 6 μL are applied.

Minisatellites (VNTR)

PCR® amplicates are diluted: 1 + 3 parts of sample buffer. 6 μL are applied.

Systems with alleles with repeats of 70 bp (e.g. YNZ 22) can be well resolved in short gels with 10 % T.

16 bp repeats (e.g. D1S80) need long distance gels.

For native PAGE of STRs see Schickle (1996).

DDRT

For optimized amplification procedures work according to Bosch and Lohmann (1996).

Separations are either performed in native long distance gels or in 15 %T short distance denaturing gels (see method 15).

Bosch TCG, Lohmann J. In Fingerprinting Methods based on PCR. Bova R, Micheli MR, Eds. Springer Verlag Heidelberg, in press.

4
Electrophoresis

In many cases enough resolution is obtained with directly used gels, when double stranded DNA has to be separated in short distances. For long distances and very high resolution it is recommended to use washed and rehydrated gels.

Rehydration of washed and dried gels in the gel buffer

Rehydration solution:
6.25 mL gel buffer + 1 mL glycerol + 1 mL ethylenglycol + 0.5 mL bromophenol blue solution, make up to 25 mL with distilled water.

The gels for short distances can be used in one piece, or – depending on the number of samples – cut into smaller portions with scissors (when they are still dry). The rest of the gel should be sealed airtight in a plastic bag and stored in a freezer.

Lay GelPool onto a horizontal table; select the appropriate reswelling chamber, pipet the rehydration solution into the chamber, for

a complete gel:	25 mL
a half gel (short):	13 mL

Set the edge of the gel-film – with the gel surface facing down – into the rehydration buffer (Fig. 5) and slowly lower it, avoiding air bubbles.

Very even rehydration is obtained when performing it on a shaker at a slow rotation rate (Fig. 5C). If no shaker is used, lift gel edges repeatedly.

Using forceps, lift the film up to its middle, and lower it again without catching air bubbles, in order to achieve an even distribution of the liquid (Fig. 5 B). Repeat this during the first 10 min.

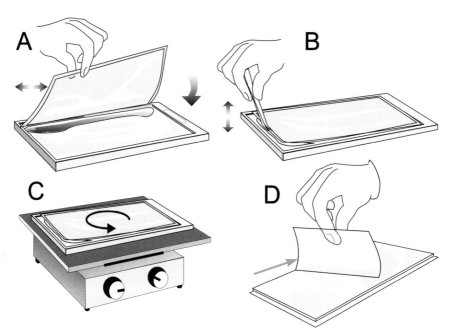

Fig. 5: Rehydration of a gel.
A Placing the dry gel into the GelPool. **B** Lifting the gel for an even distribution of the liquid. **C** Rehyd-ration on a rocking platform (not always necessary). **D** Removing the excess buffer from the gel surface with filter paper.

60 min later the gel has reswollen completely and is removed from the GelPool. Dry sample wells with clean filter paper, wipe buffer off the gel surface with the edge of a filter paper (Fig. 5D).

When the gel surface is dry enough, this is indicated by a noise like a whistle.

Preparation of the electrode wicks

Short distance gels: Lay two of the 25 × 5 cm electrode wicks into the compartments of the PaperPool. Apply 20 mL of the respective electrode buffer to each wick (Fig. 6).

Do not forget, that for one buffer system different anode and cathode buffer is used.

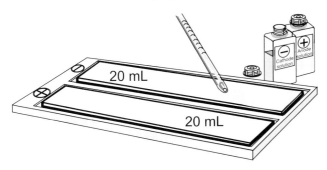

Fig. 6: Soaking the wicks with electrode buffer.

Long distance gels: Cut six electrode strips of 11.7 × 1.8 cm. Lay three of them stacked into the compartments of the PaperPool. Apply 10.5 mL of the respective buffer to each stack.

Application of the gel and the electrode wicks

Switch on the thermostatic circulator, adjusted to the 15 °C. Apply a very thin layer of kerosene (ca. 0.5 mL) onto the cooling plate with a tissue paper, in order to ensure good cooling contact (a few air bubbles do not matter).

Short distance run: Place the gel (surface up) on to the center of the cooling plate: the side containing the wells must be oriented towards the cathode (Fig. 7; Multiphor: line 5).

Place the cathodal strip onto the cathodal edge of the gel, edge of the electrode strip matching "3.5" on the cooling plate. The edge of the strip should be at least 4 mm away from the edges of the sample wells (otherwise small DNA fragments will exhibit less sharpness or hollow bands).

Place the anode strip over the anodal edge, matching "13.5" on the cooling plate. Smooth out air bubbles by sliding bent tip forceps along the edges of the wicks laying in contact with the gel.

short distance run   **long distance run**

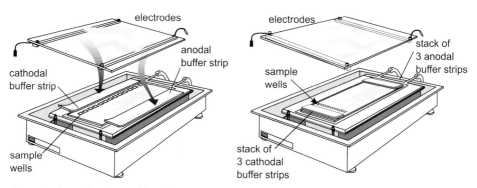

Fig. 7:   Appliance for short and long distance runs of DNA fragments.

Long distance run: Place the gel (surface up) on to the center of the cooling plate: the side containing the wells must be oriented towards the cathode.

Place the cathodal stack electrode strips to the cathodal side of the gel, the inner edge matching with line 23 of the cooling plate.

Place the other stack over the anodal side, the inner edge matching with line 2 of the cooling plate. Smooth out air bubbles by sliding bent tip forceps along the edges of the wicks laying in contact with the gel.

Sample application and electrophoresis

Apply 6 µL of each sample to the sample wells using a micropipette. Clean platinum electrode wires before (and after) each electrophoresis run with a wet tissue paper. Move electrodes so that they will rest on the outer edges of the electrode wicks. Connect the cables of the electrodes to the apparatus and lower the electrode holder plate (Fig. 7). Close the safety lid.

Running Conditions:

The suggested running conditions have to be modified for some applications: e.g. when samples of the size range 700 to 900 bp have to be well separated, at least two hours separation time is required. The dyes added to the samples are a help for the estimation of the running time: In a 10%T gel the Xylencyanol dye migrates with the same mobility like 200 bp fragments, in a 15 %T gel it migrates like 100 bp fragments.

Short distance run in a 10 % T gel,
Tris-acetate / tris-tricine buffer, 15 °C, whole gel:

Power supply settings   600 V,   25 mA,   15 W,   1 h 20 min

Short distance run in a 10 % T gel,
Tris-phosphate / tris-borate buffer, 15 °C, whole gel:

100 V,	10 mA	5 W,	20 min
600 V,	30 mA	10 W,	45 min

Power supply settings

Long distance runs:

> ■ Note:
> **High voltage and relatively low milliampere**
> **values are applied on long electrophoresis gels.**
> **This means, that with a milliampere-constant con-**
> **trol only little changes in conductivity (e.g. from**
> **25 to 28 mA) can result in severe changes of the**
> **voltage value, e.g. from 250 to 360 V.**

The electrophoresis pattern in long gels is highly dependent on the temperature and the running conditions. The conductivities of the cables and electrodes in the electrophoresis chamber have a great influence on the separation.

Different chambers have different cable and electrode lengths and diameters, which results in different conductivites inside the apparatus.

The following power supply settings are valid for one chamber type. They may be changed for a different chamber. The actual **voltage** values are most important for optimized separation patterns.

Tab. 2: Long distance run in a 10 % T gel, Tris-phosphate / tris-borate buffer, 15 °C

Phase	set	actual	set	actual	set	actual	time
1	150 V	150 V	15 mA	11 mA	5 W	2 W	30 min
2	400 V	400 V	31 mA	31 mA	10 W	10 W	10 min
3	500 V	360 V	24 mA	24 mA	10 W	7 W	45 min
4	800 V	490 V	28 mA	28 mA	13 W	13 W	2 h 20 min
Total:							3 h 45 min

Control with Xylencyanol band: at the end m_R = 230 bp should be 1 cm away from the edge of the anodal filter paper.

When a different chamber is employed, the *mA values* of the 3rd and 4th phase have to be adjusted to reach the actual voltage values of 360 V and 490 V, respectively, at the beginning of these phases: for example phase 3 → 22 mA and phase 4 → 23 mA.

It is, however, easier to employ voltage-ramping, as described in the next paragraph.

Voltage-Ramping:

In the power supply the menu "Setup" has to be switched from Volt Level=fixed to Volt Level=changing (or =gradient).

Current and power set to a maximum value.

Then the power supply will control the voltage values over the time. The power supply calculates a linear transition of the voltage from phase to phase

At least two phases have to be defined.

Further advantages of voltage-ramping:

- If the buffer system is slightly varied, the settings of the power supply do not have to be altered.
- The width of the gel does not matter.

Tab. 3: Ramping programme. Voltlevel: gradient

Phase	set	set	set	time
1	150 V	30 mA	20 W	1 min
2	150 V	30 mA	20 W	29 min
3	400 V	50 mA	30 W	1 min
4	400 V	50 mA	30 W	29 min
5	500 V	50 mA	30 W	45 min
6	800 V	50 mA	30 W	2 h 20 min
total				*3 h 45 min*

5
Silver staining

The volumes are for autopmated staining, manual staining requires at least 200 mL.

Several silver staining protocols are published. The main benefit of the procedure described: Benzene sulphonic acid in the fixing and in the silver nitrate solution is very efficient for fixing also small oligo nucleotides (<150 bp).

Tab. 4: Sensitive silver staining (20 to 50 pg per band)

Step	Solution	Volume	t [min]	
			Native gels	*Denaturing gels*
Fix	0.6 % (w/V) benzene sulphonic acid in 24 % (v/v) ethanol	150 mL	30	40
Wash	H_2O_{dist}	150 mL	–	3 × 10
Stain	0.2 % $AgNO_3$ (w/v) + 0.07 % (w/V) benzene sulphonic acid	150 mL	30	40
Wash	H_2O_{dist}	150 mL	1	2
Develop	2.5 % Na_2CO_3 + 150 µL formaldehyde + 150 µLNa-thiosulphate (2 %)	150 mL	6 (visual control)	7 (visual control)
Stop/ Preserve	2 % (v/v) acetic acid + 5 (w/v)Na-acetate + 10 % glycerol (v/v)	150 mL	30	30
Air dry				

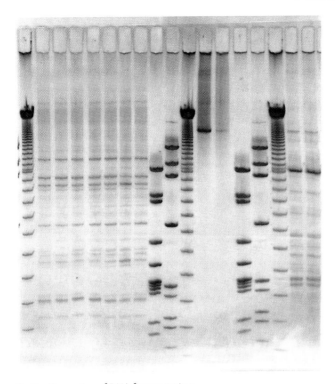

Fig. 8: Separation of DNA fragments in a
rehydrated polyacrylamide gel with a
discontinuous buffer system. Silver staining.

Ethidium bromide staining cannot be used, because, the polester
support film is fluorescent. However, fluorescent labels excitation at
red light wavelength, like Cy5 can be employed.

Method 13:
Native PAGE of single stranded DNA

The most certain and sensitive method for the detection of mutations is DNA sequencing analysis. However, this method is too costly and time-consuming for screening purposes.

A wide range of new methods in clinical diagnostics is based on mutation detection.

SSCP (single strand conformation polymorphism) analysis is a very powerful technique for mutation detection, and it can easily be applied for screening purposes.

As already mentioned on page 29, SSCP can also be employed for genetic differentiations.

The principle: Variations in the sequence as small as one base exchange alter the secondary structure of ssDNA. Changes in the sequence cause differences in the electrophoretic mobility, which are observed as band shifts (Orita et al. 1989). The mechanism of SSCP is described as: Differential transient interactions of the bent and curved molecules with the gel fibers during electrophoresis, causing the various sequence isomers to migrate with different mobilities.

The folded structure of the single strands is maintained by sequence-specific intramolecular base-pairing, which can be influenced by the running conditions.

Fig. 1 shows a SSCP analysis gel with silver stained ssDNA fragments of p53 genes (Schickle, 1996).

Schickle HP. GIT LaborMedizin 19 (1996) 159–163.

Before screening, the mutants have to be defined by direct sequencing. PCR products are denatured by heating with formamide or sodium hydroxid, chilled in an ice-water bath, and loaded on a nondenaturing polyacrylamide gel for electrophoresis.

The sequences of the two primers for the amplified fragment have to be detected.

Many samples can be screened with a considerably lower effort than direct sequencing in a relatively short time, namely within a few hours.

SSCP analysis can be performed in all types of electrophoresis chambers for polyacrylamide gels, provided that the temperature can be controlled. However, horizontal systems have several advantages over vertical systems:

- With Peltier cooling the temperature can be exactly controlled.
- The gel size can be easily modified.
- Washed and rehydrated gels can be employed.
- Experiments for the optimization of conditions are easy to perform.
- The handling of film-supported gels is very easy.

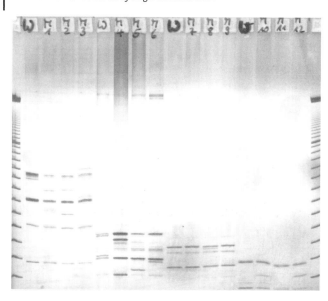

Fig. 1: SSCP analysis: Different exons in p53 mutation screening. Horizontal 0.5 mm thin polyacrylamide gel 10% *T*, 2% *C*. 10 °C; silver staining; for the other conditions see Schickle HP (1996a).

Ethidium bromide can not detect single stranded DNA.

Silver staining of the DNA fragments can replace the radio-isotopic detection methods, when this multi-step procedure can be performed without breaking of the gel. Staining of large and thin, unbacked gels, which have been run in vertical chambers are not easy to stain and handle.

This can be checked on the RNA or protein level.

SSCP – like other DNA analysis methods – can not distinguish between silent mutations which do not cause a disease, and mutations that are disease-associated.

For some exons only 60 % of mutations can be detected so far.

SSCP analysis is not a replacement but an addition to sequencing, when 100 % of defined mutations have to be detected. The band shifts do not show up automatically for all mutations and under all conditions. Unfortunately, there is not a single and unique separation condition, which can be applied to the separations of all exons.

Several parameters influence the results in SSCP analysis:

DNA-fragment-length,
denaturation procedure,
gel concentration,
crosslinking factor,
buffer system,
gel length,
temperature,
field strength,
running time,
additives (e.g. glycerol).

Hayashi and Yandell (1993); Jaksch M, Gerbitz K-D, Kilger C. J Clin Biochem. 1995.

Thus, when a mutation does not appear in the first experiment, that does not mean, that SSCP analysis can not be employed for this exon.

The following chapter should be seen as a support for selecting the optimal media and conditions for SSCP analysis.

1
Sample treatment

Fragment size
The fragments should not be longer than 200 bp. The mutation should be closer to the center of the amplified sequence than to the end.

Denaturing
In order to find the best and most reliable denaturation procedure, the author has randomly checked publications on SSCP analysis, which are listed in Tab. 1. It is sure, that the way of denaturation has a great influence on the conformation of the singel strands. Also the effectiveness of chilling, and the time for pipetting on the gel play an important role.

The author has stopped after the eighth publication.

The formamide solution should always contain 0.05% bromophenol blue and 0.05 % xylencyanol for easy pipetting and migration control.

This solution is sometimes called "stop solution".

Tab. 1: Denaturating procedures for SSCP analysis from eight randomly selected publications

PCR® product	formamide	EDTA mmol/L	NaOH mmol/L	Temp °C	t min
20 µL	–	–	–	95	5
10 µL	10 µL	–	–	85	3
10 µL	10 µL	–	–	95	3
5 µL	10 µL	–	–	95	5
3.5 µL	5.5 µL	–	–	95	5
20 µL	–	1	33	50	10
1 µL	9.0 µL	20	–	95	2
10 µL	10 µL	5	–	94	10
	+ 4.6 mol/L urea			65	5
	A three step procedure!			20	15

Chilling of the samples must be performed in ice-water, because the cooling effect of crushed ice is insufficient.

Pipetting time should be as short as possible. If there are samples with a strong tendency to form double strands and a high number of samples has to be pipetted, multi-channel pipettes might be useful.

2
Gel properties

Composition

Vidal-Puig A, Moller DE. BioTechniques 17 (1994) 492–496.

Vidal-Puig and Moller (1994) have compared different gel media with different exons, and could show differences in the sensitivity for mutation detection.

The total monomer concentration can vary from 5 to 20 %T. With a stacking gel the resolution is improved.

A gel with high crosslinking factor is more hydrophobic than a low crosslinked gel.

In general, the lower the hydrophobicity of the matrix, the higher is the sensitivity of mutation detection. This hypothesis could be proved by the discovery, that more mutations were detected in gels with low crosslinking. The gel compositions described methods 1, 4, 7 and 11 are suitable for SSCP analysis, because the resolving gels have a crosslinking factor of only 2 %C.

Also ready-made gels with 2% crosslinking are available.

For preparation and rehydration of these washed and dried gels, see methods 4 and 11.

Because temperature is an important parameter, it should be kept in mind, that the Joule heat during electrophoresis is developed by all ionic conmpounds inside the gel. The use of washed, dried and rehydrated gels is recommended.

Gel length

It has been shown in many publications, that a separation distance as short as 4 cm can be sufficient. However, minor mobility differences are better resolved in larger gels.

3
Buffers and additives

Buffer composition

SSCP analysis in vertical gels is mainly run in TBE (tris-borate EDTA) buffer. However, the composition of TBE buffer is not standardized: In the literature the content of boric acid varies from 75 to 90 mmol/L, the EDTA from 2.5 to 20 mmolL. Good results are also achieved in SDS gels with the standard buffer compositions.

For *vertical gels* the following buffer according to Maniatis *et al.* (1982) is suggested:

For the recipe see method 8.

TBE system

90 mmol/L Tris/ 75 mmol/L boric acid / 12.5 mmol/L EDTA-Na$_2$

in both, the gel and the electrode tanks.

For *horizontal gels* with electrode strips the following discontinuous buffers produce good results:

For the recipes see method 12.

Tris-acetate / Tris-tricine system

Gel:	112 mmol/L Tris-acetate pH 6.4,
	3.75 % glycerol, 3.75 ethylenglycol.
Anode:	0.45 mol/L Tris-acetate pH 8.4,
Cathode:	0.08 mol/L Tris / 0.8 mol/L tricine.

With this buffer, the storage time of the gel is not limited. Because the pH value of the gel is lower than pH 7, the matrix does not hydrolyse.

Alternative electrode buffer for washed and rehydrated or very short gels (PhastSystem®):

0.2 mol/L Tris-0.2 mol/L tricine – 0.55 % SDS

Tris-phosphate / TBE system

Gel:	90 mmol/L Tris-phosphate pH 8.4,
	3.75 % glycerol, 3.75 ethylenglycol.
Electrodes:	450 mmol/L Tris / 375 mmol/L boric acid /
	12.5 mmol/L EDTA-Na$_2$ (5 × TBE)

Additives

At least 10 % glycerol should be added to the rehydration or polymerisation solution. It is disputed, whether SDS is helpful or not.

4
Conditions for electrophoresis

The gel temperature must be exactly controlled.

Temperatures of 5, 15 or 25 °C produce different mobility shifts for single stranded DNA. Some laboratories run the same sample three times at these different temperatures to be sure not to miss a mutation.

In some cases the field strength does obviously not cause migration differences.

The field strength has to be limited to less than 20 V/cm for some samples. As a consequence, some experiments take four hours up to over night.

As already mentioned above, care should be taken, that the chilling and pipetting time does not become too long.

5
Strategy for SSCP analysis

- Use short fragments (< 200 bp).
- Denature samples by heating for 3 minutes at 95 °C with 50 % formamide (1 part PCR® product)
- + 1 part formamide) and place them in ice-water for at least 5 min prior to application on the gel.

Even standard SDS gels, as described in methods 7 and 8 can be employed.

- Try the most convenient way: ready-made gels and buffers, or the gels and buffers described above, containing 10 % (v/v) glycerol.
- The resolving gels must not have more than 2 % crosslinking.
- Try 15 °C.

See methods 7, 8, and 12 for running conditons.

- Apply the standard field strength suggested for the gels and buffers used.

If there is a problem of the single strands forming double strands during sample application, it can be tried to solve the problem by adding formamide to the stacking gel. Either cast a gel in this way or equilibrate the stacking gel area of a film-supported gel separately in a buffer solution containing 40 % (v/v) formamide like described in method 7 on page 199.

When, after the first trial, the defined mutant(s) is (are) not distinguishable from the wildtype, try the following modifications in the suggested sequence:

- Try different temperatures, first 5 °C, then 25 °C.
- Use low field strength (20 V/cm) for as long time as possible on the equipment available.
- Try higher gel concentration: 15 or 20 %T.

- Try different buffers in rehydrated gels*).
- Add more glycerol or other hydrophilic additives like mono- *Here are chances for new devel-* ethylenglycol*). *opments.*
- Modify denaturation procedure (see Tab. 1).
- Modify fragment length, try another primer pair.
- Try DGGE, TGGE or other mutation detection methods.

*) One of the benefits of washed and dried gels on film support is, that they can be cut into strips, the strips can be rehydrated in different buffers and / or additives, and they can be run side by side on the cooling plate.

Silver staining is carried out as described on page 310 for native gels.

Method 14:
Denaturing gradient gel electrophoresis

With denaturing gradient gel electrophoresis (DGGE) single base exchanges in segments of DNA can be detected with almost 100 % efficiency. The principle of DGGE is based on the different electrophoretic mobilities of partially denatured molecules caused by differences in DNA melting (see page 30 f).

This procedure needs much more efforts than SSCP, and is therefore by far less frequently employed.

Typically the 100 % denaturant gel part contains 7 mol/L urea and 40 % formamide. The gels are run at temperatures between 40 °C and 60 °C.

A denaturant gradient perpendicular to the electrophoresis direction allows the identification of the region of a point mutation. A denaturant gradient parallel to the electrophoresis run is better for screening, because many samples are run in one gel. Constant denaturing gel electrophoresis (CDGE) is employed for screening, when the denaturant concentration of differential melting of a DNA segment has been detected with DGGE.

Also with these techniques it is not possible to distinguish between silent mutations which do not cause a disease, and disease-associated mutations.

In practice, there are different technical solutions of preparing the gels and running them:

- Freshly prepared gradient gels are used in a vertical or a horizontal system.
- Washed and dried gels are rehydrated in a gradient and run horizontally.

The vertical system must have a special design because of the heating.

The latter method uses a technique, which has been introduced by Altland *et al.* (1984) for preparing immobilized pH gradient gels with a perpendicular urea gradient. Barros *et al.* (1991) have rehydrated home-made prepolymerized, washed and dried gels in a denaturant gradient, with 6 mol/L urea, 20 % formamide, and 25 % glycerol in the 100 % denaturant solution for DGGE, and run them on a horizontal chamber, the PhastSystem.

Electrophoresis in Practice, Fourth Edition. Reiner Westermeier
Copyright © 2005 WILEY-VCH Verlag GmbH & Co. KGaA, Weinheim
ISBN: 3-527-31181-5

In the following instruction very simple and reproducible procedures are described for preparing and running DGGE in perpendicular and parallel gradients, as well as CDGE. For the preparation of the washed and dried gels follow the instructions in Method 12.

1
Sample preparation

In the fragments analysed, the mutation must be in the part of the sequence with a low melting domaine. In order to detect every mutation, a GC clamp with 30 to 40 bp is frequently added to the fragment during PCR as a high melting domain. The fragment sizes are between 100 and 900 bp.

2
Rehydration solutions

Dye:
Prepare a 1 % (w/v) stock solution of bromophenol blue.

Gel buffer 0.36 mol/L Tris-phosphate pH 8.4 (4 × conc):
4.36 g Tris, dissolve in 80 mL H_2O_{dist}; titrate to pH 8.4 with phosphoric acid; make up to 100 mL with H_2O_{dist}.

Electrode buffer 450 mmol/L Tris / 75 mmol/L boric acid / 12.5 mmol/L EDTA-Na$_2$ (5 × TBE):
54.5 g Tris + 23.1 g boric acid + 4.65 g EDTA Na$_2$, make up to 1 L with H_2O_{dist}.

Tab. 1: Rehydration solutions: dense denaturing (7 mol/L urea and 40 % (v/v) formamide) and light native

	Dense denat.	Light native
Urea	4.2 g	–
Gel buffer (4 × conc)	2.5 mL	2.5 ml
Formamide	4.0 mL	–
Bromophenol blue (1%)	200 µL	–
Distilled water	–	7.5 mL
mix thoroughly	= 10 mL	= 10 mL

3
Preparing the rehydration cassette

a) Perpendicular gradient

Two glass plates 12.5 × 12.5 cm are needed. Cut a washed and dried gel in the middle (Fig. 1A). Place one glass plate on a paper towel. Pipette ca. 3 mL water on the surface. Lay the gel with the carrier film downward on the liquid, which distributes evenly between glass and film. The gel-film must jut out with the edge, which has no film margin.

Preferably a 15 %T / 2 % C gel is used here, which is prepared according to the instructions in Method 12 and 15.

Tilt the glass plate with the film up and let it stand on the paper towel for ca. 5 min (Fig. 1B). The excess water will be drawn out by gravitation and soaked off by the paper: in this way the film will adhere firmly and evenly to the glass plate.

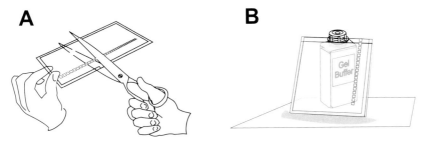

A **B**

Fig. 1: **A**: Cutting the dried gel to size. **B**: Generation of a thin water film to attach the hydrophobic side of the carrier film evenly to the glass plate.

Prepare an U-shaped gasket: Using a scalpel, cut a square of 12.5 × 12.5 cm out of a 0.5 mm silicon rubber plate or a stack of four layers of Parafilm®. Cut a rectangle of 11.5 × 12 cm out of this square leaving an U-shaped gasket of 5 mm width.

Unfortunately this gasket it not available as a ready-made product.

The glass plate with the attached dry gel-film is placed on the bench horizontally, and the gasket is laid on the three margin edges of the dry gel-film (Fig. 2). Then the second glass plate is laid on top; the sandwich is clamped together with the clamps, using two clamps at the bottom (see Figs. 3 and 4).

The gel-film should not be attached to the glass plate with a roller, because the surface is sticky and sensitive to scratching.

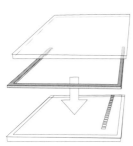

Fig. 2: Assembling the rehydration cassette.

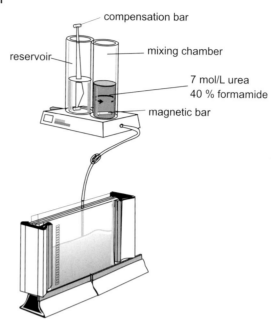

compensation bar

reservoir

mixing chamber

7 mol/L urea
40 % formamide

magnetic bar

Fig. 3: Casting a perpendicular denaturing gradient.

b) Parallel gradient

See method 12.

Complete gel: The large glass plate and the U-frame is used.

Half gel: Cut the washed and dried gel on all sides to 125 mm. The sample wells must be oriented towards the upper part of the cassette (see Fig. 4). The gel-film is laid on the glass plate with the upper edge protruding by ca. 1 mm to have a lead for the tip of the tubing of the gradient maker. All other steps are performed like described above (see also Fig. 2).

c) Constant denaturing gels

See method 12.

Complete gel: The large glass plate and the U-frame is used.

Half gel: Cut the washed and dried gel on all sides to 125 mm. The sample wells must be oriented towards the upper part of the cassette (see Fig. 4). The gel-film is laid on the glass plate with the upper edge protruding by ca. 1 mm to have a lead for the tip of the tubing of the gradient maker. All other steps are performed like described above (see also Figs. 1 and 2).

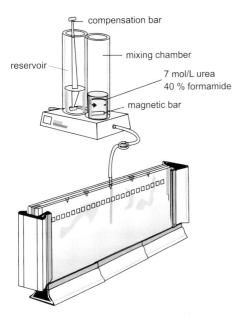

Fig. 4: Casting a parallel denaturing gradient.

4
Rehydration

The perpendicular gradient

- The valve and the clamp on the tubing must be closed.
- Pipet 4 mL of the light solution into the reservoir. The compensation bar is placed into the reservoir. The valve is shortly opened to fill the connection between reservoir and mixing chamber.
- The magnetic bar is put into the mixing chamber. Then 4 mL of the dense solution are pipetted into the mixing chamber.
- The gradient maker is placed on a magnetic stirrer motor, the outlet 5 cm above the edge of the cassette. The magnetic stirrer is started with a speed to give only a light vortex.
- The tip of the tube is placed in the middle of the cassette. *See Fig. 3.*
- First the valve of the gradient maker is opened, secondly the clamp on the tubing.
- Now the gradient is flowing into the chamber.

If it does not flow (mostly because of remains of water in the tubing from last washing), close the gradient maker valve, push a thumb or a finger into the mixing chamber, open the valve after the liquid has started to flow.

- The gradient maker must be cleaned and dried before next use.
- The cassette is left for rehydration for 2 hours.

The parallel gradient
(volumes are given for a complete gel, divide by 2 for a half gel)

- The valve and the clamp on the tubing must be closed.
- Pipet 6 mL of the light solution into the reservoir. The compensation bar is placed into the reservoir. The valve is shortly opend to fill the connection between reservoir and mixing chamber.
- The magnetic bar is put into the mixing chamber. Then 6 mL of the dense solution are pipetted into the mixing chamber.
- The gradient maker is placed on a magnetic stirring motor, the outlet 5 cm above the edge of the cassette. The magnetic stirrer is started with a speed to give only a light vortex.

See Fig. 4.
- The tip of the tube is placed in the middle of the cassette.
- First the valve of the gradient maker is opened, secondly the clamp on the tubing.
- Now the gradient is flowing into the chamber.
- If it does not flow, see perpendicular gradient.
- When the gradient maker is empty, pipette 4 mLof light solution into the cassette (the stacking gel zone should not contain a gradient!).
- The gradient maker must be cleaned and dried before the next use.
- The cassette is left for rehydration for 2 hours.

Constant denaturing gel

- Mix dense and light solution in a certain percentage according to the result of a perpendicular DGGE experiment.
- *Complete gel:* Pipet 16 mL of this solution into the cassette.
- *Half gel:* Pipet 8 mL of this solution into the cassette.
- The cassette is left for rehydration for 2 hours.
- Alternatively, rehydration of a constant denaturing gel can be performed in an horizontal tray (GelPool). See Fig. 5.

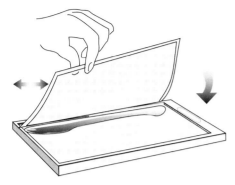

Fig. 5: Rehydration of a dried gel in the GelPool.

5
Electrophoresis

After rehydration, the cassette is dismantled, the gel is laid on a horizontal plane, with the edges of a filter paper the excess of solution is soaked off the surface (Fig. 6) – always from the nondenaturing towards the denaturing side!

Drying is performed until you can hear a squeaking.

- Switch on the thermostatic circulator, adjusted to 55 °C. Apply a very thin layer of kerosene (ca. 0.5 mL) onto the thermostatic plate with a tissue paper, in order to ensure good heating contact (small air bubbles do not matter). Place the gel (surface up) on to the center of the cooling plate: the side containing the wells must be oriented towards the cathode. Two half gels can be run together on the chamber shown in Fig. 8.

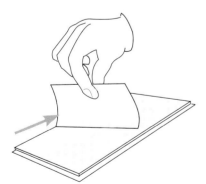

Fig. 6: Removing the excess buffer with the edge of a clean filter paper.

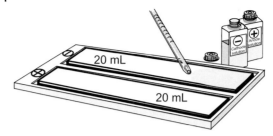

Fig. 7: Soaking the filter paper strips with electrode solutions.

- Lay two of the electrode strips into the compartments of the PaperPool. Apply 20 mL (10 mL for half strips) of the electrode buffer to each of the wicks (Fig. 7).
- Place one of the strips onto the cathodal edge of the gel, edge of the electrode strip matching line 3.5 on the cooling plate. Place the other strip over the anodal edge, matching "13.5" on the cooling plate. Smooth out air bubbles by sliding bent tip forceps along the edges of the wicks laying in contact with the gel.
- Apply 6 μL of each sample to the sample wells using a micro-pipette.
- Move electrodes so that they will rest on the outer edges of the electrode wicks.
- Connect the cables of the electrodes to the apparatus and lower the electrode holder plate. Close the safety lid.

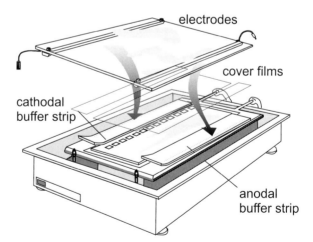

Fig. 8: Appliance for DGGE run at high temperatures. Note the cover films, which are applied after the samples have entered the gel.

Running conditions (55 °C) for two half gels or one complete gel:

100 V	15 mA	5 W	20 min
600 V	28 mA	10 W	1 h 30 min

- After 20 minutes, the first phase, stop the run, cover the gel surface and the electrode strips – between the electrode wires – with a polyester film, to prevent them from drying out, and continue the separation.
- Clean platinum electrode wires after each electrophoresis run with wet tissue paper.

6
Silver staining

The procedure for staining of these urea-containing gels is described on page 310. As indicated, the fixation and silver incubation steps have to be prolonged, also additional washing steps are necessary.

Method 15:
Denaturing PAGE of DNA

Denaturing polyacrylamide gels are employed for the following techniques:

- DNA sequence analysis
- Amplified restriction fragment length polymorphism (AFLP)
- Microsatellite analysis
- Differential display reverse transcription (DDRT)
- Ribonuclease protection assay (RPA)
- Temperature gradient gel electrophoresis (TGGE)

Most laboratories perform all these techniques – except the TGGE – in a sequencing apparatus.

In many cases it is not necessary, to run long sequencing gels and use radioactivity for detection. In Fig. 1 a silver stained 15%T gel is shown, which has been washed, dried and rehydrated in urea-buffer.

When the samples are run on thin horizontal gels polymerized on carrier films, a number of advantages is achieved:

- The gels are much easier to handle.
- Washed and dried gels can be freshly rehydrated with urea-buffer.
- The gels can be silver stained.
- When radioactivity is needed, the amount of contaminated liquid and material is much less.
- The procedure is much faster.
- A horizontal chamber can be easily modified to serve as a TGGE apparatus (Suttorp *et al.* 1996).

DNA bands can be reamplified.

In the following chapter washed and dried gels are rehydrated in urea-buffer. Gels can, of course, also be cast directly with urea and Tris-phosphate buffer as described in method 12.

Electrophoresis in Practice, Fourth Edition. Reiner Westermeier
Copyright © 2005 WILEY-VCH Verlag GmbH & Co. KGaA, Weinheim
ISBN: 3-527-31181-5

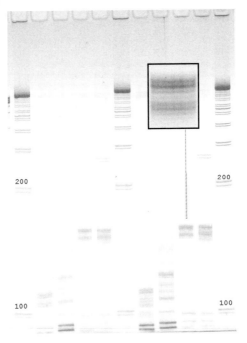

Fig. 1: Denaturing electrophoresis of microsatellites (PCR products of cancer tissues). Silver Staining. The bands in the enlargement show the resolution of one base.

1
Sample preparation

Microsatellite and AFLP samples
Denature samples by heating for 3 minutes at 95 °C with 50 % formamide (1 part PCR product + 1 part formamide) and place them in ice-water during the application procedure.

Differential display reverse transcription (DDRT):
For optimized amplification procedures work according to Bosch and Lohmann (1996). Separations are either performed in 15 %T short distance denaturing gels or in native long distance gels (see method 11).

Ribonuclease protection assay (RPA):
Prepare the samples with the help of commercial RNase protection assay kits and apply them directly on the gel.

Temperature gradient gel electrophoresis (TGGE):
The amplification products are pipetted directly on the gel.

2
Solutions

Dye:
Prepare a 1 % (w/v) stock solution of bromophenol blue.

Gel buffer 0.36 mol/L Tris-phosphate pH 8.4 (4 × conc):
4.36 g Tris, dissolve in 80 mL H_2O_{dist}; titrate to pH 8.4 with phosphoric acid; make up to 100 mL with H_2O_{dist}.

Electrode buffer 450 mmol/L Tris / 75 mmol/L boric acid / 12.5 mmol/L
EDTA-Na$_2$ (5 × TBE):
54.5 g Tris + 23.1 g boric acid + 4.65 g EDTA Na$_2$, make up to 1 L with H_2O_{dist}.

Rehydration solution (90 mmol/L Tris-phosphate pH 8.3, 7 mol/L urea,
4 % glycerol, 4 % ethylenglycol):
10.5 g urea + 6.25 mL gel buffer + 1 mL glycerol + 1 mL ethylenglycol + 0.5 mL bromophenol blue solution, make up to 25 mL with H_2O_{dist}.

3
Rehydration

The preparation of washed and dried gels on carrier films is described in Method 12. For microsatellites, DDRT, and RPA 15% *T* gels are selected, for TGGE 10 % *T* short gels and for AFLP 10 % *T* long gels are employed.

Lay GelPool onto a horizontal table; select the appropriate reswelling chamber, pipet the rehydration solution into the chamber, for

The gels for short distances can be used in one piece, or – depending on the number of samples – cut into smaller portions with scissors (when they are still dry). The rest of the gel should be sealed airtight in a plastic bag and stored in a freezer.

a complete gel:	25 mL
a half gel (short):	13 mL

Set the edge of the gel-film – with the gel surface facing down – into the rehydration buffer (Fig. 2) and slowly lower it, avoiding air bubbles.

Using forceps, lift the film up to its middle, and lower it again without catching air bubbles, in order to achieve an even distribution of the liquid (Fig. 2B). Repeat this during the first 10 min.

90 min later 7 mol/L urea and the other additives have diffused into the gel, and the gel is removed from the GelPool. Dry sample wells with clean filter paper, wipe buffer off the gel surface with the edge of a filter paper (Fig. 2D).

When the gel surface is dry enough, this is indicated by a noise like a whistle.

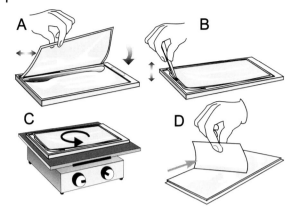

Fig. 2: Rehydration of a gel: (**A**) Placing the dry gel into the GelPool. (**B**) Lifting the gel for an even distribution of the liquid. (**C**) Rehydration on a rocking platform (not always necessary). (**D**) Removing the excess buffer from the gel surface with filter paper.

4
Electrophoresis

Preparation of the electrode wicks

Do not forget, that for one buffer system different anode and cathode buffer is used.

Short distance gels: Lay two of the 25 × 5 cm electrode wicks into the compartments of the PaperPool. Apply 20 mL of the respective electrode buffer to each wick (Fig. 3).

Long distance gels: Cut six electrode strips of 11.7 × 1.8 cm. Lay three of them stacked into the compartments of the PaperPool. Apply 10.5 mL of the respective buffer to each stack.

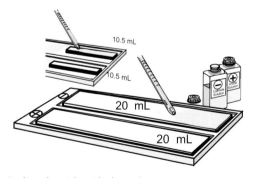

Fig. 3: Soaking the wicks with electrode buffer.

Application of the gel and the electrode wicks

Switch on the thermostatic circulator, adjusted to the 15 °C. Apply a very thin layer of kerosene (ca. 0.5 mL) onto the cooling plate with a tissue paper, in order to ensure good cooling contact (a few air bubbles do not matter).

Short distance run: Place the gel (surface up) on to the center of the cooling plate: the side containing the wells must be oriented towards the cathode (Fig. 4; Multiphor: line 5).

Place the cathodal strip onto the cathodal edge of the gel, edge of the electrode strip matching "3.5" on the cooling plate. The edge of the strip should be at least 4 mm away from the edges of the sample wells

If the cathodal strip is placed too close to the wells, small DNA fragments will exhibit less sharpness or hollow bands

Place the anode strip over the anodal edge, matching "13.5" on the cooling plate. Smooth out air bubbles by sliding bent tip forceps along the edges of the wicks lying in contact with the gel.

Long distance run: Place the gel (surface up) on to the center of the cooling plate: the side containing the wells must be oriented towards the cathode. Place the cathodal stack electrode strips to the cathodal side of the gel, the inner edge matching with line 23 of the cooling plate.

Place the other stack over the anodal side, the inner edge matching with line 2 of the cooling plate. Smooth out air bubbles by sliding bent tip forceps along the edges of the wicks laying in contact with the gel.

Sample application and electrophoresis

Apply 6 µL of each sample to the sample wells using a micropipette. Clean platinum electrode wires before (and after) each electrophoresis run with a wet tissue paper. Move electrodes so that they will rest

short distance run   **long distance run**

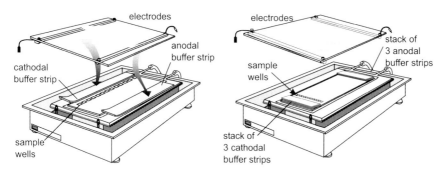

Fig. 4: Appliance for short and long distance runs of DNA fragments.

on the outer edges of the electrode wicks. Connect the cables of the electrodes to the apparatus and lower the electrode holder plate (Fig. 4). Close the safety lid.

Running Conditions:

The suggested running conditions have to be modified for some applications: e.g. when samples of the size range 700 to 900 bp have to be well separated, at least two hours separation time is required. The dyes added to the samples are a help for the estimation of the running time:

Gels, which contain 7 mol/L urea and run at 25 °C do not need to be covered with a film or a glass plate.

Short distance run in a whole 15 % T gel,
Tris-phosphate / TBE buffer pH 8.4, 7 mol/L urea, 25 °C:

400 V	17 mA	6 W	15 min
1000 V	22 mA	25 W	1 h 20 min

Long distance run in a 10 % T gel,
Tris-phosphate /TBE buffer pH 8.4, 7 mol/L urea, 25 °C:

250 V	8 mA	5 W	30 min
1000 V	15 mA	12 W	1 h 50 min

Long distance runs:

See page 308.

High voltage and relatively low milliampere values are applied on long electrophoresis gels. This means, that with a milliampere-constant control only little changes in conductivity can result in severe changes of the voltage value. In method 12 an alternative for an improved control for the running conditions are described: voltage ramping.

Higher temperatures:

When higher temperatures like 40 to 60 °C have to be applied, the gel surface has to be protected from drying. Suttorp et al. (1996) cover it with a glass plate. In Fig. 5 an alternative with cover films is shown.

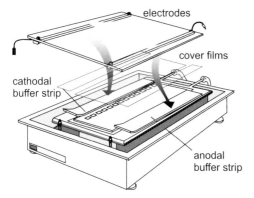

electrodes

cover films

cathodal
buffer strip

anodal
buffer strip

Fig. 5: Protection of the gel surface during high temperature runs with cover films, which are applied after the samples have entered the gel.

5
Silver staining

The procedure for staining of these urea containing gels is described on page 310. As indicated, the fixation and silver incubation steps have to be prolonged, also additional washing steps are necessary.

A1 Trouble shooting:
Isoelectric focusing

1.1
PAGIEF with carrier ampholytes

Tab. A1–1:   Gel characteristics.

Symptom	Cause	Remedy
Gel sticks to glass plate.	Glass plate too hydrophilic. Incomplete polymerization.	Clean the glass plate and coat with Repel Silane. See below.
No gel or sticky gel, insufficient mechanical stability.	No or incomplete polymerization:	
	Poor water quality.	Always use double-distilled water!
	Too much oxygen in the gel solution (radical trap).	Degas thoroughly.
	Acrylamide, Bis or APS solutions too old.	Maximum storage time in the refrigerator: Acrylamide, Bis solution, 1 week; APS solution 40%, 1 week.
	Poor quality reagents.	Only use analytical grade quality reagents.
	Photochemical polymerization with riboflavin.	Chemical polymerization with APS is much more effective.
	The pH value is too basic (narrower basic pH range).	Rehydrate the prepolymerized and dried gel in a carrier ampholyte solution.

Electrophoresis in Practice, Fourth Edition. Reiner Westermeier
Copyright © 2005 WILEY-VCH Verlag GmbH & Co. KGaA, Weinheim
ISBN: 3-527-31181-5

Symptom	Cause	Remedy
Gel peels away from the support film.	Wrong support film was used.	Only use GelBond PAG film for polyacrylamide gels not GelBond film (for agarose).
	Wrong side of the support film was used.	Only cast the gel on the hydrophilic side of the support film, test with a drop of water.
	Support film was incorrectly stored or too old.	Always store the GelBond PAG film in a cool, dry, and dark place (< 25 °C), check the expiry date.
	Insufficient polymerization.	See above.
	Gel solution contains non-ionic detergents (Triton X-100, Nonidet NP-40).	Rehydrate the prepolymerized and dried gel in a carrier ampholyte/detergent solution (method 6).

Tab. A1–2: Problems during IEF.

Symptom	Cause	Remedy
No current.	Safety turn off, "ground leakage" because of massive short circuit.	Turn off the power supply, check the separation unit and cable. Dry the bottom of the separation chamber, cooling coils and laboratory bench, turn the power on again.
Too low or no current.	Poor or no contact between the electrodes and electrode strips.	Make sure that the electrode strips are correctly placed; if a small gel or part of a gel is used, place it in the middle.
	The connecting cable is not plugged in.	Check the plug; press the plug more securely into the power supply.
Current rises during the IEF run.	Electrode strips or electrodes mixed up.	Acid solution at the anode, basic solution at the cathode.
General condensation.	The power setting is too high.	Check the power supply settings. Guide value: at most 1 W per mL of gel.

Symptom	Cause	Remedy
	Insufficient cooling.	Check temperature, if focusing is carried at a higher temperature e.g. 15 °C, reduce the power. Check the flow of the cooling fluid (bend in tubing?). Add kerosene between the cooling plate and the support film.
Condensation at the sample applicator.	Excessive salt concentration in the sample (>50 mmol/L) which causes local overheating.	Desalt the sample by gel filtration (NAP column) or dialyze against 1% glycine or 1% carrier ampholyte (w/v).
Gel swells around electrode strips.	Electroendosmosis causes a flow of water in direction of the electrodes (especially the cathode).	Normal phenomenon. It is not a problem unless it interferes with the run. It may help to occasionally blot the electrode strips.
	There is too much electrode solution in the electrode strips.	After soaking, blot the strips with filter paper.
	Electrode solution is too concentrated.	Use electrode solution at the specified concentration, dilute if necessary.
Condensation along the electrode strips.	Electrode strips reversed.	Acid solution at the anode, basic solution at the cathode.
Local condensation.	Localized hot spots due to bubbles in the insulating fluid.	Remove the air bubbles, avoid them from the beginning if possible.
Condensation over the basic half of the gel.	Electroosmotic water flow in direction of the cathode.	Use a better or fresher acrylamide solution. Reduce focusing time as much as possible, for narrow pH ranges pH > 7, blot the liquid which collects.
Lines of condensation over the whole gel.	Hot spots, conductivity gaps because of plateau phenomenon. Too long focusing time, especially for narrow pH ranges.	Fill the conductivity gaps by adding carrier ampholytes with a narrow pH range. Keep the focusing time as short as possible, or use IPG.

Symptom	Cause	Remedy
Sparking on the gel.	Same causes as for condensation, next stage (dried out gel).	Remedy as for condensation. Take measures as soon as condensation appears.
Sparking along the edge of the support film.	Electrode strips hang over the edge of the gel. High voltage and ions in the insulating fluid.	Cut the electrode strips to the size of the gel. Use kerosene or DC-200 silicone oil, not water.

Tab. A1–3: Separations.

Symptom	Cause	Remedy
The pH gradient deviates from that expected.	Gradient drift (Plateau phenomenon). Acrylic acid polymerized in the gel.	Only use analytical grade quality reagents.
	Acrylic acid polymerized in the gel because the acrylamide, Bis stock solutions were stored too long.	Maximum storage time in the refrigerator, in the dark: 1 week. The storage life can be prolongued by trapping the acrylic acid with mixed bed ion-exchanger shortly before use.
	Temperature dependence of the pH gradient (pK values!).	Check the focusing temperature.
	Too long focusing time.	Reduce the focusing time as much as possible, especially in narrow basic pH intervals; or else use IPG.
	Gel stored too long.	Gels with narrow alkaline pH intervals have a limited storage life, use rehydratable gels.
	Gel contains carbonic acid ions.	Degas the rehydration solution (removal of CO_2); avoid the effects of CO_2 during IEF (particularly in basic pH ranges): seal the separation chamber, flush with N_2, trap CO_2: add 1 mol/L NaOH to the buffer tanks.

Symptom	Cause	Remedy
Partial loss of the most basic part of the pH gradient.	Oxidation of the carrier ampholytes during the run.	Reduce the influence of CO_2 as much as possible: see above.
	Oxidation of the electrode solutions.	See above.
Wavy iso-pH bands:1. no influence of the sample. 	Too much APS was used for polymerization.	Rehydrate a polymerized, washed and dried gel in a carrier ampholyte solution. Increase the viscosity of the gel by adding 10% (w/v) sorbitol to the rehydration solution or urea (< 4 mol/L, not denaturing in most cases).
	Urea gels stored too long, urea degraded to isocyanate.	Use urea gels immediately after preparation or else rehydrate the gel shortly before use.
	Bad electrode contact.	Check the electrode contacts, especially the anode; if necessary put a weight on the electrode support.
	Unevenly or excessively wetted electrode strips.	Soak the electrode strips completely with electrode solution and blot them with filter paper.
	Wrong electrode solutions.	Use the electrode solutions recommended for the pH range in the correct concentrations.
	Gel too thin.	Ultrathin gels <200 μm are sensitive to protein overloading, varying protein concentrations, buffer and salt ions as well as diffusion of electrode solutions (compression of the gradient).
Wavy iso-pH lines. 2. induced by the sample.	Strongly varying protein concentration of the sample.	

Symptom	Cause	Remedy
(A) Protein concentration.		Either dilute the highly concentrated samples or apply in order of increasing or decreasing concentration. Prefocuse. Decrease the field strength at the beginning: (V/cm).
	Samples applied too far apart.	Apply samples closer to one another(1 – 2 mm)
	Highly concentrated samples applied at different places in the pH gradient.	It is not possible to proceed another way with the step trial test; but otherwise apply the samples as close as possible in the pH gradient. Prefocus; field strength at the beginning: < 40 V/cm.
	Overloading, – protein concentration too high.	Dilute the samples (2 to 15 μg per band) or use a thicker gel. Ultra-thin-layer IEF: use gel thickness >250 μm or IPG.
(B) Buffer, salt concentration.	The buffer or salt concentrations in the samples vary a lot.	As for (A); if necessary, desalt highly concentrated samples.
	High buffer or salt concentration in the samples.	As for (A); make sure that the samples are applied close to one another, at the same level in the gradient. Prefocus; field strength at the beginning (< 40 V/cm; at first for 30 min at < 20 V/cm let salt ions migrate out of the gel or use IPG.

Symptom	Cause	Remedy
	Buffer or salt concentrations in the samples too high – desalting too risky or not possible because of eventual protein losses.	Cast sample application strips in polyacrylamide ($T = 10\%$ ca.) or agarose (ca. 2%) containing salts in the same concentrations as the samples and place over the whole width of the gel. Apply the samples in the wells; let the salt ions migrate out for about 30 min at $E = 20$ V/cm, the salt load will then be the same over the whole width of the gel, so individual shifts in the pH gradient will be compensated. Alternatively use IPG.
Streaking or tailing of the sample.	Precipitate and/or particles in the sample.	Centrifuge the sample.
	The applicators retain the proteins and release them later.	Remove the applicators after about 30 min of IEF or use applicator strips.
	Old or denatured sample.	Check the sample preparation procedure, carry it out shortly before the separation. Store samples at $< -20\,°C$.
	High molecular weight proteins have not reached their pI yet.	Focus longer or use agarose gels.
	Poorly soluble proteins in the sample.	Focus in urea (if necessary with non-ionic or zwitterionic detergents) or 30% DMSO.
	Protein overloading.	Dilute sample or apply less.
	Protein aggregation during sample entry.	Set a lower current (mA) limit for the sample entry phase. This reduces the field strength in the beginning.

Symptom	Cause	Remedy
Diffuse bands.	Diffusion during IEF, low molecular weight peptides.	It is preferable to focus oligopeptides with molecular weight < 2 kDa in IPG where diffusion is less marked.
	Diffusion after IEF, inadequate or reversible fixing.	Check the fixing and staining methods.
	Urea IEF: Urea precipitation in the gel.	Run urea gels at 15 – 20 °C.
	Focusing time too short.	Focus for a longer time.
	Marked gradient drift.	See above.
	Influence of CO_2 on the basic bands.	See above.
Individual bands are diffuse.	See above.	See above.
	The focusing time for the individual proteins is too short (large molecules and/or low net charge).	Optimize the sample application point with a concentration test or titration curve analysis. Apply the sample on the side of the pI where the charge curve is steeper.
Missing bands.	Concentration too low or detection method not sensitive enough.	Apply more sample or concentrate the sample. Use another detection method (e.g. silver staining of the dried gel, blotting).
	The proteins are absorbed on the sample applicator.	Use sample application strips.
The proteins precipitate at the point of application.	Application too close to the pI.	Apply sample further away from its pI (step trial test, titration curve).
	The field strength is too high at the point of sample entry.	Reduce the voltage at the beginning (E< 40 V/cm).
	The molecule is too large for the pores of the gel.	Use agarose instead of polyacrylamide.
	The proteins form complexes.	Add urea (7 mol/L) to the sample and the gel; add EDTA to the sample; add non-ionic or zwitterionic detergents to the sample and the gel.

Symptom	Cause	Remedy
	Protein unstable at the pH of site of application.	Apply the sample at another point (step trial test, titration curve).
Individual bands focus at the wrong place.	The protein is unstable at the temperature used.	Change the focusing temperature.
	The proteins form complexes.	See above. If it is suspected that complexes form with the carrier ampholyte, check with IPG.
"One" protein focuses in several bands.	The proteins have lost ligands.	Check with titration curve analysis.
	The protein exists in various states of oxidation.	Check the sample preparation; eventually focus under N_2.
	The protein has dissociated into subunits.	Do not focus in the presence of urea.
	Urea IEF: Carbamylation by cyanate.	Check sample preparation and gel casting with urea.
	Different conformations of a molecule.	Focuse the protein in the presence of urea (>7 mol/L).
	Different combinations of oligomers of a protein or of subunits.	Natural phenomenon.
	Different degrees of enzymatic phosphorylation, methylation or acetylation exist.	Check the sample preparation procedure.
	Various carbohydrate moieties of glycoproteins.	Natural phenomenon. Treat the sample with neuramidase for example, to verify.
	Partial proteolytic digestion of a protein.	Check the sample preparation procedure. Add inhibitor (Pefabloc or PMSF).
	Complex formation.	If complex formation with the carrier ampholytes is suspected, verify with immobilized pH gradients.

1.2
Agarose IEF with carrier ampholytes

Only problems specific to agarose IEF are mentioned here. Consult paragraph 1.1 under PAGIEF for general problems occurring in connection with isoelectric focusing or carrier ampholytes.

When agarose IEF is used, it should be remembered that the matrix is not as electrically inert as polyacrylamide since sulfate and carboxylic groups are still bound to agarose which is of natural origin, and they give rise to electroendosmotic phenomena.

Tab. A1–4: Gel properties.

Symptom	Cause	Remedy
Insufficient gel consistency.	Incomplete solidification of the gel.	Let the gel solidify > 1 h. It is best to remove it from the cassette after 1 h and store it overnight in a humidity chamber at +4 °C (maximum storage time: 1 week).
	The agarose concentration is too low despite the fact that the agarose was precisely weighed out, the agarose has absorbed water.	Store the agarose in a dry place out of the refrigerator. Close the package well.
	Urea gel: urea disrupts the structure of agarose.	Use a higher agarose concentration (2%); let the gel solidify longer or use rehydratable agarose gels.
The gel comes off the support film.	Wrong support film used.	Only use GelBond film for agarose, not GelBond PAG film (for polyacrylamide gels).
	The wrong side of the support film was used.	Cast the gel on the hydrophilic side of the support film.
	The gel was cast at too high a temperature.	The temperature should be kept between 60 and 70 °C during casting.
	The solidification time was too short.	See above.

Tab. A1–5:   Problems during the IEF run.

Symptom	Cause	Remedy
Flooding on the surface.	The gel surface was not dried.	Always dry the surface of the gel with filter paper before IEF.
	The solidification time was too short.	See above.
	The wrong electrode solution was used.	In general it is recommended to use: at the anode: 0.25 mol/L acetic acid; at the cathode 0.25 mol/L NaOH.
	The electrode strips are too wet.	Remove the excess liquid. Blot the electrode strips so that they appear almost dry.
	Electroendosmosis.	Natural phenomenon. Blot the electrode strips every 30 min. Or replace them by new ones.
	Marked electroendosmosis.	Always use double-distilled water; use an ideal combination of chemicals: 0.8% agarose IEF with 2.7% Ampholine.
	There are no water binding additives in the gel.	Add 10% sorbitol to the gel solution.
Water build-up at the cathode.	Electroendosmosis, cathodic drift.	Dry the cathode strips more often and carefully; only focus as long as necessary.
Water build-up at the sample application site.	Electroendosmosis because of the material used for sample application.	Only use sample application strips or masks, do not use paper or Paratex for example.
	The protein or salt concentration is too high.	See polyacrylamide gels.
Formation of a ditch in the gel.	Advanced cathodic drift because of electroendosmosis.	See above. Focus at 10 to 15 °C.
	Insufficient gel consistency.	See above.

Symptom	Cause	Remedy
Formation of small hollows near the sample application site.	The power was too high during sample entrance.	Set the power at 5 to 10 W at most for the first 10 to 15 min (for a 1 mm thick gel, 25 cm wide × 10 cm separation distance, use correspondingly lower settings for smaller gels.
	Sample overloading.	See under polyacrylamide gels.
The gel dries out.	Advanced electroendosmosis.	See above.
	The gel was irregularly cast.	Position the leveling table exactly when casting horizontal gels or else use the vertical technique ("clamp" technique in prewarmed molds).
	Heat source in the proximity.	During agarose IEF do not place the separation chamber beside a thermostatic water bath.
	The air is too dry.	When the ambient humidity is too low, pour a small volume of water in the electrode tanks.
Sparking.	Advanced stage of the effects listed above.	See above, if possible take measures before this occurs.

Tab. A1–6: Separation results.

Symptom	Cause	Remedy
Bands too wide.	Too much sample solution was applied.	Reduce the sample volume.
Diffuse bands.	Focusing time too long (gradient drift) or too short (the proteins have not reached their pI yet).	See above and under polyacrylamide gel.
	Because of the larger pore size, diffusion is more marked in agarose than in polyacrylamide gels.	Check the fixing and staining procedures. Dry the gel after fixing and then stain (this is also valid for silver staining).
Missing bands.	See above.	See above.

Symptom	Cause	Remedy
Missing bands in the basic part of the gel.	Part of the gradient is lost because of a cathodic drift (more pronounced in agarose than in polyacrylamide gels).	Add a carrier ampholyte with a narrow basic range; focus for a shorter time.
Distorted bands at the edge of the gel.	Fluid has run out of the gel or the electrode strips; fluid has run along the edge of the gel and forms L-shaped "electrodes".	Blot the gel or electrode strips regularly when water oozes out.
	The samples were applied too close to the edge.	Apply the samples about 1 cm from the edge.
Wavy bands in the gel.	As for polyacrylamide gels.	See under polyacrylamide gels.
	Irregularities in the surface of the gel.	Degas the gel solution properly. Use a humidity chamber for storage.

Tab. A1–7: Problems specific to agarose IEF.

Symptom	Cause	Remedy
Bands are diffuse, disappear or do not appear.	Diffusion.	Always dry agarose gels after fixing and before staining them.
The gel comes off the support film during staining.	The fixing solution was not completely removed from the gel before drying.	Rinse the gel in twice for 20 min each time in the destaining solution containing 5% glycerol.
	A mistake was made during gel casting.	See above.

1.3
Immobilized pH gradients

Tab. A1–8: Gel properties.

Symptom	Cause	Remedy
The gel sticks to the glass plate.	The glass plate is too hydrophilic.	Clean the glass plate and coat it with Repel Silane.
	The gel was left too long in the mold.	Remove it from the cassette 1 h after the beginning of polymerization.
	The gel concentration is too low.	Do not use glass when *T* <4%, use acrylic glass (Plexiglas) instead.
	Incomplete polymerization.	See below.
No gel or sticky gel.	Poor water quality.	Always use double-distilled water!
	The acrylamide, Bis or APS solutions are too old.	Maximum storage time in the dark in the refrigerator: acrylamide, Bis solution: 1 week. 40% APS solution: 1 week.
	Poor quality reagents.	Only use reagents of analytical grade quality.
	Too little APS and/or too little TEMED were used.	Always use 1 μL of APS solution (40% w/v) per mL of gel solution and at least 0.5 μL of TEMED (100%) per mL of gel solution.
	The pH value was not optimal for polymerization.	For wide (1 pH unit) and alkaline (above pH 7.5) pH ranges: titrate both gel solutions with HCl 4 mol/L respectively NaOH 4 mol/L to pH 7 after TEMED has been added. The precision of pH paper is sufficient.
	The polymerization temperature was too low.	Let the gel polymerize for 1 h in a heating cabinet or at least an incubator at 50 °C or 37 °C respectively.

Symptom	Cause	Remedy
One half of the gel is not or insufficiently polymerized.	The APS solution has not mixed properly with the gel solution (usually the dense solution: the APS solution overlayers it because of the glycerol content).	After adding the APS solution stir vigorously for a short time. Make sure that the drops of APS solution are incorporated in the gel solution.
	One of the solutions was not titrated to pH 7.	See above.
The surface of the gel is sticky, swells during washing and detaches itself from the support film.	Oxygen has inhibited polymerization of the surface.	Overlay the surface of the gel with about 300 μL of double distilled water immediately after casting; do not use butanol.
The gel detaches itself from the support film.	The wrong support film or the wrong side of the support film were used or else the support film was stored incorrectly.	See under polyacrylamide gels.

Tab. A1–9: Effects during washing.

Symptom	Cause	Remedy
The gel has a "snake skin" structure in certain areas or all over.	This is normal. Because of the fixed buffering groups the gel possesses slight ion-exchanger properties and swells.	The gel surface will become normal again when it is dry.
The gel becomes wedge shaped.	This is normal. The buffer has different concentrations and properties which results in different swelling characteristics within the gradient.	Dry the gel after washing; rehydrate it in the reswelling cassette (the cassette prevents it from taking a wedge shape).

Tab. A1–10: Effects during drying.

Symptom	Cause	Remedy
The support film rolls up.	The gel pulls in all directions.	Add 1% to 2% of glycerol to the last washing, this makes the gel more elastic.

Tab. A1–11: Effects during rehydration.

Symptom	Cause	Remedy
The gel does not swell, or only partially.	The reswelling time is too short.	Adapt the reswelling time to the additive and the additive concentration. If the gel was stored for a long time at room temperature or if the use-by date is expired, prolong the reswelling time.
	Gel was dried too long or at too high a temperature.	Dry the gel with a fan at room temperature, the airflow should be parallel to the gel surface and the gel should be dried in a dust free atmosphere.
	The gel was stored too long at room temperature or higher.	Use the gel immediately after drying or store it hermetically sealed at < –20 °C.
The gel sticks to the reswelling cassette.	The surface of the glass is too hydrophilic.	Coat the surface of the gel within the gasket with Repel Silane.
The gel sticks to the support glass plate.	The gel surface was inadvertently applied on the wet glass plate.	Pull it away gently under water in a basin.

Tab. A1–12: Effects during the IEF run.

Symptom	Cause	Remedy
No current.	The cable is not plugged in.	Check the plug; insert the plug more securely in the power supply.
Low current.	This is normal for IPG. The gels have a very low conductivity.	Standard setting for whole IPG gels: 3500 V, 1.0 mA, 5.0 W. Regulate IPG strips with the voltage setting.
Localized condensation over specific areas.	Salt concentration in the samples too high. Salt ions form arcs when leaving the sample wells, very high salt conc. where two arcs meet.	Apply samples with high salt concentration close to one another, if samples must be applied at different areas within the pH gradient. Separate the traces by cutting strips or scraping out troughs.

Symptom	Cause	Remedy
Local sparking at specific points.	See above; next stage.	See above; if focusing is carried out overnight, do not apply more than 2500 V and turn up to 3500 V the next day.
Sparking along the edge of the gel.	High voltage and there are ions in the contact fluid.	Use kerosene as contact fluid between the cooling plate and the film.
Sparking at an electrode.	The gel has dried out because of electroendos-mosis. This occurs in narrow pH gradients at extreme pH intervals (< 4.5; > pH 9).	Either add glycerol (25%) or 0.5% non-ionic detergent to the reswelling solution.
	The electrode solutions are too concentrated.	Soak both electrode strips in double-distilled water. The conductivity is suffi-cient; in addition, the field strength decreases at the beginning of IEF for improved sample entrance.
	Gel insufficiently polymerized.	See above.
A narrow ridge develops over the whole width of the gel and slowly migrates in direction of an electrode.	Normal phenomenon during IPG: ion front at which a jump in the ionic strength and a reversal of the electroend-osmotic effect occur.	Wash the gel thoroughly. Add 2 mmol/L acetic acid to reswelling solution for samples applied at the anode and 2 mmol/L Tris for samples applied at the cathode.
The ridge does not migrate any further.	The gel contains too many free ions, the difference in conductivity within the gel is so large that the voltage is not sufficient to carry the ions further.	Wash the gel thoroughly. Use a power supply with a high voltage (3500 V are sufficient). Focus for a long time, overnight if necessary.

Tab. A1–13: Separation results.

Symptom	Cause	Remedy
The bands and iso-pH lines form arcs in the gel.	The gel polymerized before the concentration gradient had finished leveling.	Cool the casting cassette in the refrigerator before casting (this delays the onset of the polymerization). Use glycerol and not sucrose (its viscosity is too high) to make the acid solution denser.
	The catalyst was not properly washed out.	See above.
The bands are diffuse.	Focusing time too short.	Focus for a longer time, overnight for example.
	The field strength is not sufficient when the pH range is narrow or the separation distance is long (10 cm).	High voltages are necessary for narrow pH ranges and long separation distances: use a 3500 V power supply.
	There are problems with polymerization, for example the acrylamide or Bis solutions are old; see above.	Use fresh stock solutions.
The bands in the basic part of the gel are diffuse.	Influence of CO_2.	Trap CO_2 during IEF: seal the chamber, add soda lime or 1 mol/L NaOH to the buffer tanks.
No bands are visible.	The pH gradient is wrongly orientated.	Place the gel on the cooling plate with the acid side towards the anode and the basic side towards the cathode; the basic side has an irregular edge and the support film sticks out.
The proteins have stayed at the site of application.	The field strength was too high at first.	Do not prefocus (the pH gradient already exists). Keep the field strength low at the beginning.
	The proteins have aggregated at the site of application because their concentration was too high.	Dilute the sample with water or water/non-ionic detergent; it is preferable to apply a large sample volume than a concentrated solution.

Symptom	Cause	Remedy
	Some proteins have formed complexes and obstructed the pores.	Add EDTA to the sample. Add urea to the sample and rehydration solution; complex formation is prevented by a urea concentration of 4 mol/L but most enzymes are not denatured yet.
	Conductivity problems.	Apply the sample to the other side, or direct the ionic front as described above.
	The salt concentration in the sample is too high.	Dilute the sample with water and apply a larger sample volume.
	High molecular weight proteins are unstable when the ionic strength is low.	Prepare a gel matrix with large pores so that the protein can penetrate the gel before it has completely separated from the low molecular substances. As emergency measure it is recommended to add 0.8% (w/v) carrier ampholyte to the sample and 0.5% (w/v) carrier ampholyte from the corresponding pH range to the rehydration solution.
The pI of the proteins lies outside of the immobilized pH gradient.	Narrow pH range: the focusing was performed at wrong temperature.	Focus at 10 °C and/or widen the pH range.
	The pI obtained by carrier ampholyte IEF is shifted compared to the one obtained by IPG.	Use a wider or different pH range.
The immobilized pH range is not correct or not present.	Immobiline was not stored correctly. The acrylamide or Bis solutions are too old. A mistake was made when the Immobiline was pipetted.	Follow the recipes for Immobiline and casting instructions exactly; otherwise: see above.
	The focusing time is not sufficient.	Lengthen the focusing time, if necessary focus overnight.

Symptom	Cause	Remedy
Some bands are missing, are diffuse or are at the wrong place.	Oxygen sensitive proteins have oxidized in the gel (Immobiline gels trap oxygen from the air during drying).	Add a reducing agent to the rehydration solution when working with proteins which are sensitive to oxygen.
The separation lanes are curved and run from one another.	The conductivity of the gel is much lower than the conductivity of the sample (proteins, buffer, salts).	Direct the ionic front as described above; apply the samples beside one another; separate the lanes by cutting the gel or scraping out troughs.

Table A1–14: Specific staining problems with IPG gels.

Symptom	Cause	Remedy
There is a blue background after Coomassie Blue staining.	Basic Immobiline groups tend to bind Coomassie Blue.	Use a solution with 0.5% Coomassie; or, even better, use colloidal staining: no background staining!

A2 Trouble shooting:
SDS electrophoresis

Tab. A2–1: Gel casting.

Symptom	Cause	Remedy
Incomplete polymerization.	Poor quality chemicals.	Only use analytical grade quality reagents.
	The acrylamide and/or APS solutions were kept too long.	Always store the stock solutions in the refrigerator in the dark; the 40% APS solution can be stored for one week, solutions of lower concentration should be freshly prepared every day; in case of doubt make new stock solutions.
	The water is of poor quality.	Always use double-distilled water.
The gel sticks to the glass plate.	The glass plate is too hydrophilic.	Wash the glass plate and coat it with Repel Silane.
Leakage from the gradient mixer.	The rubber gasket is dry.	Open the gradient mixer, coat the gasket with a thin layer of CelloSeal®.
Gradient gels: the gel solution already polymerizes in the gradient mixer.	Too much APS was used.	Reduce the amount of APS. Open the gradient mixer and clean it.
Air bubbles are trapped in the cassette.	This cannot always be prevented.	Carefully pull them out with a strip of film.
Gradient gels: one half of the gel is not or is incompletely polymerized.	The APS solution has not mixed with the gel solution (it has stayed on the sides or on the dense solution).	Pipette carefully; stir vigorously for a short time so that the APS solution is drawn into the dense solution.

Electrophoresis in Practice, Fourth Edition. Reiner Westermeier
Copyright © 2005 WILEY-VCH Verlag GmbH & Co. KGaA, Weinheim
ISBN: 3-527-31181-5

Symptom	Cause	Remedy
The gel separates from the support film.	Wrong support film was used.	Only use GelBond PAG film for polyacrylamide gels not GelBond film (for agarose).
	The wrong side of the support film was used.	Cast the gel on the hydrophilic side; test with a drop of water.
	The support film was wrongly stored or too old.	Always store the GelBond PAG film in a cool, dry and dark place.
There is a liquid film on the surface.	The polyacrylamide matrix has hydrolysed because the buffer is too alkaline (pH 8.8).	Alkaline polyacrylamide gels should not be stored longer than 10 days in the refrigerator.
Holes in the sample wells.	Air bubbles were incorporated during casting.	Cut out the slot former with a sharp scalpel. Press the cut edge down with a pair of curved tweezers; use "crystal clear" Tesafilm or Dymo tape with a smooth surface; otherwise small air bubbles which inhibit polymerization in their vicinity can form.
Holes in the gel ("Swiss cheese effect").	Many very small bubbles in the gel solution.	Degas the gel solution; when casting gradient gels do not stir too fast, because SDS solutions foams.
The edges of the gel have not polymerized enough.	The gel solution was not overlayed.	Overlay the gel solution.
	Oxygen from the air has diffused through the seal.	Polymerize the gel at a higher temperature (37 to 50 °C, ca. 30 min).
	Polymerization is too slow.	Degas the gel solution; add a little more TEMED and APS.

Tab. A2–2: Effects during electrophoresis.

Symptom	Cause	Remedy
Droplets on the surface of the gel.	Buffer dripped on the surface of the gel when the electrode wicks soaked in buffer were placed on the surface of the gel. High ambient temperature and high humidity, water condensation on the surface of the cooled gel. (summer time !).	Make sure that the electrode strips are never held over the surface of the gel; if this happens, carefully remove the drops with a bit of filter paper. Apply gel and samples on the cooling plate before the cooling waterbath is connected to the cooling plate. Do not connect to the cooling system before electrodes and safety lid cover the gel.
Sample applicator strip: the samples merge together.	The sample applicator strip is not applied well enough.	Press the sample applicator down properly, do not touch it anymore, do not touch the strip with the tip of the pipette when applying the sample; use sample application pieces.
Sample wells: the samples leave the wells and spread over the gel surface.	Glycerol oder sucrose in the sample, osmotic distribution.	Prepare samples without glycerol or sucrose; those are only necessary for vertical PAGE.
	High protein content, sample contains proteins with low surface tension.	Add 8 mol/L urea to each sample (**after** heating!), the urea does not influence the electrophoresis pattern.
No current.	The electrode cable is not plugged in.	Check whether all the cables are properly connected.
The power supply switches itself off and exhibits ground leakage.	Electricity is leaking from the chamber.	Make sure that the laboratory bench is dry; if the ambient temperature and humidity are high, regularly wipe off the condensation water from the tubings: ideally use a foam rubber tubing to cover the cooling tubing.

Symptom	Cause	Remedy
The current decreases quickly, the voltage increases quickly.	The system runs out of buffer ions, because the electrodes are placed too close together.	Place electrodes as far as possible to the outer edges of the buffer strips, in order to include the complete buffer between the electrodes.
The front migrates into the wrong direction.	The plugs are inverted: the gel is wrongly oriented.	Check whether the cable is plugged in properly; place the gel so that the sample application point lies near the cathode.
The front migrates too slowly; the separation takes too long.	The current flows under the support film.	Use kerosene as contact fluid, not water.
The electrophoresis takes too long.	There are chloride ions in the cathode buffer.	The cathode buffer (Tris glycine) must not be titrated with HCl even if the pH value given in the recipe (usually pH 8.3) is not reached with glycine; the pH usually sets itself at 8.9 which is all right.
Condensation.	The power is too high.	Check the setting of the power supply: at most 2.5 W/mL of gel.
	The cooling is insufficient.	Check the cooling temperature (10 to 15 °C are recommended); check the flow of the cooling fluid (bend in the tubing?); add contact fluid (kerosene) between the cooling plate and the support film.
The front is crooked.	Irregular electrical contact.	See above.
	The buffer concentration in the electrode strips is irregular because they were not held straight when they were soaked or placed on the gel.	Make sure that the electrode wicks are always held straight.

Symptom	Cause	Remedy
	The casting mold was not levelled when the disk or gradient gels were cast.	Set the mold level with a spirit-level.
The front is curved.	The gel polymerized before the density gradient had settled or before the density discontinuity was finished.	Delay the onset of the polymerization: either by reducing the quantity of APS and/or by cooling the casting cassette in the refrigerator.
The front is uneven or wavy.	The gel surface has dried out in places under the holes in the lid.	Only for Multiphor I: Place a glass plate or electrode holder over the electrode wicks and the gel.
Formation of white precipitates and an irregular gel surface.	There is dirt in the electrode wicks which forms a precipitate with SDS.	Use electrode paper of the best quality; only handle the paper with rubber gloves.
The gel dries along one edge of the paper and burns through there.	This is an electroendosmotic effect due to poor quality of the chemicals and/or an old acrylamide solution.	Only use chemicals of analytical grade quality; store the acrylamide stock solution in the refrigerator in the dark for a short time (at most 2 weeks for SDS gels).
The slots dry out and burn through after a while; at the same time water collects at the cathodic side of the slots.	The samples were wrongly prepared; free SH groups lead to the formation of disulfide bridges between the various polypeptides; these aggregates are too large for the gel pores and highly negatively charged (SDS), → electro-osmotic flow of water towards the cathode.	Protect DTT from oxidation with EDTA; after reducing, heating and recooling the samples, add the same quantity of DTT as for the reduction; or else alkylate with iodoacetamide.

Tab. A2–3: Separation results.

Symptom	Cause	Remedy
The bands are not straight but curved.	Part of the gradient was shifted by heat convection during polymerization.	Add a little more APS to the light solution than to the dense one so that the polymerization starts at the top and proceeds towards the bottom.
The bands are close together at the buffer front.	The concentration of the resolving gel is too low.	Increase the concentration of the resolving gel.
	Low molecular weight peptides are poorly resolved.	Use gradient gels (if necessary concave exponential pore gradients); or use the buffer system of Schägger and von Jagow (1987).
The low molecular weight proteins are missing.	The post-polymerization is not finished.	Polymerize the gel at least one day ahead.
The bands are "fuzzy" and diffuse.	The post-polymerization is not finished. The resolving gel is not properly polymerized.	Polymerize the resolving gel at least one day ahead. Always prepare the resolving gel at least one day before use, because a slow post-polymerization takes place in the gel matrix.
The bands are diffuse.	The proteins were applied too close to the cathode.	Apply the samples at least 1 cm from the edge of the cathode wick or the buffer strips.
	The samples are old.	Prepare fresh samples; it is helpful to boil the samples again (adding reducing agent before and after); or else alkylate them.
	A homogeneous gel system was used or else the buffer system was not optimal for the samples.	Try disc PAGE and/or gradient gel PAGE or another buffer system.
	Insufficient gel polymerization.	See above.

Symptom	Cause	Remedy
Artifactual double bands.	Partial refolding of the molecules because the SH groups are not sufficiently protected.	Alkylate the samples after reduction.
The sample concentration is irregularly distributed in the bands.	Very small sample volumes were applied so the filling of the sample wells was irregular.	Dilute the samples with buffer and apply a correspondingly larger volume or apply the sample directly on the surface of the gel (for < 5 µL).
Crooked bands.	There is too much salt in the samples which leads to marked conductivity differences at the beginning.	Desalt the samples by gel filtration (NAP column) or by dialyzing against sample buffer.
Precipitates at the edge of the slots.	Gradient gel: the gradient was poured the wrong way round and the samples applied in the part with narrow pores.	Make sure that the slots are in the area with the lower acrylamide concentration.
	Electro-decantation, the slots are too deep.	Make the slots less deep (decrease the number of layers of Tesa film), they should not be more than 1/3 of the thickness of the gel. If larger sample volumes must be applied, it is preferable to increase the surface of the slots.
	A few proteins are too large to penetrate the gel pores.	Use gradient gels which allow separate of a broader molecular weight spectrum.
	Mistake during sample application; the free SH groups were not protected and disulfide bridges have formed between the polypeptides which are now too large for the pores.	Protect DTT from oxidation with EDTA; after reduction, heating and cooling, add the same amount of DTT as for the reduction to the samples; or else alkylate them, with iodoacetamide for example.
High molecular weight "ghost bands".	Mistake during sample preparation, see above.	See above.

Symptom	Cause	Remedy
Protein precipitate at the edge of the sample well; a narrowing then widening trace stretches below the sample well through the separation trace.	Overloading; the polypeptides form complexes and aggregate when entering the gel matrix; protein dissolves intermittently and migrates towards the anode.	Dilute the sample, apply less.
The bands streak.	There are particles in the sample.	Centrifuge or filter the sample.
There are individual streaks which sometimes begin in the middle of the gel.	Dust particles or dandruff etc. have fallen on the surface.	Do not leave the lid of the chamber open too long; apply the samples as quickly as possible; do not lean over the gel
Smudged bands, the bands tail.	There is grease in the sample.	Remove lipophilic substances completely during sample preparation.
	Insufficient loading with SDS.	The sample buffer should contain at least 1% SDS; the SDS/sample ratio should be higher than 1.4:1.
	Incomplete stacking of some highly concentrated protein fractions.	Use a complete discontinuous buffer system either by casting a resolving and a stacking gel in the traditional way, or by equilibrating the stacking zone of a film supported gel selectively in a vertical buffer chamber (see page 228).
	Very acidic and basic proteins, nucleoproteins do not react well in the SDS system.	Try CTAB electrophoresis (cationic detergent in an acid buffer system) (Eley *et al.* 1979, Atin *et al.* 1985).
The molecular weights do not agree with other experimental results.	The molecular weights of non-reduced samples were determined with reduced marker proteins.	An estimation of the molecular weights is possible by comparison with non-reduced globular polypeptides.

Symptom	Cause	Remedy
	Glycoproteins migrate more slowly than polypeptides with the same molecular weight, because the sugar moieties are not charged with SDS.	Use a Tris-borate-EDTA buffer; borate binds to the sugar moieties and the mobility increases because of the additional negative charge. Also use a gradient gel.
Background in the separation lanes.	Protease activity in the sample.	Proteases are also active in the presence of SDS; add inhibitor if necessary (e.g. PMSF).

Tab. A2–4:   SDS-PAGE specific staining problems.

Symptom	Cause	Remedy
Coomassie Blue: insufficient staining power.	SDS was not properly removed from the sample.	Only use pure SDS, low percentages of 14-C and 16-C sulfate bind more strongly to proteins; wash out SDS with 20% TCA; stain for longer than native electrophoresis.
	The alcohol content of the destaining solution is too high.	Avoid alcohol containing staining and destaining solutions; or else: use colloidal development methods.
Silver staining: negative bands.	Impure SDS.	See above.

Tab. A2–5: Drying.

Symptom	Cause	Remedy
The gel comes off the support film during staining, tears during drying, the dried gel rolls up.	The bonds between the gel and the support film have been partly hydrolyzed by the strong acids (TCA, H_3PO_4) in some staining solutions, they do not resist the high mechanical tensions of the concentrated gels.	Prepare highly concentrated gels with a degree of crosslinking $C=2\%$ instead of $C=3\%$ as usual. But less concentrated plateaus or stacking gels should be crosslinked with $C=3\%$ otherwise they are not stable enough; add 10% (v/v) glycerol to the last destaining solution, 15 min.

A3 Trouble shooting:
Vertical PAGE

Only problems specific to vertical gels are listed here. Consult paragraphs A2 and A5 for general problems occuring in SDS and Two-dimensional electrophoresis.

Tab. A3–1: Gel preparation.

Symptom	Cause	Remedy
Refractive lines in different directions are visible in the gel.	Dirty plates.	Wash and dry glass and ceramics plates thoroughly.
	CelloSeal® on the ceramics plate.	Do not coat the gaskets with CelloSeal®.
Sample well arms are too short.	Poor polymerization effectiveness	Use stacking gel buffer pH 6.8 for the low acrylamide part; degas monomer solution, add more APS solution.
Stacking gel and well arms have dried.	Gels have not been kept in sufficient humidity.	Remove gels after 1 hour after casting from the stand and place them immediately with enough gel buffer into completely sealed plastic bags.

Tab. A3–2: Running the gels.

Symptom	Cause	Remedy
Cathodal buffer leaking out.	A crinkle in the gasket has formed.	Take the gasket out and place it back into the groove. Do not coat it with CelloSeal®.

Electrophoresis in Practice, Fourth Edition. Reiner Westermeier
Copyright © 2005 WILEY-VCH Verlag GmbH & Co. KGaA, Weinheim
ISBN: 3-527-31181-5

Symptom	Cause	Remedy
Sample (visible because of added bromophenol blue) does not deposit at the bottom of the well.	No density difference. The samples must contain sucrose or glycerol.	Add 25 % (v/v) glycerol to the sample buffer.
Bromophenol blue dye does not migrate.	Incomplete filling of the cathodal compartment.	Fill 75 mL in, check buffer level, check gasket for leaking.
	Wrong buffer in the cathodal compartment.	Pour the Tris-glycine buffer into the cathodal compartment.
	One gel is run. Short circuit in the anode buffer.	Clamp a glass or a ceramics plate to the opposite side of the core.
"Smiling front".	Insufficient dissipation of Joule heat.	Reduce current setting. When direct cooling is employed: check, whether the Parafilm® has been removed before clamping the cassette to the core.

Tab. A3–3:   Separation results.

Symptom	Cause	Remedy
The bands are close together at the front.	Poor sieving properties of the gel. Gel has partially hydrolized due to: too long storage. too warm storage.	Store gels in a refrigerator, however no longer than 2 months.

Tab. A3–4:   Preservation.

Symptom	Cause	Remedy
Dried with cellophane clamped in two frames: Gels are cracking.	Air bubbles caught between gel and cellophane.	Carefully remove all air bubbles.
	Cellophane dries too fast, because it has been soaked in water.	Soak cellophane in the same glycerol solution like the gel.

A4 Trouble shooting: Semidry blotting

Tab. A4–1: Assembling the blotting sandwich.

Symptom	Cause	Remedy
Air bubbles between the filter papers.	Air bubbles are trapped between the layers because the sandwich was not made up under buffer.	After assembling the sandwich carefully and slowly push out the air bubbles with a roller, trying not to expel too much buffer.
Gels which are polymerized on support films are difficult to handle.	It is difficult to remove the support film: the gels tear or warps.	Use the Film Remover.
Bits of the gel remain on the support film.	The metal wire was not pulled through regularly, the process was interrupted a few times.	Pull the wire completely through using a single action with constant speed.
The gel warps when the support film is removed.	Very thin gels or gels with large pores, as well as gels containing glycerol, often stick to the film.	Spray a a few drops of transfer buffer between the gel and the film with a Pasteur pipette.

Tab. A4–2: Problems during blotting.

Symptom	Cause	Remedy
The power supply does not work properly, it jumps between different current settings or switches itself off.	Some power supplies do not work at the very low voltages required for blotting: 3 to 6 V.	Use a power supply with a low resistance.

Electrophoresis in Practice, Fourth Edition. Reiner Westermeier
Copyright © 2005 WILEY-VCH Verlag GmbH & Co. KGaA, Weinheim
ISBN: 3-527-31181-5

Symptom	Cause	Remedy
High power (ca. 20 W).	The surface of the filter paper is too large, the current flows around the blot stack.	Cut the filter paper to the size of the gel and the blotting membrane, or else cut a plastic mask and place it under the blotting membrane. (The window should be the same size as the gel and the blotting membrane).
The voltage increases during blotting.	Electrolysis gas pockets have formed between the graphite plates and the filter paper and this causes the conductivity to decrease.	Place a 1 kg weight on the graphite plate so that the trapped gas escapes from the sides.
The power increases during blotting, the blot sandwich warms up.	The buffer is too concentrated, the current is too high.	Only use the recommended buffer concentration, if possible do not set the current at more than 0.8 mA/cm2.

Tab. A4–3: After blotting.

Symptom	Cause	Remedy
It is difficult to remove the blotting membrane.	The gel surface is too sticky.	Use a blotting membrane with0.45 µm pores, no smaller, for gels with large pores (focusing gels, agarose gels); soak the gels briefly (3 to 4 min) in the transfer buffer before assembling the blotting sandwich; for IEF gels place the side of the gel that was previously bound to the support film on the blotting membrane (not the other surface).

Tab. A4–4: Results.

Symptom	Cause	Remedy
No transfer, there is nothing on the blotting membrane.	The current supply was wrongly connected.	Check the connections: the current should flow towards the anode in basic buffers and towards the cathode in acid buffers.
	The blotting sandwich was not correctly assembled, the blotting membrane is on the wrong side.	Check the assembling order of the blotting sandwich; see above.
	The surface of the filter paper is too large, the current flows around the blot.	Cut the filter paper and the blotting membrane to the size of the gel or else use a plastic mask, see above.
	The urea concentration of the focusing gel is too high: SDS does not bind to the proteins so the transfer does not occur because of the missing electrical charge.	Soak focusing gels which contain urea in the cathode buffer for a few minutes so that the urea diffuses out, then assemble the blot.
	The focusing gel contains non-ionic detergents which prevent the isoelectric proteins from becoming charged.	Native electrophoresis should be carried out in the absence of non-ionic detergents; as yet no satisfying solution exist for IEF (subject for a thesis).
	The proteins lie within the blotting membrane and cannot be seen.	Use a general stain to check, e.g. Amido Black or Indian Ink, then scan the blot with a laser densitometer.
The transfer is incomplete (the molecules stay in the gel).	The gel concentration is too high.	Use a lower gel concentration.
	The molecular weight of a few proteins is too high.	Carry out a limited proteolysis of the high molecular weight proteins before the transfer in the gel.

Symptom	Cause	Remedy
	The speeds of migration of the proteins are too variable: the high molecular weight ones are still in the gel, the low molecular weight ones have already passed through the membrane.	Use pore gradient gels.
	The methanol in the transfer buffer causes the gel to shrink too much.	Remove methanol or reduce its concentration.
Irregular transfer.	The current is irregular because the graphite plates are too dry.	Always wet the graphite plates with distilled water before assembly.
Incomplete transfer.	The transfer time was too short.	The recommended transfer time (1 h) should be prolonged (e.g. by 1/2 h) for thick gels, e.g. 3 mm and/or gels with very narrow pores.
	The charge/mass ratio of a few of the proteins is not favorable.	Equilibrate the gel in the cathode buffer for 5 to 10 min before blotting.
	Part of the current has flowed beside the gel and the blotting membrane.	Always cut the filter paper and blotting membrane to exactly the same size as the gel; or else use a plastic mask with openings to isolate the area around the gel.
	When several "trans units" are transferred simultaneously, the transfer efficiency decreases.	In such a case, only blot one transfer unit at a time.
	A continuous buffer system was used; this is less effective than a discontinuous buffer.	Use a discontinuous buffer system.

Symptom	Cause	Remedy
	The blotting with a discontinuous buffer system was performed at too low a temperature, this causes the pK value of the terminating ion to change and it migrates faster than the protein.	If 6-aminohexanoic acid is used as terminating ion, the blotting should be carried out at room temperature (20 to 25 °C); if the blotting of temperature sensitive enzymes has to be carried out at a lower temperature, use another buffer system.
Poor transfer (the pattern is distorted).	Contact problem: there are air bubbles in the blotting sandwich.	Make sure that there are no air bubbles; carefully press them out with a hand roller.
	The polyacrylamide gel swells during blotting.	Add 20% methanol to the transfer buffer; soak the gel in the transfer buffer beforehand (5 min).
	Diffusion.	Reduce the time for sandwich assembly and equilibration; do not shift the gel and the blotting membrane once they have come into contact.
	A lot of buffer in the graphite plates after long and /or intensive use of the semidry blotter.	Always wet the graphite plates intensively with distilled water before assembling the blot. From time to time the buffer ions must be eluted from the graphite plates. Soak a stack of filter paper in blotting buffer and make a 30 min run with reversed polarity (change anode and cathode plug in the power supply sockets. Repeat this procedure two times. Soak tissue paper with dist. water and lie them on the graphite plates for a few min. Repeat this.

Symptom	Cause	Remedy
The transfer was not effective (the molecules migrate out of the gel, but too few are found on the membrane).	The binding is not strong enough.	After the transfer let the blot lie overnight or for 3 h in a heating cabinet at 60 °C so that the bonds are reinforced; only then stain or block.
	The low molecular weight peptides have been washed out during detection.	Use another blotting membrane: e.g. nitrocellulose with a smaller pore size, cyanobromide activated NC, PVDF or nylon film. Or else place several sheets of blotting membrane behind one another so that blotted molecules are trapped. Use nylon membranes and fix with glutaraldehyde after the transfer (Karey and Sirbasku, 1989)
	The binding capacity has been reduced by detergents.	Do not use detergents or use agarose-intermediate-gels (Bjerrum *et al.* 1987).
	The pH was too high or too low.	Modify the pH accordingly.
	The transfer time was too long.	Shorten the transfer time.
There are additional bands on the blotting membrane.	A few proteins were blotted through when several trans units were blotted simultaneously and have migrated on to the next membrane.	When several trans units are blotted simultaneously insert a dialysis membrane between them.
The transfer was not effective – the proteins have left the gel but are not on the blotting membrane.	The transfer time is too long; the proteins have been blotted through.	Shorten the transfer time; guideline: 1 h for IEF gels. Less time should be used for native gels.

Tab. A4–5: Detection.

Symptom	Cause	Remedy
Fast Green cannot be washed out.	The development was too long.	Stain for a shorter time.
Too little protein was detected.	The detergent (e.g. Tween, NP-40) in the blocking buffer has eluted part of the proteins.	Reduce the detergent concentration or do without detergent.
	The sensitivity is too low.	The sensitivity of immuno detection can be increased by alkaline treatment; perform immuno reactions and enzyme reactions at 37 °C.
	The sensitivity of the peroxidase reaction is too low.	Use another peroxidase substrate, e.g. the tetrazolium method; use the immuno gold method.
	The sensitivity of the immuno gold reaction is too low.	Enhance it by silver staining.
	The sensitivity is very poor because the antigen concentration is extremely low.	Use an amplifying enzyme system: avidin-biotin-alkaline phosphatase method; or use an enhanced chemiluminescent detection method; or else use autoradiography.
Dark background.	Ineffective blocking.	Block for a longer time and at a higher temperature (37 °C).
	Cross-reactions with the blocking reagent.	Use another blocking reagent e.g. fish gelatin or skim milk powder.
The immuno detection is not specific.	SDS was not properly washed out, antibodies also bind to SDS protein micelles.	Wash longer.
	Poor quality antibodies.	Use other antibodies.
	The wrong secondary antibodies were used. For example: investigation of antigens of plant origin. The secondary antibody animal nourishes itself with plants.	Use other secondary antibodies.

Symptom	Cause	Remedy
Indian Ink staining: too low sensitivity, some bands in the middle are not stained.	The Indian Ink solution was too alkaline.	Add 1% acetic acid to the Indian Ink solution.
The background is not completely white.	There are particles in the Indian Ink solution.	Filter the Indian Ink solution, do not touch it with bare hands.
No bands.	The wrong dye was used.	Try with TBS-Tween instead of PBS-Tween.

A5 Trouble shooting:
2-D Electrophoresis

Very comprehensive trouble shooting guides on 2-D electrophoresis are found in the handbook by Berkelman and Stenstedt (2002) and at the website of Angelika Görg's group: found on this website: http://www.weihenstephan.de/blm/deg/
A few important points are listed here:

Tab. A5–1: First dimension: IEF in immobilized pH gradients.

Symptom	Cause	Remedy
The urea in the IPG strips crystallizes out.	The IEF temperature is too low.	Focus at 20 °C.
	The surface dries out.	Add 0.5% glycerol to the rehydration solution.
Sparking.	High voltage and ions in the contact fluid.	Use kerosene.
Formation of strong precipitates at the sample application point.	There are aggregates and complexes in the sample.	Check the sample preparation procedure: see above. The addition of at most 0.8% (w/v) carrier ampholyte to the sample increases the solubility.
	The protein and/or salt concentrations in the sample are too high.	Dilute the sample, it is preferable to apply a higher sample volume. The contact surface between the sample and the gel should be as small as possible. Apply the samples with pieces of tubing or an applicator strip.

Symptom	Cause	Remedy
	The field strength is too high at first.	Do not prefocus (the pH gradient already exists); when IEF is carried out in individual strips, regulate the field strength with the voltage: 1 h at max. E=15 V/cm, then turn up the setting.
The bands in the basic area are blurred.	Influence of CO_2.	Run the strips under paraffin oil.
Horizontal streaks.	A few proteins have not focused.	Lengthen the focusing time, see above.
	IEF temperature too low.	Run the first dimension at 20 °C.
	A few proteins have precipitated on the surface of the IEF gel but dissolved again after a while.	Minimize precipitate formation (see above).
Horizontal streaking at the basic end.	Deprotonation of reducing agent.	Use DeStreak (disulphide) in the rehydration solution instead of DTT or DTE. Apply sample with cup-loading at anodal end of the pH gradient.
There are precipitates on the surface.	The proteins are not properly solubilized.	Increase the urea concentration (up to 9 mol/L); add a non-ionic or a zwitterionic detergent to the sample and the gel; add carrier ampholyte and DTT to the solubilizing solution. If this does not help, try nondetergent sulfobetains (Vuillard *et al.* 1995).
	The proteins concentrate before entering the gel and suddenly aggregate.	Use a low field strength at first.
	Nucleic acids in the sample have precipitated with basic carrier ampholytes.	Apply the sample at the anode.

Symptom	Cause	Remedy
	High molecular weight nucleic acids form strongly ionic precipitates which bind to the protein.	Treat the sample with RNAse or DNAse.
	The basic part of the gradient is overbuffered by 2-mercaptoethanol (pK 9.5) from the sample.	Apply the sample at the anode, use DTT.

Tab. A5–2: Second dimension: SDS-electrophoresis.

Symptom	Cause	Remedy
The IPG strip becomes thinner at the anodic end and swells at the cathodic end and then turns up, the gel burns through.	Electroendosmosis: the IPG gel becomes negatively charged by equilibration in SDS, this induces transport of water towards the cathode.	Equilibration for 2 × 15 min under constant agitation in a modified equilibration buffer (add 6 mol/L urea, 30% glycerol) compensates the electro-osmotic effect; remove the IPG strips after 75 min, displace the cathodic electrode strips.
Spots are missing, loss of protein in the second dimension.	Electroendosmosis (see above): electroosmotic flow carries part of the proteins (up to 2/3) towards the cathode.	Use a modified equilibration system as described above for horizontal as well as vertical systems.
Horizontal streaks on the gel.	The proteins are not completely focused.	Focuse for a longer time; this is no problem during IPG because the gradient cannot drift.
	The equilibration was not effective enough.	Use an equilibration system as described above; the times (2 × 15 min) must absolutely be held.
	Artifacts due to the reducing agent occur.	Add iodoacetamide (2.5 times the amount of DTT) to the second equilibration step (to trap the excess of reducing agent).

Symptom	Cause	Remedy
Horizontal streaks.	There are air bubbles between the first– and second-dimensional gels.	Ensure that the contact between the first– and second-dimensional gels is free from air bubbles.
There are three streaks over the whole width of the gel.	Artifacts due to the reducing agent.	Equilibrate in two steps; during the second step trap the excess of reducing agent with iodoacetamide.
Vertical streaks	Problems with protein solubility.	Prepare a fresh urea solution to prevent the formation of isocyanate; use urea (6 mol/L) to equilibrate; increase the SDS content to 2%.
	Artifacts due to the reducing agent.	See above.
There are vertical streaks and a distorted spot pattern.	Micelles have formed between the non-ionic detergent from the IEF gel and the anionic detergent.	Only use 0.5% non-ionic detergent in the gel instead of 2% as usually used for IEF, or else use narrower strips for the first dimension.
There are vertical streaks in the high molecular weight range.	The equilibration step was not sufficient for several of the proteins.	Lengthen the equilibration time, increase the SDS concentration in the equilibration buffer (up to 4%), increase the temperature (up to 80 °C).
	The protein concentrating effect ("stacking") is not sufficient.	Use a discontinuous gel system.
There are vertical streaks in the low molecular weight range.	The protein concentration effect ("stacking") is not sufficient.	Use the method of Schägger and von Jagow (1987) for the second dimension.
There is a dark background in different areas of the gel.	There is protease activity in the sample.	Check the sample preparation method; if necessary add protease inhibitor (Pefabloc or PMSF).
There is a conspicuous row of spots with the same moleular weight.	A few proteins were carbamylated by isocyanate.	Check the sample preparation procedure; prepare a fresh urea solution, avoid high temperatures; only use very pure urea.

Symptom	Cause	Remedy
Spots are missing.	The first dimension was stored too long or not correctly.	Equilibrate immediately after the first dimension and carry out the second dimension; or store the IPG strips in liquid nitrogen or at $< -80\,°C$.
	Some proteins are not soluble any more after the first-dimension run has been stained or fixed.	Perform intermediate staining or fixing only of those proteins which are easily soluble.

A6 Trouble shooting:
DNA electrophoresis

Tab. A6–1: Preparing the samples.

Symptom	Cause	Remedy
Inadequate resolution.	Sample too concentrated.	Thumb rule: When a sample can be detected with Ethidiumbromide, dilute it 1 : 5 with sample buffer.
SSCP analysis: Only double bands in the gel.	Chilling of the sample not efficient.	Use an ice-water bath instead of crushed ice. Apply samples quickly and start electrophoresis as fast as possible.
Mutation can not be detected.	SSCP analysis is influenced by many factors.	For a strategy to achieve optimal results check method 13 on page 313 ff.
Heteroduplex analysis: Only homoduplex bands and single strands in the gel.	Sample was too diluted to form heteroduplexes.	Samples must have the concentration to be detectable with Ethidiumbromide during heating and cooling. They are diluted to silver staining concentrations afterwards.
Denaturing run: Blurred bands an inadequate resolution.	Sample has not been denatured before application.	Heat sample with formamide before application.

Electrophoresis in Practice, Fourth Edition. Reiner Westermeier
Copyright © 2005 WILEY-VCH Verlag GmbH & Co. KGaA, Weinheim
ISBN: 3-527-31181-5

Tab. A6–2: Running the gels.

Symptom	Cause	Remedy
Inadequate separation, poor resolution, no polymorphism expressed.	Wrong buffer system applied: different buffer systems can result in different patterns. Important for SSCP!	Test for optimal buffer system, apply: Tris-acetate / tricine; Tris-phosphate / borate, or else.
	Wrong temperature applied. Different temperatures can result in different patterns. Important for SSCP!	Test for optimal temperatures, run gel at different temperatures: 5 °C, 15 °C, 25 °C.
	Electrode solutions have been mixed up.	Add bromophenol blue to the anodal buffer to avoid this mistake.
	Electrodes are too close together; this leads to a lack of ions.	Set the electrodes on the outer edges of the electrode wicks, see drawing:
	Gel surface too wet.	Dry the gel surface with the edge of a filter paper until you can hear a "squeaking".
Curved front and band distribution.	Uneven rehydration of the gel.	Lift the edges of the gel-film repeatedly , or rehydrate on a rocking platform, use sufficient buffer volume.
Smiling effect.	Too much kerosene applied.	Apply only a very thin layer of kerosene with tissue paper to the cooling plate. A few air bubbles do not matter!
Uneven front and band distribution.	Air bubbles under electrode wick, uneven contact of gel edge and electrode wick.	Slide bent tip foreceps aong the edges of the wicks laying in contact with the gels.
Slanted front and band distribution.	Uneven buffer concentration in the wicks.	After soaking electrode wicks in buffer, hold them horizontally with two foreceps during the transfer from the PaperPool to the gel.

Symptom	Cause	Remedy
Sample does not stay in sample well, spreads out over stacking gel.	Water condensation on the gel due to high humidity and / or high room temperature.	Apply the sample on the gel, when cooling plate has still ambient temperature. Connect cooling plate to thermostatic circulator after sample application.
	Glycerol or sucrose, or organic solvent in sample, osmotic distribution.	Prepare sample without glycerol or sucrose; those are only necessary for submarine agarose gels and vertical PAGE. Remove organic solvent with a SpeedVac.
	Water condensation on the gel due to high humidity and / or high room temperature.	Apply the sample on the gel, when cooling plate has still ambient temperature. Connect cooling plate to thermostatic circulator after sample application.
Less than 6 µL fit into sample well of a gel.	Buffer in sample wells: due to insufficient drying after rehydration.	Dry also the sample wells with filter paper.
	Buffer in sample wells: due to flooding with buffer from the cathodal electrode wick.	After soaking the cathodal wick: flap it upright along the long edge with two forceps, leave it for a few seconds to drain excess buffer off. Lay this wick on the cathodal edge of the gel with the formerly upper edge oriented towards the sample wells.

Tab. A6–3: Silver staining.

Symptom	Cause	Remedy
No bands detected.	Formaldehyde has precipitated.	Store formaldehyde at room temperature, never in a refrigerator.
Gel shows dark background, and no or faint bands.	Compositon and/or quality of chemicals and/ or water is inadequate.	Use only high purity chemicals. Try another supplier for $AgNO_3$.

Symptom	Cause	Remedy
Dark brown background.	Plastic trays are used, which contain softener compounds, which reduce silver.	Use only glass or stainless steel trays. Clean these trays thoroughly after each silver staining.
	Temperature of solutions to high	Perform silver staining with solutions cooled down below + 15 °C.
Gel shows darker areas and white areas after silver staining.	Uneven rehydration of the gel.	Lift the edges of the gel-film repeatedly, or rehydrate on a rocking platform, use sufficient buffer volume.
Silver stained fragment can not be reamplified.	Wrong staining protocol. TCA and/or glutardialde-hyde has destroyed the complete DNA in a band.	Use only DNA staining protocols, not those for proteins.

References

Aebersold RH, Teplow D, Hood LE, Kent SBH. J Biol Chem. 261 (1986) 4229–4238.

Aebersold RH, Pipes G, Hood LH, Kent SBH. Electrophoresis. 9 (1988) 520–530.

Alba FJ, Bermudez A, Daban J-R. BioTechniques 21 (1996) 625–626.

Alban A, David S, Bjorkesten L, Andersson C, Sloge E, Lewis S, Currie I. Proteomics 3 (2003) 36–44.

Allen RC, Budowle B, Lack PM, Graves G. In Dunn M, Ed. Electrophoresis '86. VCH, Weinheim (1986) 462–473.

Altland K, Banzhoff A, Hackler R, Rossmann U. Electrophoresis. 5 (1984) 379–381.

Altland K, Hackler R. In: Electrophoresis'84. Neuhoff V, Ed.Verlag Chemie, Weinheim (1984) 362–378.

Altland K. Electrophoresis. 11 (1990) 140–147.

Alwine JC, Kemp DJ, Stark JR. Proc Natl Acad Sci USA. 74 (1977) 5350–5354.

Anderson UN, Colburn AW, Makarov AA, Raptakis EN, Reynolds DJ, Derrick PJ, Davis SC, Hoffman AD, Thomson S. Rev Sci Instrum 69 (1998) 1650–1660.

Anderson L, Seilhamer J.A. Electrophoresis 18 (1997) 533–537.

Anderson NG, Anderson NL. Anal Biochem. 85 (1978) 331–340.

Anderson NL, Anderson NG. Mol Cell Proteomics 1 (2002) 845–867.

Andrews AT. Electrophoresis, theory techniques and biochemical and clinical applications. Clarendon Press, Oxford (1986).

Ansorge W, De Maeyer L. J Chromatogr. 202 (1980) 45–53.

Ansorge W, Sproat BS, Stegemann J, Schwager C. J Biochem Biophys Methods. 13 (1986) 315–323.

Atin DT, Shapira R, Kinkade JM. Anal Biochem. 145 (1985) 170–176.

Baldo BA, Tovey ER. Ed. Protein blotting. Methodology, research and diagnostic applications. Karger, Basel (1989).

Baldo BA. In Chrambach A, Dunn M, Radola BJ. Eds. Advances in Electrophoresis 7. VCH, Weinheim (1994) 409–478.

Banks R. Dunn MJ, Forbes MA, Stanly A, Pappin DJ, Naven T, Gough M, Harnden P, Selby PJ. Electrophoresis 20 (1999) 689–700.

Barros F, Carracedo A, Victoria ML, Rodriguez-Calvo MS. Electrophoresis 12 (1991) 1041–1045.

Bassam BJ, Caetano-Annollés G, Gresshoff PM. Anal Biochem. 196 (1991) 80–83.

Bauer D, Müller H, Reich J, Riedel H, Ahrenkiel V, Warthoe P, Strauss M. Nucleic Acid Res. 21 (1993) 4272–4280.

Baumstark M, Berg A, Halle M, Keul J. Electrophoresis. 9 (1988) 576–579.

Bayer EA, Ben-Hur H, Wilchek M. Anal Biochem. 161 (1987) 123–131.

Electrophoresis in Practice, Fourth Edition. Reiner Westermeier
Copyright © 2005 WILEY-VCH Verlag GmbH & Co. KGaA, Weinheim
ISBN: 3-527-31181-5

Beavis RC, Chait BT. Rapid Commun Mass Spectrom. 3 (1989) 432–435.

Beavis RC. Org Mass Spectrom. 27 (1992) 653–659.

Béguin P. Anal Biochem 131 (1983) 333–336.

Beisiegel U. Electrophoresis. 7 (1986) 1–18.

Berg DE, Akopyants NS, Kersulyte D. Meth Molec Cell Biol. 5 (1994)13–24.

Berggren KN, Schulenberg B, Lopez MF, Steinberg TH, Bogdanova A, Smejkal G, Wang A, Patton WF. Proteomics 2 (2002) 486–498.

Berkelman T, Stenstedt T. 2-D electrophoresis using immobilized pH gradients. Principle & methods. Amersham Biosciences Handbook 80–6429–60 (2002).

Berndt P, Hobohm U, Langen H.Electrophoresis 20 (1999) 3521–3526.

Berson G. Anal Biochem. 134 (1983) 230–234.

Bjellqvist B, Ek K, Righetti PG, Gianazza E, Görg A, Westermeier R, Postel W. J Biochem Biophys Methods. 6 (1982) 317–339.

Bjellqvist B, Linderholm M, Östergren K, Strahler JR. Electrophoresis. 9 (1988) 453–462.

Bjellqvist B, Sanchez J-C, Pasquali C, Ravier F, Paquet N, Frutiger S, Hughes GJ, Hochstrasser D. Electrophoresis. 14 (1993) 1375–1378

Bjerrum OJ, Selmer JC, Lihme A. Electrophoresis. 8 (1987) 388–397.

Bjerrum OJ. Ed. Paper symposium protein blotting. Electrophoresis. 8 (1987) 377–464.

Blake MS, Johnston KH, Russell-Jones GJ. Anal Biochem. 136 (1984) 175–179.

Blakesley RW, Boezi JA. Anal Biochem. 82 (1977) 580–582.

Blomberg A, Blomberg L, Norbeck J, Fey SJ, Larsen PM, Roepstorff P, Degand H, Boutry M, Posch A, Görg A. Electrophoresis. 16 (1995) 1935–1945.

Böckelmann R, Bonnekoh B, Gollnick H. Skin Pharmacol Appl Skin Physiol.12 (1999) 54–63.

Bøg-Hansen TC, Hau J. J Chrom Library. 18 B (1981) 219–252.

Bosch TCG, Lohmann J. In Fingerprinting methods based on PCR. Bova R, Micheli MR, Eds. Springer Verlag Heidelberg, (1997).

Brada D, Roth J. Anal Biochem. 142 (1984) 79–83.

Brewer JM. Science. 156 (1967) 256–257.

Brown RK, Caspers ML, Lull JM, Vinogradov SN, Felgenhauer K, Nekic M. J Chromatogr. 131 (1977) 223–232.

Burgess R, Arthur TM, Pietz BC. Meth Enzymol. 328 (2000) 141–157.

Burnette WN. Anal Biochem. 112 (1981) 195–203.

Bussard A. Biochim. Biophys Acta. 34 (1959) 258–260.

Caetano-Annollés G, Bassam BJ, Gresshoff PM. Bio/Technology 9 (1991) 553–557.

Chapman JR Ed. Mass spectrometry of proteins and peptides. Methods in Molecular Biology 146. Humana Press, Totowa, NJ (2000).

Chiari M, Casale E, Santaniello E, Righetti PG. Theor Applied Electr. 1 (1989a) 99–102.

Chiari M, Casale E, Santaniello E, Righetti PG. Theor Applied Electr. 1 (1989b) 103–107.

Chrambach A. The practice of quantitative gel electrophoresis. VCH Weinheim (1985).

Cohen AS, Karger BL. J Chromatogr. 397 (1987) 409–417.

Davis BJ. Ann NY Acad Sci. 121 (1964) 404–427.

Comisarow MB, Marshall AG. Chem Phys Letters. 25 (1974) 282–283.

Damerval C, DeVienne D, Zivy M, Thiellement H. Electrophoresis 7 (1986) 53–54.

Denhardt D. Biochem Biophys Res Commun. 20 (1966) 641–646.

Desvaux FX, David B, Peltre G. Electrophoresis 11 (1990) 37–41.

Dieefenbach CW, Dveksler GS, eds. PCR primer. A laboratory manual. Cold Spring Harbour Laboratory Press (1995).

Diezel W, Kopperschläger G, Hofmann E. Anal Biochem. 48 (1972) 617–620.

Dockhorn-Dworniczak B, Aulekla-Acholz C, Dworniczak B. Pharmacia LKB Sonderdruck A37 (1990).

Dunn MJ, Burghes AHM. Electrophoresis. 4 (1983) 97–116.

Dunn MJ, Ed. From Genome to proteome. Advances in the practice and application of proteomics. WILEY-VCH, Weinheim (1999).

Eckerskorn C, Mewes W, Goretzki H, Lottspeich F. Eur J Biochem. 176 (1988) 509–519.

Eckerskorn C, Lottspeich F. Chromatographia. 28 (1989) 92–94.

Eckerskorn C, Strupat K, Karas M, Hillenkamp F, Lottspeich F. Electrophoresis 13 (1992) 664–665.

Eley MH, Burns PC, Kannapell CC, Campbell PS. Anal Biochem. 92 (1979) 411–419.

Everaerts FM, Becker JM, Verheggen TPEM. Isotachophoresis, Theory, instrumentation and applications. J Chromatogr Library Vol 6. Elsevier, Amsterdam (1976).

Ferguson KA. Metabolism. 13 (1964) 985–995.

Fischer SG, Lerman LS. Proc Natl Acad Sci. 60 (1983) 1579–1583.

Fountoulakis M, Takacs B, Langen H. Electrophoresis 19 (1998) 761–766.

Friedman DB, Hill S, Keller JW, Merchant NB, Levy SE, Coffey RJ, Caprioli RM. Proteomics 4 (2004) 793–811.

Fujimura RK, Valdivia RP, Allison MA. DNA Prot Eng Technol. 1 (1988) 45–60.

Gharahdaghi F, Weinberg CR, Meagher D, Imai BS, Mische SM. Electrophoresis 20 (1999) 601–605.

Gershoni JM, Palade GE. Anal Biochem. 112 (1983) 1–15.

Giaffreda E, Tonani C, Righetti PG. J Chromatogr. 630 (1993) 313–327.

Gianazza E, Chillemi F, Gelfi C, Righetti PG. J Biochem Biophys Methods. 1 (1979) 237–251.

Gianazza E, Chillemi F, Righetti PG. J Biochem Biophys Methods. 3 (1980) 135–141.

Gianazza E, Chillemi F, Duranti M, Righetti PG. J Biochem Biophys Methods. 8 (1983) 339–351.

Görg A, Postel W, Westermeier R. Anal Biochem. 89 (1978) 60–70.

Görg A, Postel W, Westermeier R, Gianazza E, Righetti PG. J Biochem Biophys Methods. 3 (1980) 273–284.

Görg A, Postel W, Günther S, Weser J. Electrophoresis. 6 (1985) 599–604.

Görg A, Postel W, Weser J, Günther S, Strahler JR, Hanash SM, Somerlot L. Electrophoresis. 8 (1987) 45–51.

Görg A, Postel W, Günther S. Electrophoresis. 9 (1988a) 531–546.

Görg A, Postel W, Weser J, Günther S, Strahler JR, Hanash SM, Somerlot L, Kuick R. Electrophoresis. 9 (1988b) 37–46.

Görg A, Boguth G, Obermaier C, Posch A, Weiss W. Electrophoresis. 16 (1995) 1079–1086.

Görg A, Obermaier C, Boguth G, Csordas A, Diaz J-J, Madjar J-J: Electrophoresis. 18 (1997) 328–337.

Görg A, Boguth G, Obermaier C, Weiss W. Electrophoresis. 19 (1998) 1516–1519.

Görg A, Boguth G, Obermaier C, Harder A, Weiss W. Life Science News 1 (1998) 4–6.

Görg A, Obermaier C, Boguth G, Weiss W. Electrophoresis. 20 (1999) 712–717.

Görg A, Obermaier C, Boguth G, Harder A, Scheibe B, Wildgruber R, Weiss W. Electrophoresis. 21 (2000) 1037–1053.

Görg A, Boguth G, Köpf A, Reil G, Parlar H, Weiss W. Proteomics 2 (2002) 1652–1657.

Grabar P, Williams CA. Biochim Biophys Acta. 10 (1953) 193.

Günther S, Postel W, Weser J, Görg A. In: Dunn MJ, Ed. Electrophoresis '86. VCH Weinheim (1986) 485–488.

Hanash SM, Strahler JR, Somerlot L, Postel W, Görg A. Electrophoresis. 8 (1987) 229–234.

Hancock K, Tsang VCW. Anal Biochem. 133 (1983) 157–162.

Handmann E, Jarvis HM. J Immunol Methods. 83 (1985) 113–123.

Hannig K. Electrophoresis. 3 (1982) 235–243.

Hardy E, Santana H, Sosa AE, Hernandez L, Fernandez-Patron C, Castellanos-Serra L. Anal Biochem. 240 (1996) 150–152.

Hayashi K, Yandell DW. Hum Mutat. 2 (1993) 338 - 346.

Haynes PA, Gygi SP, Figeys D, Aebersold R. Electrophoresis 19 (1998) 1862–1871.

Hedrick JL, Smith AJ. Arch Biochem Biophys. 126 (1968) 155–163.

Hellman U, Wernstedt C, Góñez J, Heldin C-H. Anal Biochem. 224 (1995) 451–455.

Heukeshoven J, Dernick R. In: Radola BJ, Ed. Electrophorese-Forum '86. (1986) 22–27.

Heukeshoven J, Dernick R. Electrophoresis. 6 (1985) 103–112.

Himmelreich R, Hilbert H, Plagens H, Pirkl E, Li B-C, Herrmann R. Nucl Acids Res. 24 (1996) 4420–4449.

Hjalmarsson S-G, Baldesten A. In: CRC Critical Rev in Anal Chem. (1981) 261–352.

Hjertén S. Arch Biochem Biophys Suppl 1 (1962) 147.

Hjertén S. J Chromatogr. 270 (1983) 1–6.

Hoefer Protein Electrophoresis Applications Guide (1994) 18–54.

Hoffman WL, Jump AA, Kelly PJ, Elanogovan N. Electrophoresis. 10 (1989) 741–747.

Hsam SLK, Schickle HP, Westermeier R, Zeller FJ. Brauwissenschaft 3 (1993) 86–94.

Hsu D-M, Raine L, Fanger H. J Histochem Cytochem. 29 (1981) 577–580.

Jackson P, Thompson RJ. Electrophoresis. 5 (1984) 35–42.

Jaksch M, Gerbitz K-D ,Kilger C. J Clin Biochem. 1995.

Janson JC, Rydén L, Eds. Protein purification. Principles, high-resolution methods, and applications. WILEY-Liss, New York (1998).

Jenne DE, Denzel K, Blätzinger P, Winter P, Obermaier B, Linke RP, Altland K. Proc Natl Acad Sci USA. 93 (1996) 6302–6307.

Jeppson JO, Franzen B, Nilsson VO. Sci Tools. 25 (1978) 69–73.

Johansson K-E. Electrophoresis. 8 (1987) 379–383.

Johnson DA, Gautsch JW. Sportsman JR. Gene Anal Technol. 1 (1984) 3–8.

Jorgenson JW, Lukacs KD. Anal Chem 53 (1981) 1298–1302

Jovin TM, Dante ML, Chrambach A. Multiphasic buffer systems output. Natl Techn Inf Serv. Springfield VA USA PB(1970)196 085–196 091.

Karas M, Hillenkamp F. Anal Chem. 60 (1988) 2299–2301.

Karey KP, Sirbasku DA. Anal Biochem. 178 (1989) 255–259.

Kaufmann R, Kirsch D, Spengler B. Int J Mass Spectrom Ion Processes 131 (1994) 355–385.

Kellner R, Lottspeich F, Meyer H, Eds. Microcharacterization of proteins. Second edition. WILEY-VCH Weinheim (1999).

Keen JD, Lester D, Inglehearn C, Curtis A, Bhattacharya. Trends Genet. 7 (1991) 5.

Keough T, Youngquist RS, Lacey MP. Proc Natl Acad Sci USA. 96 (1999) 7131–7136.

Kerenyi L, Gallyas F. Clin Chim Acta 38 (1972) 465–467.

Kittler JM, Meisler NT, Viceps-Madore D. Anal Biochem. 137 (1984) 210–216.

Kleine B, Löffler G, Kaufmann H, Scheipers P, Schickle HP, Westermeier R, Bessler WG. Electrophoresis 13 (1992) 73–75.

Kohlrausch F. Ann Phys. 62 (1987) 209–220.

Kohn J. Nature 180 (1957) 986–988.

Krause I, Elbertzhagen H. In: Radola BJ, Ed. Elektrophorese-Forum'87. (1987) 382–384.

Kyhse-Andersen J. J Biochem Biophys Methods. 10 (1984) 203–209.

Laemmli UK. Nature. 227 (1970) 680–685.

Laing P. J Immunol Methods. 92 (1986) 161–165.

Lamond AI, Mann M. Trends Cell Biol. 7 (1997) 139–142.

Lane LC. Anal Biochem. 86 (1978) 655–664.

Landegren U, Ed. Laboratory Protocols for Mutation Detection. Oxford University Press (1996).

Langen H, Röder D, Juranville J-F, Fountoulakis M. Electrophoresis.18 (1997) 2085–2090.

Langen H, Röder D. Life Science News 3 (1999) 6–8.

Langen H, Takács B, Evers S, Berndt P, Lahm H-W, Wipf, B, Gray C, Fountoulakis M. Electrophoresis. 21 (2000) 411–429.

Laurell CB. Anal Biochem. 15 (1966) 45–52.

Leifheit H-J, Gathof AG, Cleve H. Ärztl Lab. 33 (1987) 10–12.

Liang, P, Pardee AB. Science 257 (1992) 967–971.

Link AJ. Ed. 2-D Proteome Analysis Protocols. Methods in Molecular Biology 112. Humana Press, Totowa, NJ (1999).

Loessner MJ, Scherer S. Electrophoresis.13 (1992) 461–463

Lohmann J, Schickle HP, Bosch TCG. BioTechniques 18 (1995) 200–202.

Mackintosh JA, Choi H-Y, Bae S-H, Veal DA, Bell PJ, Ferrari BC, Van Dyk DD, Verrills NM, Paik Y-K, Karuso P. Proteomics 3 (2003) 2273–2288.

Maniatis T, Fritsch EF, Sambrook J. Molecular cloning. A laboratory manual. Cold Spring Laboratory (1982).

Matsudaira P. J Biol Chem. 262 (1987) 10035–10038.

Matsui NM, Smith DM, Clauser KR, Fichmann J, Andrews LE, Sullivan CM, Burlingame AL, Epstein LB. Electrophoresis. 18 (1997) 409–417.

Maurer RH. Disk- Electrophorese - Theorie und Praxis der diskontinuierlichen Polyacrylamid-Electrophorese. W de Gruyter, Berlin (1968).

Maxam AM, Gilbert W. Proc Natl AcadSci USA. 74 (1977) 560–564.

Méchin V, Consoli L, Le Guilloux M, Damerval C, Proteomics 3 (2003) 1299–1302.

Merril CR, Switzer RC, Van Keuren ML. Proc Nat Acad Sci 76 (1979) 4335–4339.

Merill CR, Goldman D, Sedman SA, Ebert MH. Science. 211 (1981) 1437–1438.

Moeremans M, Daneels G, De Mey J. Anal Biochem. 145 (1985) 315–321.

Moeremans M, Daneels G, Van Dijck A, Langanger G, De Mey J. J Immunol Methods. 74 (1984) 353–360.

Moeremans M, De Raeymaeker M, Daneels G, De Mey. J. Anal Biochem. 153 (1986) 18–22.

Möller A, Wiegand P, Grüschow C, Seuchter SA, Baur MP, Brinkmann B. Int J Leg Med 106 (1994) 183–189.

Mørtz E, Saraneva T, Haebel S, Julkunen I, Roepstorff P. Electrophoresis 17 (1996) 925–931.

Montelaro RC. Electrophoresis. 8 (1987) 432–438.

Mosher RA, Saville DA, Thormann W. The Dynamics of Electrophoresis. VCH Weinheim (1992).

Neuhoff V, Stamm R, Eibl H. Electrophoresis. 6 (1985) 427–448.

O'Farrell PH. J Biol Chem. 250 (1975) 4007–4021.

O'Farrell PZ, Goodman HM, O'Farrell PH. Cell. 12 (1977) 1133–1142.

Olsson BG, Weström BR, Karlsson BW. Electrophoresis. 8 (1987) 377–464.

Olsson I, Larsson K, Palmgren R, Bjellqvist B. Proteomics 2 (2002) 1630–1632.

Olszewska E, Jones K. Trends Gen. 4 (1988) 92–94.

Orita M, Iwahana H, Kanazewa H, Hayashi K, Sekiya T. Proc Natl Acad Sci USA. 86(1989) 2766–2770.

Ornstein L. Ann NY Acad Sci. 121 (1964) 321–349.

Ouchterlony Ö. Allergy. 6 (1958) 6.

Pandey A, Mann M. Nature 405 (2000) 837–846.

Patestos NP, Fauth M, Radola BJ. Electrophoresis. 9 (1988) 488–496.

Pennington SR, Wilkins MR, Hochstrasser DR, Dunn MJ. Trends cell biol. 7 (1997) 168–173.

Pennington SR, Dunn MJ, Eds. Proteomics from protein sequence to function. BIOS Scientific Publishers Limited, Oxford (2001).

Perella M, Heyda A, Mosca, A, Rossi-Bernardi L. Anal Biochem 88 (1978) 212–224.

Perlman D, Chikarmane H Halvorson HO. Anal Biochem. 163 (1987) 247–254.

Pflug W, Laczko B. Electrophoresis. 8 (1987) 247–248.

Pharmacia LKB Development Technique File No 230 PhastSystem (1989).

Poduslo JF. Anal Biochem. 114 (1981) 131–139.

Pretlow TG, Pretlow TP Methods: A Companion to Methods in Enzymology 2 (1991)183–191.

Prieur B, Russo-Marie F. Anal Biochem. 172 (1988) 338–343.

Puers C, Hammond HA, Jin L, Caskey CT, Schumm JW. Am J Hum Genet 53 (1993) 953–958.

Rabilloud T, Gianazza E, Cattò N, Righetti PG. Anal Biochem 185 (1990) 94–102.

Rabilloud T, Valette C, Lawrence JJ. Electrophoresis. 15 (1994) 1552–1558.

Rabilloud T. Electrophoresis 19 (1998) 758–760.

Rabilloud T, Chalmont S. In Rabilloud T, Ed. Proteome research: Two-dimensional gel electrophoresis and identification methods. Springer Berlin Heidelberg New York (2000) 107–126.

Radola BJ. Biochim Biophys Acta. 295 (1973) 412–428.

Ramsby ML, Makowski GS, Khairallah EA. Electrophoresis 15 (1994) 265–277.

Raymond S. Weintraub L. Science. 130 (1959) 711–711.

Rehbein H. Electrophoresis. 16 (1995) 820–822.

Rehbein H, Mackie IM, Pryde S, Gonzales-Sotelo C, Perez-Martin R, Quinteiro J, Rey-Mendez M. Inf. Fischwirtsch. 42 (1995) 209–212.

Rehbein H, Kündiger R, Pineiro C, Perez-Martin RI. Electrophoresis 21 (2000) 1458–1463.

Reiser J, Stark GR. Methods Enzymol. 96 (1983) 205–215.

Renart J, Reiser J, Stark GR. Proc Natl Acad Sci USA. 76 (1979) 3116–3120.

Rickwood D, Hames BD. Gel electrophoresis of nucleic acids. IRL Press Ltd (1982).

Riesner D, Steger G, Wiese U, Wulfert M, Heibey M, Henco K. Electrophoresis 10 (1989) 377–389.

Righetti PG, Drysdale JW. Ann NY Acad Sci. 209 (1973) 163–187.

Righetti PG. J. Chromatogr. 138 (1977) 213–215.

Righetti PG, Tudor G, Gianazza E. J Biochem Biophys Methods 6 (1982) 219–227

Righetti PG. In:Work TS, Ed. Burdon RH. Isoelectric focusing: theory, methodology and applications. Elsevier Biomedical Press, Amsterdam (1983).

Righetti PG, Gelfi C. J Biochem Biophys Methods. 9 (1984) 103–119.

Righetti PG, Wenisch E, Faupel M. J Chromatogr. 475 (1989) 293–309.

Righetti PG. In: Burdon RH, van Knippenberg PH. Ed. Immobilized pH gradients: theory and methodology. Elsevier, Amsterdam (1990).

Rimpilainen M, Righetti PG. Electrophoresis. 6 (1985) 419–422.

Robinson HK. Anal Biochem. 49 (1972) 353–366.

Roepstorff P, Fohlmann J. Biomed Mass Spectrom. 11 (1984) 601.

Rosengren A, Bjellqvist B, Gasparic V. In: Radola BJ, Graesslin D. Ed. Electrofocusing and isotachophoresis. W. de Gruyter, Berlin (1977) 165–171.

Rossmann U, Altland K. Electrophoresis. 8 (1987) 584–585.

Rothe GM, Purkhanbaba M. Electrophoresis. 3 (1982) 33–42.

Rothe GM. Electrophoresis of Enzymes. Springer Verlag, Berlin (1994).

Salinovich O, Montelaro RC. Anal Biochem. 156 (1986) 341–347.

Sanchez J-C, Appel RD, Golaz O, Pasquali C, Ravier F, Bairoch A, Hochstrasser DF. Electrophoresis. 16 (1995) 1131–1151.

Sanger F, Coulson AR. J Mol Biol. 94 (1975) 441–448.

Schägger H, von Jagow G. Anal Biochem. 166 (1987) 368–379.

Schägger H, von Jagow G. Anal Biochem. 199 (1991) 223–231.

Schägger H. Chapter 4: Native electrophoresis. In: A Practical guide to membrane protein Purification (Von Jagow G, Schägger H, eds.) Academic Press, New York (1994) 81–103.

Scherz H. Electrophoresis. 11 (1990) 18–22.

Schickle HP. GIT LaborMedizin. 19 (1996) 159–163.

Schickle HP. GIT LaborMedizin. 19 (1996) 228–231.

Schickle HP, Lamb B, Hanash SM. Electrophoresis. 20 (1999)1233–1238.

Schumacher J, Meyer N, Riesner D, Weidemann HL. J Phytophathol. 115 (1986) 332–343.

Schwartz DC, Cantor CR. Cell. 37 (1984) 67–75.

Serwer P. Biochemistry. 19 (1980) 3001–3005.

Seymour C, Gronau-Czybulka S. Pharmacia Biotech Europe Information Bulletin (1992)

Shapiro AL, Viñuela E, Maizel JV. Biochem Biophys Res Commun. 28 (1967) 815–822.

Shevchenko A, Wilm M, Vorm O. Mann M. Anal Chem. 68 (1996) 850–858.

Simpson RJ, Moritz RL, Begg GS, Rubira MR, Nice EC. Anal Biochem. 177 (1989) 221–236.

Simpson RJ. Proteins and proteomics. A laboratory manual. Cold Spring Harbor Laboratory Press, Cold Spring Harbor, New York (2003).

Simpson RJ, Ed. Purifying proteins for proteomics: A laboratory manual. Cold Spring Harbor Laboratory Press, Cold Spring Harbor, New York (2003).

Sinha PK, Bianchi-Bosisio A, Meyer-Sabellek W, Righetti PG. Clin Chem. 32 (1986) 1264–1268.

Sinha P, Köttgen E, Westermeier R, Righetti PG. Electrophoresis. 13 (1992) 210–214

Smith MR, Devine CS, Cohn SM, Lieberman MW. Anal Biochem. 137 (1984) 120–124.

Siuzdak G. The expanding role of mass spectrometry for biotechnology. MCC Press, San Diego (2003).

Smithies O. Biochem J. 61 (1955) 629–641.

Southern EM. J Mol Biol. 98 (1975) 503–517.

Speicher DW, Zuo X. Anal Biochem. 284 (2000) 266–278.

Stasyk T, Hellman U, Souchelnytskyi S. Life Science News 9 (2001) 9–12.

Steinberg TH, Hangland RP, Singer VI. Anal Biochem. 239 (1996).

Strahler JR, Hanash SM, Somerlot L, Weser J, Postel W, Görg A. Electrophoresis. 8 (1987) 165–173.

Strupat K, Karas M, Hillenkamp F, Eckerskorn C, Lottspeich F. Anal Chem. 66 (1994) 464–470.

Susann J. The Valley of the Dolls. Corgi Publ. London (1966).

Sutherland MW, Skerritt JH. Electrophoresis. 7 (1986) 401–406.

Suttorp M, von Neuhoff N, Tiemann M, Dreger P, Schaub J, Löffer H, Parwaresch R, Schmitz N. Electrophoresis 17 (1996) 672–677.

Svensson H. Acta Chem Scand. 15 (1961) 325–341.

Taketa K. Electrophoresis. 8 (1987) 409–414.

Terabe S, Otsuka K, Ichikawa K, Tsuchiya A, Ando T. AnalChem.64 (1984) 111–113

Terabe S, Chen N, Otsuka K. In Chrambach A, Dunn M, Radola BJ. Eds. Advances in electrophoresis 7. VCH, Weinheim (1994) 87–153.

Tiselius A. Trans Faraday Soc. 33 (1937) 524–531.

Tovey ER, Baldo BA. Electrophoresis. 8 (1987) 384–387.

Towbin H, Staehelin T, Gordon J. Proc Natl Acad Sci USA. 76 (1979) 4350–4354.

Towbin H, Özbey Ö, Zingel O. Electrophoresis 22 (2001) 1887–1893.

Ünlü M, Morgan ME, Minden JS. Electrophoresis. 18 (1997) 2071–2077.

Vandekerckhove J, Bauw G, Puype M, Van Damme J, Van Montegu M. Eur J Biochem. 152 (1985) 9–19.

Vesterberg, O. Acta Chem. Scand. 23 (1969) 2653–2666.

Vidal-Puig A, Moller DE. BioTechniques 17 (1994) 492–496.

Vuillard L, Marret N, Rabilloud T. Electrophoresis. 16 (1995) 295–297.

Wagner H, Kuhn R, Hofstetter S. In: Wagner H, Blasius E. Ed. Praxis der elektrophoretischen Trennmethoden. Springer-Verlag, Heidelberg (1989) 223–261.

Washburn MP, Wolters D, Yates JR III. Nature Biotechnol 19 (2001) 242–247.

Wasinger VC, Cordwell SJ, Cerpa-Poljak A, Yan JX, Gooley AA, Wilkins MR, Duncan MW, Harris R, Williams KL, Humphery-Smith I. Electrophoresis 16 (1995) 1090–1094.

Weber K, Osborn M. J Biol Chem. 244 (1968) 4406–4412.

Weiss W, Vogelmeier C, Görg A. Electrophoresis 14 (1993) 805–816.

Welsh J, McClelland M. Nucleic Acids Res. 18 (1990) 7213–7218

Wenger P, de Zuanni M, Javet P, Righetti PG. J Biochem Biophys Methods 14 (1987) 29–43.

Werhahn W, Braun H-P. Electrophoresis 23 (2002) 640–646.

Wessel D, Flügge UI. Anal. Biochem. 138 (1984) 141–143.

Westermeier R, Postel W, Weser J, Görg A. J Biochem Biophys Methods. 8 (1983) 321–330.

Westermeier R, Naven T. Proteomics in Practice. A laboratory manual of proteome analysis. WILEY-VCH, Weinheim (2002).

Westermeier R. In Cutler P, Ed. Protein purification protocols. Second edition. Methods in molecular biology. Volume 244. Humana Press, Totowa, NJ (2004) 225–232.

Wildgruber R, Harder A, Obermaier C, Boguth G, Weiss W, Fey SJ, Larsen PM, Görg A. Electrophoresis. 21 (2000) 2610–2616.

Wilkins MR, Williams KL, Appel RD, Hochstrasser DF. Eds. Proteome research: New frontiers in functional genomics. Springer, Berlin (1997).

Williams JGK, Kubelik AR, Livak KJ, Rafalski JA, Tingey SV. Nucleic Acids Res. 18 (1990) 6531–6535

Willoughby EW, Lambert A. Anal Biochem. 130 (1983) 353–358.

Wilm M, Mann M. Int J Mass Spectrom Ion Proc. 136 (1994) 167–180.

Wilm M, Shevchenko A, Houthaeve T, Breit S, Schweigerer L, Fotsis T, Mann M. Nature 379 (1996) 466–469.

White MB, Carvalho M, Derse D, O'Brien SJ, Dean M. Genomics 12 (1992) 301–306.

Wurster U. Isoelectric focusing of oligoclonal IgG on polyacrylamide gels (PAG plates pH 3.5 – 9.5) with automated silver staining. Amersham Pharmacia Biotech Application Note (1998).

Yamashita MJ, Fenn B. J Phys. Chem 88 (1984) 4451–4459.

Zischka H, Weber G, Weber PJA, Posch A, Braun RJ, Bühringer D, Schneider U, Nissum M, Meitinger T, Ueffing M, Eckerskorn C. Proteomics 3 (2003) 906–916.

Index

1-D gel evaluation   87 f
2-D electrophoresis   42, 92 ff, 100 ff, 275 ff, 379 ff
  s. a. Two-dimensional electrophoresis
2-D gel evaluation   88 f

a
A, C, G, T s. adenine, cytosine, guanine, thymine
Acid violet 17 staining   271 f
Acid violet staining   272, 290
Acrylamide   16, 339, 352, 359
  formula   16
  gel recipes, DNA gels   243, 300 f
    IEF gel   200, 259
    Immobiline gel   259, 279 f
    native PAGE   177
    SDS gradient gel   218, 236, 284
    titration curve gel   164
    ultrathin layer electrophoresis   148
    vertical gels   233
  gradient   36
  immobilized pH gradient   59
  polymerization   148, 164, 180, 200, 221, 235, 266, 285
Additive gradient   30, 62, 321 ff
Adenine   25
Adsorption
  on blotting membranes   68 ff, 109, 247 ff
  on capillaries   11
A/D transformer s. analog digital transformer
Affinity electrophoresis   20, 151
AFLP   331
Agarose gel   15, 151, 189, 348 ff
  blotting   68
  clinical routine   17
  electrophoresis   15
  immunoelectrophoresis   18
  native immunoblotting   69

PFG electrophoresis   23
preparation   152 ff, 190 ff
protein detection   159 ff, 195 f
rehydratable   54
submarine electrophoresis   12, 22, 125
Agarose IEF   54
  method   189 ff, 348 ff
Albumin   94
Alkaline blotting   74
Alkaline phosphatase   20, 76
Alkylation   37, 101, 116, 214, 363
Allelic ladder   32
Amido Black staining   75, 253
Ammonium nitrate   162, 171, 187, 196, 206
Ammonium persulfate   16, 164, 339, 343, 352
  gel recipe, DNA gels   243, 303
    IEF gel   200
    Immobiline gel   259, 281
    native electrophoresis   177
    SDS gradient gel   218, 236, 284
    titration curve gel   164, 166
    ultrathin layer gel electrophoresis   148
Amphoteric buffer   4, 42, 175, 187
Amphoteric substance   5, 51, 65
Amplification products   27 ff, 299 ff, 385
Amplifying enzyme   77
Analog digital transformer   111
Analysis, quantitative   5, 48 ff, 81 ff
Anion   5, 47
Antibody   18, 67, 151 ff
  detection an blots   76 f
  immunoblotting   76, 109, 253
  quantity   82
  reactivity   75
  solution   155, 158, 196, 209
Antigen   18, 67, 159
Antigen-antibody titer   82, 159
API   114

Electrophoresis in Practice, Fourth Edition. Reiner Westermeier
Copyright © 2005 WILEY-VCH Verlag GmbH & Co. KGaA, Weinheim
ISBN: 3-527-31181-5

APS s. ammonium persulfate
ARDRA   28
Automation   12, 25, 78, 128 f
Autoradiography   13, 24, 32, 76, 85,
    103, 377

b
BAC s. bisacryloylcystamine
β-Alanine   58
Base pair ladder   299
Base pairs   27, 32
Basic pH gradients   61, 101
Bi-directional electrophoresis   24
Bi-directional transfer   68, 71
Bioinformatics   92, 118
Biotin-avidin   24, 77
Bis s. NN′-methylenebisacrylamide
Bisacryloylcystamine   17
Blocking   75, 253, 377 f
Blotting   4, 13, 67 ff, 104
    blotting equipment   69 ff, 127, 129, 247 ff
    electrophoretic   24, 70, 247 ff, 371 ff
    of IEF   69
    of SDS electropherograms   250
    semi dry   70, 247 ff
    trouble-shooting   371 ff
Blotting membranes   67 ff, 72, 249
    blocking of –   75
    staining of –   75, 253
Blue Native PAGE   41, 99
Blue toning   226 f
Bovine serum albumin   75, 90
bp s. base pairs
BSA s. bovine serum albumin
Buffer   7, 34 ff, 73 f, 300 f, 317
    agarose strips   128
    amphoteric   6, 42, 175 ff
    strips, paper   14, 127, 149, 183, 222, 292,
        307 f, 328, 334
    strips, polyacrylamide   40
    system, continuous   248
    system, discontinuous   5, 22, 34, 45, 74,
        218, 228, 235, 238, 248 f, 300 f, 362

c
Calibration curve   39, 85, 89
Capillary blotting   68
Capillary electrophoresis   3, 7, 10 ff, 27, 48,
    53, 82
Capillary isotachophoresis   48
CAPS s. 3-(cyclohexylamino)-propanesulfonic
    acid
Carboxymethyl   73

Carrier ampholytes   55 ff, 62, 96, 100, 163,
    189, 197, 268, 277, 339 ff
    recipe   166, 191, 201, 283
Catalyst   16, 42, 167, 175, 189
Cathodic drift   58, 100, 208, 349
    s. a. plateau phenomenon
Cation   5, 41, 47
Cationic detergent   41, 42
cDNA   32
Cell disruption   276
Cellulose acetate   13, 84
Cerebrospinal fluid   16, 206
Cetyltrimethylammonium bromide   42, 366
CHAPS s.
    3-(3-cholamidopropyl)dimethylammonio-1-
    propane sulfate
Charge heterogeneity   21, 81
Chemically assisted fragmentation   113
Chemiluminescence   24, 77, 84
3-(3-Cholamidopropyl)dimethylammonio1-
    propane sulfate   277, 283
Chromatography
    of peptides   92
    of proteins   99
Chromosome separation   23
Clinical diagnostic   13, 17, 151
Clinical laboratory   17
CM s. carboxymethyl
Co-analytical modifications   96
Collagen Molecular Weight   214
Collision-induced dissociation   115
Colloidal gold   75, 77
Colloidal staining, Coomassie   170, 185,
    204 f, 224 f, 294, 358
Complex formation   7, 82
Composite proteins   184
Computer
    for DNA sequencing   26
    for evaluation   83 ff, 92, 118
Concentration effect   5, 47
    s. a. zone sharpening effect
Configurational changes   6, 52, 53, 96
Confocal scan head   86
Constant denaturing gel electrophoresis   30,
    321 ff
Continuous buffer   39, 74, 146, 151, 244
Continuous buffer system   249
Continuous freeflow electrophoresis   10
Cooling   14, 29, 54, 122 ff, 318
    thermoelectric   128 f
Coomassie Blue staining   13, 41, 103
    agarose gels   159, 195

colloidal   170, 185, 204 f, 224 f, 294, 358, 367
   fast   170, 185, 205, 224
   immunoelectrophoresis   161
Counter immunoelectrophoresis   19
Counter-ion   45
Cross-linking   16, 27, 297, 315
Cryo-isoelectric focusing   54, 209
Cryoproteins   54
CTAB s. cetyltrimethylammonium bromide
Current   122
Cy3 and Cy5 label   26, 104, 311
3-(Cyclohexylamino)-propanesulfonic acid   73
Cystein gels   101
Cysteins   37, 101
Cytosine   25

d

Da s. Dalton
Dalton   20, 38
Database   89, 108, 118
Deep Purple™   104, 295
Denaturing conditions
   DNA   24, 30, 313 ff, 322 ff, 331 ff
   proteins   37, 57, 96, 208, 211 ff, 268, 277
Densitometric evaluation   83 ff
Densitometry   83 ff
Density gradient   36, 60, 219, 236, 240, 265 f, 269, 281, 324 ff
Desoxyribonucleic acid   116
   blotting   68 ff
   Separation methods   12, 15, 22 ff, 299 ff
Detection   12, 22, 29, 49, 76 ff, 84 ff, 102
   s. a. autoradiography
   s. a. blotting
   s. a. staining
Detector   48, 112
Detergent   12
   anionic   211
   cationic   31, 366
   non-ionic   41, 53, 96
   zwitterionic   53, 277 ff
Dextran gel   62
DGGE s. Denaturing gradient gel electrophoresis
Dialysis   278
Diazobenzyloxymethyl   72
Diazophenylthioether   72
Diethylaminoethyl   73
Differential display reverse transcription   32, 299, 331
Diffusion   5, 13, 17, 52, 56, 346, 375

immunodiffusion   19, 158
Diffusion blotting   68
DIGE   93, 98, 104 ff, 105 ff
Digestion, tryptic   110
Dimethylsulfoxide   209, 345
Discontinuous buffer System   5, 22, 34 ff, 45, 175, 218, 228, 235, 238, 362
   blotting   74, 248 f
   DNA separations   300 ff
   proteins   34 ff, 218, 235, 284
Dissociation constant   3, 54, 59, 230, 260 f, 281
Disulfide bond   17, 37, 96, 212, 277
Disulfides   101, 380
Dithioerythreitol   96, 268
Dithiothreitol   37, 96, 214, 268, 273, 277, 363 ff, 380 ff
DMSO s. dimethylsulfoxide
DNA s. desoxyribonucleic acid
DNA sequence   24 ff, 91 ff, 116
DNA sequencing   24, 91, 116, 125 f
   automated   12, 25
Double Replica Blotting   71
DPT s. diazophenylthioether
Drying of gels   298
DSCP   30
DTE s. dithioerythreitol

e

EDTA s. ethylendinitrilotetracetic acid
Electric field   3, 23, 55, 121
   strength   25, 45, 158, 236
Electrode buffer   7, 14, 22, 34 f, 70 ff, 131, 146
   precipes for DNA   244, 300 f, 322, 333
   recipes for DNA   317
   recipes for proteins   152, 183, 215, 228, 233, 245, 248 f, 280
Electrode solutions   57, 193, 208, 286, 347
Electroendosmosis   7, 13, 15, 54, 175, 189, 280
Electroendosmotic effect   290, 341, 348, 380
   in capillaries   11
Electro-osmotic movement   7, 19, 22
Electrophoresis   4, 9, 122
   2D electrophoresis   42, 92 ff, 275 ff, 379 ff
   automated   12, 25, 128 f
   bidirectional   24
   continuous free flow zone   10
   discontinuous   34 f
   in pulsed electric field   23, 126
Electrophoretic blotting   70, 127, 247 ff, 371 ff
Electrophoretic mobility   3, 15, 21, 23, 45
Electrospray ionization   114, 119

Elution, electrophoretic 63, 110
Emission 26, 85, 104
Enzymatic digestion 92, 110 ff
Enzyme
  enzyme blotting 76
  enzyme detection 13, 18, 20, 76
  enzyme detection System 76
  enzyme inhibitor 20
  enzyme substrate complex 54, 209
  enzyme substrate reaction 76
  separation 13, 18, 20, 209
Equilibration 280, 290, 381
Equipment 123 f
Equivalence point 19
Escherichia coli 277 f
ESI 115
Ethidium bromide 22, 27, 84, 132, 244, 299, 311
Ethylendinitrilotetraacetic acid
  buffer 24, 75, 225, 244, 277, 317, 333
  sample preparation 277, 316
  staining 225
Evaluation 81 ff, 107
Excitation 26, 85, 104

f

Far western blotting 70
Ferguson plot 21
Field strength 3, 25, 45, 158, 270
Flatbed systems 14, 126 f, 145 ff, 292 f
Fluorescence
  detection 85 f
  labeling 25, 104
  staining 104
Fluorography 13, 25
Focusing, isoelectric s. isoelectric focusing
Free flow electrophoresis 10, 99 f

g

GC s. group specific components
GC clamps 322
Gel 3, 13
  non-restrictive 17
  rehydrated 41
    for DNA 300 ff, 314 ff, 323 f, 333
    for proteins 42, 53, 61, 62, 163, 175, 197 ff
  restrictive 21, 32
Glassfiber membrane 63, 73, 78
Glycoproteins 41, 72, 73, 78, 367
Gold, colloidal 75, 77
Gradient
  additive 269, 321

pH 4 f, 45 ff, 163, 189 ff, 197, 265, 281
  porosity 219, 236 f, 240, 284 f
Gradient gel 36, 129
  preparation 219 f, 236 f, 240, 265, 269, 281, 284 f, 323
Group specific component 268
Guanine 25

h

Henderson-Hasselbalch equation 60
HEPES s. N-(2-hydroxyethyl)piperazine-N'-2-ethanesulfonic acid
Heteroduplex 30, 299
High Molecular Weight 79, 214
High-resolution 2D electrophoresis 42, 92 ff, 275 ff
  trouble shooting 379 ff
Histones 41, 61
HMW s. High Molecular Weight
Hybridization 76
Hydrolysis 40, 102, 259

i

Identification 13, 76, 76 f, 109
IEF s. isoelectricfocusing
IEF marker 54, 65
IgG s. immunoglobulin G
Image analysis 87 ff, 107 ff
Immobilized pH gradients 59 ff, 63, 100, 257 ff, 279 ff, 352 ff
  recipes 261 f, 281
Immobilizing membranes 67 ff
Immunoblotting 76, 109, 253, 377
  native 79
Immunodiffusion 19, 158
Immunoelectrophoresis 19 ff, 42, 82, 151 ff
Immunofixation 18, 151, 160, 189, 196, 209
Immunoglobulin G 19, 76, 151
Immunoglobulin M 54
Immunoglobulins 94
Immunoprecipitation 13, 18 ff
Immunoprinting 13, 18
Indian Ink staining 75, 254
Internal standard 93, 107
Ion-exchange chromatography 172
Ionic strength 6, 34, 39, 60, 228
Ion trop 115
IPG s. immobilized pH gradients
IsoDalt system 276
Isoelectric focusing 4, 51 ff, 100 f, 173, 189 ff, 197 ff, 257 ff
  blotting of – 69, 73, 372

current during – 122
equipment 127, 130
in agarose gel 189 ff
in immobilized pH gradients 257 ff
in polyacrylamide gel 197 ff
principles 51 ff
sample application 82, 194, 204, 269 f,
 286 ff
trouble-shooting 339 ff
Isoelectric membranes 63, 100
Isoelectric point 5, 42, 65, 87 f, 93
Isotachophoresis 5, 35, 45 ff
Isotachophoresis effect 35, 71
ITP s. isotachophoresis

j
Joule heat 6, 10, 122, 288

l
Labelling
 of nucleic acids 25 ff
 of proteins 98, 105 ff
Laboratory equipment 133 ff
Lactate dehydrogenase 18, 21
Laser 26, 78, 85 f, 107, 111
Laser capture universdissection 94
LDAO, Lauryldimethylamine-N-oxide 209
Leading ion 4, 45, 248
Lectin 20, 67
Lectin blotting 78
Linear polyacrylamide 12, 27, 54
Lipopolysaccharide 13, 73
Low molecular Weight 56, 116, 196
 peptides 40, 81, 210, 230, 242 ff
 substances 20, 145, 182
Lysis buffer 96, 277

m
MALDI 78, 110 ff, 111
Marker
 DNA 299
 protein 39, 54, 82, 90
Mass spectrometry 78, 91 ff, 92, 102, 297 f
Medium, stabilizing 6, 7, 12 ff
MEKC 12
 s. a. Micellar electrokinetic chromatography
 Membrane
Membrane
 cellulose acetate 13
 immobilizing 67 ff
 isoelectric 63 f
Membrane proteins 41, 268, 277

MES, 2-(N-Morpholino)ethanesulfonic acid
 42
Micellar electrokinetic chromatography 12
Microsatellites 31
Migration, electrophoretic 7, 19, 121
Minimum labelling 98, 105
Mixed disulfides 101
Mobility 30, 45, 52
 relative electrophoretic 4, 12, 45
Molecular diameter, radius 21, 36
Molecular weight 12, 37, 41, 42, 87, 214
MOPS, 3-(N-morpholino)propanesulfonic
 acid 42
Mouse liver 95
Moving boundary electrophoresis 5, 9
MuDPIT 92
Mutation detection 28 ff

n
N-(2-Hydroxyethyl)piperazine-N′-
 ethanesulfonic acid 42, 175
NAP s. Nucleic Acid Purifier
Native electrophoresis DNA 27, 76
 electrophoresis proteins 22, 175
 immunoblotting 79
Negative spots 295
Negative staining 103, 225, 295
NEPHGE 100
Net charge 5, 17, 35, 39
 isoelectric focusing 51
 speed of migration 21, 121
Net charge curve 51, 64, 172
Neutral blotting 74
Nitrocellulose 72, 253
NN′-Methylenebisacrylamide
 formula 17
 gel recipe, DNA gels 303
  IEF gel 200, 259
  Immobiline gel 258, 279
  native PAGE 177
  SDS gradient gel 215, 233
  titration curve gel 164
  ultrathin layer gel 145
Nonidet s. non-ionic detergent
Non-ionic detergent 42, 53, 79, 96, 197, 268,
 340
Non-restrictive medium 17
Northern blotting 70
Nucleic Acid Purifier 151, 163, 176, 190, 198,
 258
Nucleoprotein 41, 52
Nylon membrane 72, 75

o

O.D. s. optical density
Oligoclonal IgG   206
Oligo nucleotides   310
Oligo peptides   185, 210, 230, 236, 242 ff
    peptides   204
Optical density   84
Oxidation   96, 116, 209, 343, 363

p

Paper electrophoresis   12
Paper replica   62
PBS s. Phosphate Buffered Saline
PCR® s. Polymerase Chain Reaction
Pefabloc   97, 277
Peptide gels   40, 229 ff, 242 ff
Peptide massfingerprinting   116
Peptides   40, 63, 92, 102, 229 ff, 242 ff
Peptides detection   170, 204, 210, 245, 271
Peroxidase, horseradish   76, 77
PFG s. Pulsed-Field Gel Electrophoresis
Phenylmethyl-sulfonylfluoride   97, 258
pH gradient   5, 52 ff, 55, 65, 122, 339 ff
    2D electrophoresis   42, 96, 379 ff
    IEF method   202 ff, 259 ff, 281
Phosphatase, alkaline   20, 76, 77
Phosphate Buffered Saline   254, 276
Phosphoglucomutase   268
Phosphor imager   85
Photographing   83
PI s. Protease Inhibitor
pK value s. dissociation constant
Plasma, human   94
Plasma proteins   94
Plateau phenomenon   58, 100, 341
PMSF s. phenylmethyl-sulfonylfluoride
Polyacrylamide gel   16 ff, 20, 24 ff, 227
    DNA   24 ff, 244, 300 ff, 385 ff
        proteins   175
    electrophoresis   145
    isoelectric focusing   53, 197 ff, 257 ff, 279 ff
    linear   12, 27
    rehydratable   41, 163, 175, 197 ff, 257 ff,
        281 f, 299 ff
Polymerase Chain Reaction   3, 27 ff, 244,
    299 ff, 385 ff
Polymerization   17
Polyphenols   145
Polysaccharide   15
Polysaccharide separation   13
Polyvinyledendifluoride   72, 376
Pore gradient gels   36, 129, 219, 236 f, 240,
    284 f

preparation   219 f, 236 f, 240, 284 f
Post-source decay   113
Post-translational modifications   91 ff, 113,
    116, 118
Power, electric   122
Precipitation   18, 34, 97, 277, 278
    line   18, 82
Prefractionation   98
Pressure blotting   68
Protease Inhibitor   61, 97, 258, 277
Protein A, radioiodinated   76
Protein sequencing   78, 109
Proteome analysis   5 f, 43, 92 ff, 275
Pulsed-Field Gel Electrophoresis   23
Purity control   81
Purity test   5
PVDF s. polyvinylidenfluoride

q

Quantification   5, 81 ff, 89 ff
Quantitative analysis   5, 48

r

r s. molecular radius
Radioactive labeling   13, 24, 32, 76, 85, 103,
    377
Radioiodinated protein A   76
RAPD, Random amplified polymorphic
    DNA   28, 299
Reflectron   112
Rehydratable agarose gel   54
Rehydrated polyacrylamide gels   300 ff
    proteins   41, 175, 197 ff, 230, 257 ff, 281 f
Relative mobility   12, 21
Resolution   13, 52, 60, 92, 112, 228, 242, 257
Resolving gel   34 ff, 180, 218, 235 ff,
    243, 284, 303
Restriction Fragment Length Polymorphism
    22
Restrictive medium   21, 34
Retardation coefficient   21
Reversible staining   13, 75, 103, 225, 253, 295
RFLP s. Restriction Fragment Length
    Polymorphism
Ribonucleic acid   20, 32 ff, 68, 76, 299
RNA s. ribonucleic acid
RNase protection assay   332
Rocket technique   19, 155
Routine, clinical   13, 84

s

Sample   6
Sample preparation

DNA   305, 315 f, 322, 332
   proteins   82, 96 ff
      denaturing   212 ff, 276 ff
      immunoelectrophoresis   151
      Isoelectric focusing   189 f, 198, 258
      native electrophoresis   151, 176
      SDS treatment   212 ff
      titration curve analysis   163
   vertical gels   232
Saturation labelling   98, 105
Scanner   85 f, 104
Scanning   83 ff
SDS s. sodium dodecyl sulfate
SDS electrophoresis   37 ff, 101
SDS treatment   212 ff, 290
Secondary antibody   77, 377
Semi dry blotting   70 f, 74, 127, 129, 247 ff, 371 ff
Separator IEF   58
Sephadex   62, 100
Sephadex IEF   62, 100
Sequencing
   DNA   24 ff, 116
   proteins   109, 113, 118
Silver amplifying   77
Silver staining   16, 29, 27 ff, 103, 128, 129, 367
   recipe, protein PAGE   226
   recipes, agarose gels   162, 196
      PAGIEF   206 ff
      protein PAGE   186 f, 295
Single nucleotide polymorphism   31
Single strand conformation polymorphism   29, 313 ff, 385
Slotformer   146, 152 f, 178, 216, 301
SNP Analysis   31
Sodium dodecyl sulfate   35 ff, 73, 78, 101, 211, 231, 248, 250, 279, 284
Software, image analysis   87 f
Spot
   identification   109
   picker   104, 107 f
SSCP   29
   s. a. Single strand conformation polymorphism
Stabilizing medium   12
Stable isotope labeling   103
Stacking gel   34 ff, 101, 180, 218, 228, 235 ff, 284, 303
Staining   13
   2-D gels   103 f, 294 ff
   agarose gels   159, 195
   blotting membranes   75 ff, 253 f

DNA gels   22, 310
IEF gels   195, 204 ff, 271 f, 290
native gels   185 f
SDS gels   224 ff, 294 ff
titration curve gels   170 f
Starch gel   15
Storage phosphor plates   85, 103
Storage phosphor screen   85
"Submarine" chamber, gel   22, 125
SYPRO Ruby®   102, 104

t
T s. total acrylamide concentration [%]
Tank blotting   70, 73
TBE s. Tris borate EDTA
TBP   101
TCEP   101
TEMED, N,N,N′,N′-tetramethylethylendiamine   16, 42, 175, 189, 303
   gel recipes   148
      DNA gels   235 ff, 303
      Immobiline gel   264, 281
      native electrophoresis   180
      SDS gradient gel   218, 235 ff, 284
      titration curve gel   167
Temperature gradient gel electrophoresis   31, 331
Terminating ion   4 f, 47, 248
TGGE s. Temperature gradient gel electrophoresis
Thermoblotting   68
Thiourea   97, 279
Thymine   25, 299
Titration curve analysis   64, 163 ff, 175, 346
TMPTMA   254
Transferrin   268
Transmission   107
Trichloroacetic acid, fixing   159, 170, 185, 195, 205, 272
Tris, tris(hydroxymethyl)-aminomethane   74
   acetate   41, 228, 245, 300, 317
   ascorbic acid   273
   borate-EDTA   24, 41, 244, 301, 317, 322, 333, 367
   glycine   34 ff, 74, 101, 215, 233, 248, 280
   HCl   34 ff, 74, 101, 164, 177, 200, 215, 233, 280
   phosphate   301, 317, 322, 333
   tricine   41, 152, 228, 245, 300, 317
   tricine lactate   152
Triton X-100 s. detergent, non-ionic

Trouble-shooting guide   339 ff
Two-dimensional electrophoresis   42, 60, 62,
    92, 100 ff, 275 ff, 379 ff
    DNA   33

u
Ultraviolet light   11, 48, 75, 81, 254
Urea   30 ff, 53, 57, 74, 79, 189, 208, 268
    2D electrophoresis   42, 96 ff, 268, 277
    DNA   24, 322 f, 331 ff
UV s. ultraviolet light

v
Vacuum   111, 115
    blotting   69

Vertical systems   14, 124 f, 231 ff, 291
Viroids   33
Viscoelastic relaxation time   23
VLDL, very low density lipoproteins   269
Voltage   122

w
Western blotting   70

z
Zone electrophoresis   17 ff, 19, 36,
    52, 170
Zone sharpening effect   5, 35, 36, 45